Quaternary Quadratic Forms

Gordon L. Nipp

Quaternary Quadratic Forms
Computer Generated Tables

With a diskette

Springer-Verlag
New York Berlin Heidelberg London
Paris Tokyo Hong Kong Barcelona

Gordon L. Nipp
Department of Mathematics
California State University
Los Angeles, CA 90032
USA

AMS Classification: 10C05

Library of Congress Cataloging-in-Publication Data
Nipp, Gordon L.
    Quaternary quadratic forms : computer generated tables / Gordon L.
  Nipp.
      p.   cm.
    Includes bibliographical references (p.   )
    ISBN 0-387-97601-9 (Springer-Verlag New York Berlin Heidelberg :
  acid-free paper). — ISBN 3-540-97601-9 (Springer-Verlag Berlin
  Heidelberg New York : acid-free paper)
      1. Forms, Quadratric — Tables.   2. Forms, Quaternary — Tables.
  I. Title.
  QA243.N56   1991
  512'.74 — dc20   91-16275

Printed on acid-free paper.

Camera-ready copy provided by the author.
Printed and bound by Edwards Brothers, Inc., Ann Arbor, MI.
Printed in the United States of America.

9 8 7 6 5 4 3 2 1

ISBN 0-387-97601-9 Springer-Verlag New York Berlin Heidelberg
ISBN 3-540-97601-9 Springer-Verlag Berlin Heidelberg New York

Preface

This book of tables includes a reduced representative of each class of integral positive definite primitive quaternary quadratic forms through discriminant 1732. The classes are grouped into genera; also included are Hasse symbols, the number of automorphs and the level of each such form, and the mass of each genus. An appendix lists p-adic densities and p-adic Jordan splittings for each genus in the tables for p=2 and for each odd prime p dividing the discriminant.

The book is divided into several sections. The first, an introductory section, contains background material, an explanation of the techniques used to generate the information contained in the tables, a description of the format of the tables, some instructions for computer use, examples, and references. The next section contains a printed version of the tables through discriminant 500, included to allow the reader to peruse at least this much without the inconvenience of making his/her own hard copy via the computer. Because of their special interest, we include tables of discriminants 729 and 1729 at the end of this section. Limitations of space preclude publication of more than this in printed form. A printed appendix through discriminant 500 and for discriminants 729 and 1729 follows. The complete tables and appendix through discriminant 1732 are compressed onto the accompanying 3.5 inch disk, formatted for use in a PC-compatible computer and ready for research use particularly when uploaded to a mainframe. Documentation is included in the Introduction.

Computer work was done on the California State University Cyber 760-960 and on the California State University, Bakersfield, Cyber 730. Thanks must go to the Computer Center at California State University, Los Angeles, for their help in expediting the work and, particularly, to Mike Fleming at the California State University, Bakersfield, Computer Center for his help with the system during the early stages of the work. Support from the California State University by way of large amounts of computer time was always forthcoming. To those many unknown Cyber 760-960 time-sharers across California whose runs I slowed, I express my appreciation for their patience.

I would like to offer my deep gratitude to Professor John Hsia who suggested that I compile these tables and whose insight and encouragement were a necessary condition for their completion. Special appreciation must go to Professor Dennis Estes who has been very generous in sharing his knowledge and interest. I would also like to thank Professor Meinhard Peters for his comments on early versions of the tables. Of course, any errors in the tables are the sole responsibility of the author.

Los Angeles
December 1990

Gordon L. Nipp

Contents

Introduction

1. Background. The following computer-generated tables of reduced regular primitive positive definite quaternary quadratic forms over the rational integers are inspired by and are an outgrowth of the remarkable Brandt-Intrau tables of reduced positive ternary forms [2], published in pre-computer days after what must have been an incredible amount of effort. The Brandt-Intrau tables serve not only as a model of accuracy but also as a starting point for generating classes of reduced forms with an additional variable by computer. The technique used for this was essentially that of Germann [13], who began with the ternary forms through discriminant 16 from the Brandt-Intrau tables and computed all classes of reduced positive quaternary forms through discriminant 61. While Germann had few enough forms to allow separation into classes on an individual basis, our task was somewhat less manageable. Beginning with the ternary forms of discriminant $\delta \leq 320$ from the Brandt-Intrau tables (with the imprimitive ones adjoined), we generated a large file of quaternary forms containing representatives of all possible classes. Directed by a conjecture of John Hsia [14] that no two inequivalent such forms with the same discriminant would have the same theta series, two forms with the same discriminant were considered "equivalent" as a preliminary sorting measure if their first twenty theta coefficients were the same (the number 20 being chosen arbitrarily). This was followed by a computer search to find (a significant number of) classes wrongfully eliminated; i.e., to find those inequivalent forms of a given discriminant with the same first twenty theta coefficients. As a curiosity, it might be noted that the first such example occurred at discriminant 352; up to that point positive quaternary quadratic forms are classified by the number of times they represent the integers 1 through 20. Furthermore, Alexander Schiemann [25] has since produced a counterexample to Hsia's conjecture. The following two forms of discriminant 1729 are in distinct classes but have the same theta series:

$$2x_1^2 + 4x_2^2 + 4x_3^2 + 5x_4^2 + x_1x_2 + x_2x_3 + x_1x_4 + 3x_2x_4 + 4x_3x_4$$
$$2x_1^2 + 4x_2^2 + 5x_3^2 + 5x_4^2 + 2x_1x_2 + x_1x_3 - 2x_2x_3 + x_1x_4 + x_2x_4 + 5x_3x_4.$$

It should be remarked that the author has verified by a computer search of these tables that no examples of this phenomenon exist in discriminants less than 1729. Further details on generating quaternary forms from ternary forms and on separating them into classes are contained in section 3 below.

Separation into genera was accomplished using the standard local techniques as contained in O'Meara's book [22;92:2 and 93:29], programmed into a computer. Previously, Hoefliger [30] had begun with Germann's tables and used exactly these methods to derive a small table of positive quaternary forms grouped into genera through discriminant 64. Our tables agree with Hoefliger's except for one instance in discriminant 64, where two of Hoefliger's genera should be combined into a single genus. We remark that for all discriminants contained in our tables except $d=729$, the genus and spinor genus coincide. For $d=729$, separation into spinor genera is easily accomplished using known techniques. These and other details about separation into genera are included in section 4 below.

The tables were then checked by computing the mass of each genus via the Siegel mass formula and comparing this value with the sum of the reciprocals of the number of automorphs of each class in the genus. This check revealed some discrepancies at various stages of the computations. For example, it showed the necessity of including imprimitive ternary forms in generating the original quaternary forms, an aspect originally overlooked because the Brandt-Intrau tables do not include imprimitive forms. As a means of dealing with possible approximation errors, we used Minkowski's result [20;p.216,v.1] that no class of positive quaternary forms has more than 5760 automorphs. The mass itself, involving an infinite product, was computed in each instance through at least 2600 factors, sufficient to ensure that the approximation error would be less than $1/5760$ and enough to allow one to determine any genus with a missing class. Details of the mass computations are given in section 5.

In addition to the efforts of Germann and Hoefliger, there are two other earlier small tables of reduced positive quaternary quadratic forms. The tables published by Charve [4] in 1883 contain those classes of positive quaternary forms with (our) discriminant divisible by 16 through discriminant 320 and with even product terms (even f_{ij}, $i \neq j$, with notation and discriminant as below). While similarly limited, Townes' tables [28] from 1940 expand the Charve tables to discriminant 400, separate the classes into genera, and correct some errors. There are several inaccuracies in the Townes tables as well, one of which we use as an example of separation of forms into genera in section 4. In spite of their limitations, these tables were quite useful in serving as an early means of checking the accuracy of our computer-generated results. In 1973, Furter [12] extended Germann's tables through discriminant 128. While Furter's tables are not separated into genera and contain one error (the two forms with equal diagonal coefficients in discriminant 97 are in the same class), they were used as additional verification for our tables. In his dissertation (see [25]), Schiemann includes tables of positive quaternary forms through discriminant 200 separated into classes and containing the level and the number of automorphs for each class.

In the ternary case, precursors of the Brandt-Intrau tables date back to Seeber's work [26] in 1831. In 1851, Eisenstein [10] published a set of tables of reduced positive ternary forms with even product terms, and these were extended in 1890 by Borisow [1]. In 1930, Dickson [7] compiled short tables of both definite and indefinite ternary forms, and tables of positive forms computed by Jones [16] appeared in 1935.

Several papers dealing with linear independence of certain theta series associated with positive definite forms have recently appeared. The works of Hsia and Hung [15] and of Kramer [18] contain results of this nature as well as a listing of those positive quaternary forms of discriminant 389 where a linear relation among the corresponding theta series occurs. This example was calculated first by Kitaoka and served as a check on our tables for this discriminant and as a means of verifying the techniques used in generating them.

The reduction theory necessary in these tables to pick a representative reduced form from each class is based on an article by van der Waerden [29]. The papers of van der Waerden, Germann, Hoefliger, and others have been assembled into book form [30]. Computations were performed on the California State University Cyber 760-960 and on the California State University, Bakersfield, Cyber 730.

2. **Definitions and Format.** The discriminant d of a quaternary quadratic form

$$f = f_{11}x_1^2 + f_{22}x_2^2 + f_{33}x_3^2 + f_{44}x_4^2 + f_{12}x_1x_2 + f_{13}x_1x_3 + f_{23}x_2x_3 +$$

$$+ f_{14}x_1x_4 + f_{24}x_2x_4 + f_{34}x_3x_4$$

is $d = 16(\det F)$ where

$$F = \begin{bmatrix} f_{11} & \tfrac{1}{2}f_{12} & \tfrac{1}{2}f_{13} & \tfrac{1}{2}f_{14} \\ \tfrac{1}{2}f_{12} & f_{22} & \tfrac{1}{2}f_{23} & \tfrac{1}{2}f_{24} \\ \tfrac{1}{2}f_{13} & \tfrac{1}{2}f_{23} & f_{33} & \tfrac{1}{2}f_{34} \\ \tfrac{1}{2}f_{14} & \tfrac{1}{2}f_{24} & \tfrac{1}{2}f_{34} & f_{44} \end{bmatrix}$$

is the matrix associated with f. We assume throughout that the f_{ij}, $i,j = 1,2,3,4$, are rational integers and that the form f is positive definite.

Then d is an integer congruent to 0 or 1 modulo 4. In keeping with the Brandt-Intrau tables, the discriminant δ of a ternary form will be 4 times the determinant of the analogous 3x3 coefficient matrix.

If D is either the ring of rational integers or the ring of p-adic integers, we say that two forms with coefficients in D are in the same class (local class, if $D = Z_p$) if one can be transformed into the other by a change of variables whose matrix has entries in D and unit determinant. Such a transformation will be called an isometry, and two forms in the same class will be called isometric or equivalent (p-adically equivalent). An isometry of a form to itself will be called an automorph of the form.

The level N of f is the smallest positive integer such that $N(2F)^{-1}$ is an integral matrix with even diagonal entries. The level for each form in the tables was computed using standard computer techniques and is listed in the column headed "LEVEL".

For each discriminant, the tables list one representative form from each class, each row (under the heading) containing the coefficients of one such form as well as other information about the class and genus. Two forms are in the same genus if they are associated with the same number in the column headed "GENUS#". In order to distinguish among the quadratic spaces associated with the (classes of) forms, the Hasse symbols at p = 2 and at each odd prime dividing the discriminant are listed, the values being given for these primes in ascending order. (We use O'Meara's definition for the Hasse symbol. See the beginning of section 4.) For example, in discriminant d = 21 , the values of the Hasse symbol for the form in the first genus are -1 at p = 2 and at p = 3 , and the value is 1 at p = 7 . The columns headed "AUTOS" and "MASS" give the number of automorphs of each class of forms and the exact value of the mass of the genus of each form. The latter was computed both via the mass formula and as the sum over each class in the genus of the reciprocals of the number of automorphs of each class; comparing the values served as an accuracy check.

To facilitate computer use, the tables and the appendix are all contained on the enclosed 3.5 inch disk formatted and designed for use in a PC-compatible computer with MS-DOS 2.0 or higher. The tables are subdivided into files each of a size appropriate to be copied to a 5.25 inch disk. For example, the first two files on the disk are labeled D457 (containing the tables from discriminant 4 through 457) and D641 (containing the tables from discriminant 460 through 641). You can list the file names to the screen of your PC by inserting the disk into the appropriate disk drive (probably drive A) and, after the prompt, typing

    dir a: <ENTER>.

The files on the disk are executable. They have been compressed (so that everything fits onto one disk) using a commercial program called PKZIP, a product of PKWARE, Inc., and they each contain a program to make them "self-extracting". That is, this program will automatically expand the files to normal, uncompressed form. A good system for extracting the tables from the disk would include a PC with a hard disk (drive C), a 3.5 inch disk drive (drive A), and a 5.25 inch disk drive (drive B). As an example, to extract (uncompress) the information in file D641, insert the 3.5 inch disk into drive A . Type

C: <ENTER>

to select drive C as the default drive. After the prompt, type

copy a:d641.exe <ENTER> .

The computer will respond that it has copied a file. After the prompt, type

d641 <ENTER> .

The file d641 is now in normal, uncompressed form on the hard disk. At this point, you may want to upload the file onto a mainframe. Uploading procedures vary widely from system to system; consult your local computer specialists for the login and uploading procedures specific to your system. We remark that some protocols are tediously slow-paced; you would be well-advised to use one of the more efficient protocols. If you wish to copy the file from the hard disk to a 5.25 inch disk in drive B, type

copy d641 b: <ENTER>   .

Since you probably don't want to consume space on the hard disk with these files, you can delete them by typing

del d641 <ENTER>

and

del d641.exe <ENTER>   .

On the disk itself the tables are arranged somewhat differently from the printed copy so that each line contains all the above information relating to a single class of forms. Written according to the FORTRAN FORMAT statement

FORMAT(12I4,1X,5I2,I4,3I5),

the first entry in each line lists the discriminant and the second gives the genus # (two forms are in the same genus if and only if they have the same genus #). The next ten entries are the coefficients

$$f_{11}, f_{22}, f_{33}, f_{44}, f_{12}, f_{13}, f_{23}, f_{14}, f_{24}, f_{34},$$

respectively, of the reduced form chosen as a representative of the class. Then in ten spaces come the Hasse symbols at all primes $p \mid 2d$ (in ascending order of these primes). This is followed by the level and the number of automorphs, and the last two numbers are the numerator and denominator, respectively, for the mass written as a fraction. For example, from the line

$$84 \quad 2 \quad 1 \quad 1 \quad 1 \quad 6 \quad 0 \quad 0 \quad 0 \quad 1 \quad 1 \quad 1 \quad 1\text{-}1\text{-}1 \quad 42 \quad 96 \quad 5 \quad 96$$

we see that the form

$$x_1^2 + x_2^2 + x_3^2 + 6x_4^2 + x_1x_4 + x_2x_4 + x_3x_4,$$

in genus #2 of discriminant 84, has Hasse symbols 1, -1, and -1 at the primes 2, 3, and 7, respectively. It has level 42, it has 96 automorphs, and its mass is 5/96.

For each genus in the tables and for each prime p dividing 2d, the appendix lists respectively the p-adic density and a p-adic Jordan splitting for the form representing the first class listed in the genus. The densities are the results of lengthy computations and arise as part of the mass formula. The p-adic Jordan splittings given arise specifically when one applies the procedure in [22; section 94] and, for $p = 2$, performs some simplifications. As O'Meara notes in this reference, "the Jordan splitting is all that is needed in the non-dyadic theory. The Jordan invariants are also needed in the 2-adic theory, but these can be read off from the Jordan splitting." Thus, concise information determining the genus can be most readily specified via the Jordan splittings. Details are contained in section 4.

In the appendix notation, the modular component lattices of the Jordan splittings are separated by brackets [and] and the numbers given are coefficients of the forms associated with them. For $p \neq 2$, the forms are diagonal so that, for example, if $d = 5$ and $p = 5$, each form in the only genus is 5-adically equivalent to

$$L_5^1 \perp L_5^2$$

where L_5^1 has associated form

$$x_1^2 + (3/4)x_2^2 + x_3^2$$

and L_5^2 has associated form $(5/12)x_4^2$. Here L_5^1 and L_5^2 are clearly 5-adically equivalent to

$$x_1^2 + 3x_2^2 + x_3^2 \quad \text{and} \quad 15x_4^2,$$

respectively. We retain the fractions so that one can easily check discriminants.

If $p=2$, a component of the splitting may have a two-dimensional sublattice whose form

$$\theta = ax_1^2 + bx_1x_2 + cx_2^2$$

is not diagonal. We can assume that a, b, and c are integers and that b is odd, in which case the number $\Delta = b^2-4ac$ is either 1 or -3 (mod 8). As Watson [32] notes, it follows that if $\Delta \equiv 1$ (mod 8), then θ is 2-adically equivalent to

$$H = x_1x_2,$$

and if $\Delta \equiv -3$ (mod 8), then θ is 2-adically equivalent to

$$A = x_1^2 + x_1x_2 + x_2^2.$$

For example, if $d=8$ and $p=2$, each lattice in the only genus is 2-adically equivalent to

$$A \perp L_2^2 \perp L_2^3$$

where L_2^2 has associated form x_3^2 and L_2^3 has associated form

$$(2/3)x_4^2,$$

2-adically equivalent to

$$6x_4^2.$$

As another example, if $d=5$ and $p=2$, the Jordan splitting consists of a single modular component which is composed of the orthogonal sum of the two 2-dimensional sublattices A and H. Over the 2-adic integers, the associated form is

$$x_1^2 + x_1x_2 + x_2^2 + x_3x_4 .$$

3. **Separation into Classes.** The problem of systematically choosing in some natural way a single quadratic form (a "reduced" form) as a representative of each class of positive forms has a long history. For positive binary forms, Lagrange gave the accepted definition in 1773 [19]:

$$f = f_{11}x_1^2 + f_{12}x_1x_2 + f_{22}x_2^2$$

is reduced when

$$|f_{12}| \leq f_{11} \leq f_{22} .$$

The transition to a widely-accepted definition of a reduced form in n variables involved contributions by Seeber, Gauss, Dirichlet, Eisenstein, Hermite, Minkowski, Siegel, and many others. Dickson [6] gives many details and references up to 1919. Additional history and references are given in a comprehensive article by van der Waerden [29] based on Minkowski's definition of a reduced form. Cassells also considers Minkowski's definition to be most suitable and devotes a complete chapter of his book [3; Chapter 12] to the topic. It is van der Waerden's exposition that we follow, and it is in his article that proofs of many of the following results can be readily found.

A positive quadratic form

$$f(x_1,x_2,\ldots,x_n) = \sum_{i\le k} f_{ik}x_ix_k$$

is reduced (in the sense of Minkowski) if

$$f_{kk} \le f(s_1,s_2,\ldots,s_n) \tag{1}$$

for all integers $s_1,s_2,\ldots,s_n$ with $\gcd(s_k,\ldots,s_n)=1$. Every class of positive forms contains at least one reduced form. In the quaternary case, reduction condition (1) implies that

$$f_{11} \le f_{22} \le f_{33} \le f_{44} \tag{2}$$

and that

$$|f_{jk}| \le f_{jj} \tag{3}$$

for $j<k$. Furthermore, $f_{11}, f_{22}, f_{33}, f_{44}$ are successive minima of the quaternary form f in the sense that f_{11} is the least positive integer represented by f (where $f(x)=f_{11}$), f_{22} is the least positive integer value of $f(y)$ where x and y are linearly independent integral 4-vectors, f_{33} is the least positive integer value of $f(z)$ where x, y, and z are linearly independent, and f_{44} is the least positive integer value of $f(w)$ where x, y, z, and w are linearly independent. As van der Waerden notes, if d is the discriminant of the quaternary form f, the "fundamental inequality of reduction theory" is

$$f_{11}f_{22}f_{33}f_{44} \le d/4. \tag{4}$$

Additionally, for ternary forms with discriminant δ

$$\delta \le 4f_{11}f_{22}f_{33} \tag{5}$$

By setting $x_4=0$, one can associate a ternary form g with each positive quaternary form f. Then, reduction conditions (1) may be reformulated as:

(a) g is reduced.
(b) $f_{33} \leq f_{44}$ (6)
(c) $g(s_1, s_2, s_3) \geq f_{14}s_1 + f_{24}s_2 + f_{34}s_3$ for all $s_i = -1, 0,$ or 1,

the simplification on the s_i's being justified by Minkowski [20;p.78, v.II].

For discriminant d bounded above by some fixed value, conditions (2) and (4) imply that there are only a finite number of positive choices for $f_{11}, f_{22}, f_{33},$ and f_{44}. Inequality (5) then yields an upper bound for the discriminant δ of the ternary form g. For example, for quaternary forms with d<2000, one need examine relatively few cases to see that the corresponding ternary form g will have discriminant $\delta \leq 320$. Then, beginning with each reduced ternary form with $\delta \leq 320$ from the Brandt-Intrau tables (or an imprimitive multiple thereof with $\delta \leq 320$), condition (3) limits the choices of $f_{14}, f_{24},$ and f_{34}. For each such choice we determine whether (6c) holds for all $s = -1, 0, 1$ and, if so, we choose f_{44} satisfying (6b) such that the discriminant d is appropriately small. While some simplifications are possible (see [13;section 3]), the large but finite collection of reduced positive quaternary forms generated in this way will contain at least one representative from each class with discriminant less than the fixed value noted above. Germann's article [13] gives details and examples.

As a preliminary sorting method, we considered two forms to be "equivalent" if they had the same number of representatives of n for integral n from 1 through 20. The representative of each such equivalence class was chosen somewhat arbitrarily, although we attempted to pick one with simple coefficients. Since this procedure does not classify the forms in the usual sense, we then had the computer search the original tables for missing non-isometric forms. This procedure was simplified by the fact that no two such reduced forms can be isometric if their diagonal coefficients $f_{11}, f_{22}, f_{33}, f_{44}$ disagree. The first instance of two non-isometric forms with the same first twenty theta coefficients occurs for discriminant d=352. Here, the two forms

$$x_1^2 + 2x_2^2 + 2x_3^2 + 6x_4^2 + 2x_2x_4$$

and

$$x_1^2 + 2x_2^2 + 2x_3^2 + 6x_4^2 + x_1x_4 + x_2x_4 + x_3x_4$$

disagree first in the 22nd theta coefficient. As another example, for positive integer $n \geq 2$ the forms

$$f = x_1^2 + x_2^2 + 2x_3^2 + nx_4^2 + x_1x_4 + x_2x_4 \text{ and}$$
$$g = x_1^2 + x_2^2 + 2x_3^2 + nx_4^2 + 2x_3x_4$$

both of discriminant $16(2n-1)$, differ first in the n-th theta coefficient. Here f has four more representations of n than does g, and their theta coefficients agree up to n (with $x_4 = 0$).

The computer program used to determine the number of representations of an integer n by a positive quaternary form f relies on standard Gram-Schmidt orthogonalization to diagonalize f over the rational numbers. It is then easy to determine bounds on the four variables and count the finite number of integral vectors $x = (x_1, x_2, x_3, x_4)$ satisfying $f(x) = n$.

The problem of determining when two positive quaternary forms f and g with the same discriminant and with matrices F and G are isometric is equivalent to that of finding a 4x4 integral matrix T such that $T' FT = G$. Here T' denotes the transpose of T. By using the computer program described in the previous paragraph, we compute for each $i = 1, 2, 3, 4$, the finite number of integral column vectors $t_i = (t_{1i}, t_{2i}, t_{3i}, t_{4i})$ satisfying $f(t_i) = g_{ii}$ and form all possible matrices $T = (t_{ij})$. f and g are in the same class if and only if $T' FT = G$ for at least one such T. We remark that if $F = G$, one need only count the number of distinct such matrices T satisfying $T' FT = F$ to find the number of automorphs of F (a result which is listed in the tables for each class).

4. **Separation into Genera.** Unless otherwise noted, we will use the notation of O'Meara's book [22], and we assume that the reader is familiar with O'Meara's development.

Let V be the rational quadratic space associated with an integral positive definite quaternary quadratic form, and let V_p be the localization of V at a rational prime p. If we take a splitting

$$V_p = [\alpha_1] \perp [\alpha_2] \perp [\alpha_3] \perp [\alpha_4]$$

then the Hasse symbol is defined by

$$S_p(V_p) = \prod_{1 \leq i \leq j \leq 4} (\alpha_i, \alpha_j)$$

where (α_i, α_j) is the Hilbert symbol at p.

From [22;63:20] the local quadratic spaces associated with two integral positive quaternary forms with the same discriminant d are isometric if they

have the same Hasse symbols at p. It follows from the Hasse-Minkowski
Theorem that if the Hasse symbols agree at each $p|2d$ (information listed
in the tables), then the two forms are in the same (or isometric) rational
quadratic spaces. Since it is necessary for two such forms to be in the same
quadratic space if they are to be in the same genus, determination of the
Hasse symbol at each $p|2d$ yields a preliminary separation.

Two integral forms in the same rational quadratic space are in the same
genus if they are locally in the same class for each prime p. By [31;p.69] we
need only test this condition for primes p dividing the discriminant d. For
this finite set of primes, we use the conditions of [22;92:2] if $p{\neq}2$ and
[22;93:29] if $p=2$. A brief explanation of these procedures follows.

If L_p is the localization at the prime p of the lattice L corresponding
to an integral positive quaternary form, a Jordan splitting for L_p is a
decomposition

$$L_p = L_p^1 \perp L_p^2 \perp \ldots \perp L_p^t$$

where $1{\leq}t{\leq}4$, L_p^i is an orthogonal sum of one- or two-dimensional modular
sublattices of $L_p (1{\leq}i{\leq}4)$ and

$$sL_p^1 \supset sL_p^2 \supset \ldots \supset sL_p^t.$$

Here sL_p^i is the scale of the sublattice L_p^i. The procedure for actually
computing a Jordan splitting rests heavily on the Gram- Schmidt process and
is thoroughly described in [22;section 94]. If

$$L_p^{(i)} = L_p^1 \perp \ldots \perp L_p^i,$$

then

$$L_p^{(1)} \subset L_p^{(2)} \subset \ldots \subset L_p^{(t)}$$

is called the Jordan chain associated with the given splitting.

In the case $p=2$ we compute the Jordan splitting

$$L_2 = L_2^1 \perp \ldots \perp L_2^t$$

and the quantities t, dim $L_2^{(i)}, sL_2^{(i)} = s_i$, and $nL_2^i = n_i$ (the norm of L_2^i), the Jordan
invariants of L_2. We can determine when two forms with the same Jordan
invariants are in the same 2-adic class by applying the following theorem
[22;93:29]:

Theorem. Let L and K be lattices on the same regular quadratic space over
the rational numbers, and suppose that L_2 and K_2 have the same Jordan
invariants. Consider the Jordan chains for L_2 and K_2

$$L_2^{(1)} \subset L_2^{(2)} \subset \ldots \subset L_2^{(t)} \text{ and } K_2^{(1)} \subset K_2^{(2)} \subset \ldots \subset K_2^{(t)}.$$

Then $\mathrm{cls}L_2 = \mathrm{cls}K_2$ if and only if the following conditions hold:

(a) $dL_2^{(i)}/dK_2^{(i)} \cong 1 \mod n_i n_{i+1}/s_i^2$ for $1 \le i \le t-1$,

(b) For $1 \le i \le t-1$, $FL_2^{(i)} \twoheadrightarrow FK_2^{(i)} \perp <2^{u_i}>$ when $n_{i+1} \subseteq 4n_i$.

Here, $dL_2^{(i)}/dK_2^{(i)}$ is the quotient of the determinants of coefficient matrices for forms corresponding to the appropriate lattices, and, given the modularity of the components in the Jordan splitting, this value is a 2-adic unit. Condition (a) reduces to determining when

$$dL_2^{(i)}/dK_2^{(i)} \equiv 1 \mod n_i n_{i+1}/s_i^2$$

for $n_i n_{i+1}/s_i^2 = 4$ or 8 and $1 \le i \le t-1$. When appropriate in the context, n_i and s_i will be understood to be the power of 2 which generates the ideal.

From [22; 63:20 and 63:21], it is easy to see that condition (b) holds if and only if

$$S_2(FL_2^{(i)})(n_i dK_2^{(i)}, -dL_2^{(i)})_2 = S_2(FK_2^{(i)})(n_i, -dK_2^{(i)})_2$$

for $1 \le i \le t-1$ when $n_{i+1} \ge 4n_i$. This condition, involving Hasse and Hilbert symbols, is easily checked on the computer.

In the case $p \ne 2$, $n_i = s_i$ and the situation is somewhat less complicated. For $p \ne 2$, two integral lattices on the same regular rational quadratic space and with the same Jordan invariants are in the same p-adic class if and only if the discriminants (in O'Meara's sense) of corresponding components in their Jordan splittings are equal [22;92:2].

As an example, we consider the forms

$$f = x_1^2 + x_2^2 + x_3^2 + 8x_4^2 + x_1x_2 + x_1x_3 \quad \text{and}$$

$$g = x_1^2 + x_2^2 + 2x_3^2 + 3x_4^2 + x_1x_2 + x_1x_4$$

of discriminant 64 representing classes in genus #2 and genus #5 of the tables. As noted in the appendix, f and g have the 2-adic Jordan splittings

$$[A] + [2/3] + [8] \quad \text{and} \quad [A] + [2] + [8/3] ,$$

respectively. So f is 2-adically equivalent to

$$x_1^2 + x_1 x_2 + x_2^2 + 6x_3^2 + 8x_4^2,$$

and g is 2-adically equivalent to

$$x_1^2 + x_1 x_2 + x_2^2 + 2x_3^2 + 24x_4^2,$$

The Jordan invariants are the same; so we test condition (a) above. This condition holds for $i=1$. For $i=2$, $s_2 = n_2 = 2$ and $n_3 = 8$. Since

$$n_2 n_3 / s_2^2 = 4 \quad \text{and} \quad dL_2^{(2)} / dK_2^{(2)} = 2/(2/3) = 3,$$

condition (a) is violated for $i=2$, and the forms are not in the same genus.

In [28], Townes asserts that the forms

$$f = x_1^2 + x_2^2 + 4x_3^2 + 4x_4^2 \quad \text{and}$$
$$g = 2x_1^2 + 2x_2^2 + 3x_3^2 + 3x_4^2 + 2x_1 x_3 + 2x_2 x_3 + 2x_1 x_4 + 2x_2 x_4 + 2x_3 x_4$$

both of discriminant 256 and inequivalent, are in the same genus. From the appendix, f and g have the respective 2-adic Jordan splittings

$$[(1) + (1)] + [(4) + (4)] \quad \text{and} \quad [(3) + (5/3)] + [12/5) + (4/3)].$$

Thus f is already a diagonal form, while g is 2-adically equivalent to

$$3x_1^2 + 15x_2^2 + 60x_3^2 + 12x_4^2$$

The Jordan invariants are the same, and, since $n_1 = s_1 = 1$, $n_2 = 4$, and

$$dL_2^{(1)} / dK_2^{(1)} = 5,$$

condition (a) holds. But, since the 2-adic Hasse symbol for the rational quadratic space with form

$$3x_1^2 + 15x_2^2$$

is -1, condition (b) fails to hold. These two forms are therefore in separate genera.

To illustrate the use of the p-adic case with $p \neq 2$, consider the forms

$$f = x_1^2 + x_2^2 + x_3^2 + 5x_4^2 + x_1 x_2 + x_1 x_3 + x_1 x_4 \quad \text{and}$$
$$g = x_1^2 + x_2^2 + x_3^2 + 3x_4^2 + x_1 x_4 + x_2 x_4 + x_3 x_4$$

both of discriminant 36, inequivalent and having the same Hasse symbols for $p=2$ and 3. Both f and g have 2-adic Jordan splittings equivalent to

$$[A] + [2A];$$

so they are 2-adically equivalent. For $p=3$, they have respective Jordan splittings

$$[(1)+(19/4)+(14/19)]+[9/14] \quad \text{and}$$

$$[(1)+(1)+(1)]+[9/4] .$$

Clearly, the Jordan invariants are the same. The discriminants (in O'Meara's sense) of the first components of the Jordan splittings are 7/2 and 1, respectively. Since these discriminants are defined only up to squares of units, we may use the values 14 and 1 instead, and, since 14 is not a 3-adic square, these discriminants are not equal. Hence, the two forms are not 3-adically equivalent and are in different genera.

Once the classes of forms have been grouped into genera by these procedures, separation into spinor genera is almost always trivial. From [8] or [31;p. 111] the genus and spinor genus will coincide in these tables whenever d is not divisible by the sixth power of any integer greater than 2. Therefore in our case (with $d \leq 1732$), the spinor genus and genus coincide except possibly for $d=729$. For $d=729$ it is well-known (see [31;p. 114] or [21;section 6]) that genus #5 in our tables with three classes has more than one spinor genus. Specifically the form

$$x_1^2+3x_2^2+3x_3^2+7x_4^2+3x_2x_3+x_1x_4$$

is in a spinor genus of one class. Since the number g of spinor genera in a genus must be a power of 2 [22;102:8], the other two forms in this genus are in the same spinor genus. In all other cases for $d=729$, the genus and spinor genus coincide.

To verify the latter statement, we consider the mass of a spinor genus (equal to the sum of the reciprocals of the number of automorphisms of a representative form, one from each class in the spinor genus). It is known that the masses of any two spinor genera in a genus of positive integral quaternary forms must be equal [17]. For $d=729$, the mass of genus #2 in our tables is 27/16. If the number g of spinor genera in this genus is 2, then the mass of each spinor genus must be 27/32; this is impossible since no combination of sums of the numerators of 2/32, 4/32, 8/32, and 16/32 (reciprocals of the numbers of automorphisms for the classes in this genus) can be odd. Similarly $g \neq 4$. With the exception, of course, of genus #5, analogous arguments show that $g=1$ for all other genera in this discriminant.

5. **Mass Computations.** For a positive quaternary form f with matrix F as in section 2, let $A = 2F$ (so that $d = 16(\det F) = \det A$). Following Pall [24] and Watson [32], we define the local (p-adic) density of f for prime p to be

$$\alpha_p = \frac{\nu_p}{2p^{6r}}$$

where ν_p is the number of integral 4x4 matrices X which satisfy

$$X' AX \equiv A(\text{mod } p^r) .$$

Siegel proved that for large enough values of the positive integer r, α_p is independent of r. The mass of the genus of f is

$$M = M(f) = \frac{d^{5/2}}{\pi^4 \cdot \prod_p \alpha_p} \tag{1}$$

where the product is taken over all prime integers p. This formula follows easily from Siegel's basic result [27;p. 568].

For $p = 2$, determination of the local density was based on Watson's paper [32], computing his 2-adic canonical form and, from it, the various values needed in the calculation. While the many details are very tedious and, in themselves, unrewarding, Watson's algorithm is remarkably precise and effective. To compute the correct value of α_2, the interested reader need only do as the present author did - slavishly and stubbornly follow Watson's guidelines. The resulting computer-calculated values are listed in the appendix. Conway and Sloane use a somewhat different slant in [5], an article which also contains a history of the development of the mass formula and references to the pertinent literature.

The local densities α_p for $p \neq 2$ were computed via Pall's results [24;p. 101], noting Watson's correction [32;p. 94]. These computations are somewhat less complicated than those for $p = 2$, and the simplest cases are for p not dividing d. Here

$$\begin{aligned}
\alpha_p &= (1 - p^{-2})^2 \quad \text{if } (d \mid p) = 1 \\
\\
\alpha_p &= 1 - p^{-4} \quad \text{if } (d \mid p) = -1
\end{aligned} \tag{2}$$

for $p \neq 2$ and p not dividing d, where $(d \mid p)$ is the Legendre symbol.

If $p \neq 2$ and $p \mid d$, a p-adic Jordan splitting with t components can be written in the form

$$p^{e(1)}\phi_1 + p^{e(2)}\phi_2 + \ldots + p^{e(t)}\phi_t$$

where, for $i=1,2,...,t$, ϕ_i is a diagonal form in r_i variables with determinant (product of the diagonal entries) d_i, $p^{e(i)}$ is the highest power of p that can be factored from the i-th component, and $0=e(1)<e(2)< ... <e(t)$. For $i=1,2,...,t$, let

$$m_i = r_i + r_{i+1} + .. . + r_t$$

$$v = \sum_{i=1}^{t-1} e(i)r_i[m_i - (r_i - 1)/2] + e(t)r_t(r_t+1)/2$$

$$\alpha_p^{(i)} = 1 \quad \text{if} \quad r_i=1$$

$$\alpha_p^{(i)} = 1 - (d_i\mid p)p^{-1} \quad \text{if} \quad r_i=2$$

$$\alpha_p^{(i)} = 1 - p^{-2} \quad \text{if} \quad r_i=3$$

$$\alpha_p^{(i)} = (1 - p^{-2})(1 - (d_i\mid p)p^{-2}) \quad \text{if} \quad r_i=4 .$$

Then the p-adic density is

$$\alpha_p = 2^{t-1}p^v\alpha_p^{(1)}\alpha_p^{(2)}...\alpha_p^{(t)}$$

As an example, consider the form of discriminant 25

$$f = x_1^2 + x_2^2 + 2x_3^2 + 2x_4^2 + x_1x_2 + x_1x_3 + x_1x_4 + x_2x_4 + 2x_3x_4$$

A 5-adic Jordan splitting for f is

$$x_1^2 + (3/4)x_2^2 + (5/3)x_3^2 + (5/4)x_4^2 ,$$

5-adically equivalent to

$$x_1^2 + 3x_2^2 + 5(3x_3^2 + x_4^2) .$$

Here, $t=2$, $\phi_1 = x_1^2 + 3x_2^2$, $\phi_2 = 3x_3^2 + x_4^2$, $e(1)=0$, $e(2)=1$, $r_1=r_2=2$, $d_1=d_2=3$, $v=0+3=3$,

$$\overset{(1)}{\alpha_5} = 1- (-3\,|\,5)\,(1/5) = 1+ 1/5 = 6/5$$

$$\overset{(2)}{\alpha_5} = 6/5 , \quad \text{and}$$

$$\alpha_5 = 2^{(2-1)}(5^3)(6/5)(6/5)= 360.$$

From the appendix, $\alpha_2 = 9$. If $p \neq 2$ and $p \neq 5$,

$$\alpha_p = (1 - p^{-2})^2 ,$$

and since

$$\prod_p (1 - p^{-2})^2 = \zeta(2)^{-2} = \frac{36}{\pi^4} ,$$

then

$$\prod_p \alpha_p = (2^3)(3^2)(5^5)/\pi^4$$

Therefore,

$$M(f) = \frac{(25)^{5/2}}{(\pi^4)(2^3)(3^2)(5^5)/\pi^4} = \frac{1}{(2^3)(3^2)} = \frac{1}{72} .$$

Since f is in a genus of one class with 72 automorphs, this verifies the result in the tables.

Let $w(g)$ be the number of automorphs of the positive quaternary form g (or, equivalently, of its class). If the genus of such a form contains h classes, represented by the forms $g_1, g_2, \ldots, g_h$, a key check on our tables follows from the fact that

$$M(f) = 1/w(g_1) + 1/w(g_2) + \ldots + 1/w(g_h). \tag{3}$$

Since we can compute the number of automorphs of a form, the use of two independent methods of calculating the mass of a genus yields a test for the number of classes in the genus of each form and, hence, a check on the accuracy of our tables.

The mass formula (1) involves an infinite product which was truncated in our actual calculations. In order to deal with approximation error, let

$$M' = M'(f) = \frac{d^{5/2}}{\pi^4 \cdot \prod_{p \times p(k)} \alpha_p}$$

where, for fixed k, $p(k)$ is the k-th prime.
From (2), $\alpha_p \geq (1 - p^{-2})^2$ for large enough primes p and

$$\prod_p (1 - p^{-2})^2 = \zeta(2)^{-2} = \frac{36}{\pi^4}$$

implies that

$$0 \leq M - M' = M' \left(\frac{1}{\prod_{p > p(k)} \alpha_p} - 1 \right) \leq M \left(\frac{\pi^4}{36} \cdot \prod_{p \times p(k)} (1 - p^{-2})^2 - 1 \right)$$

It follows from a result of Minkowski [20;p.216,v.1] that $w(g) \leq 5760$ for positive quaternary forms g. Noting that the mass of any such form with $d \leq 1000$ is no greater than 22 (the largest value occurs for $d = 937$), it is a simple computation to see that if we make our choice of $k = 2600$ we can be assured that $M-M'$ will be less than $1/5760$ (in most cases this holds for much smaller k). We remark that the smaller the mass, the more accurate is the approximation M' (with $k = 2600$) to the actual value of M. For those relatively few discriminants greater than 1000 with mass greater than 22, an appropriately larger choice of k was made. In all cases the computer found that M' and M (computed as in (3)) differed by less than $1/5760$, so that no genus could be missing a class. The mass column in the tables is the actual value of the mass, computed as in (3) and written as a fraction.

6. **Further Remarks.** In pre-publication form, these tables have served as a starting point in establishing several results. In [11], Estes and Nipp began in the tables with the maximal quaternion orders whose class and spinor genus coincide. Then, using a computer algorithm based on a result of Pall [23], they found all quaternion orders which admit certain natural factorizations, including those in a spinor genus of one class. In [9], Earnest and Nipp found that the class number is one for all the quaternary genera for which the associated theta series lie in spaces of modular forms where the dimension of the subspace of cusp forms is zero. This could be checked from the tables for small discriminants, but for larger discriminants a minor variant of Pall's algorithm was used to generate the necessary classes from the known ones of smaller discriminant. This technique can serve as an alternative means of generating sublattices from the maximal lattices containing them and has been used to verify and extend the tables for certain discriminants.

REFERENCES

[1] E. Borisow, *Reduction of positive ternary quadratic forms by Selling's method, with a table of the reduced forms for all determinants from 1 to 200*, St. Petersburg, 1890. (Russian)

[2] Brandt-Intrau, *Tabellen reduzierter, positiver ternaerer quadratischer Formen*, Akademie Verlag, Berlin, 1958.

[3] J.W.S. Cassels, *Rational quadratic forms*, Academic Press, New York, 1978.

[4] L. Charve, "Table des formes quadratiques quaternaires positives reduites dout le determinant est egal ou inferieur a 20", *Comptes Rendus*, Academie des Sciences, Paris, 96 (1883), 773-775.

[5] J.H. Conway and N.J.A. Sloane, "Low-dimensional lattices. V. The mass formula", *Proc. R. Soc. Lond.* A 419 (1988), 259-286.

[6] L.E. Dickson, *History of the theory of numbers*, vol.3, Carnegie Institution of Washington, 256 (1919); Chelsea, New York, 1966.

[7] L.E. Dickson, *Studies in the theory of numbers*, Chelsea, New York, 1930.

[8] A.G. Earnest and J.S. Hsia, "Spinor norms of local integral rotations", II, *Pacific J. Math.* 61 (1975), 71-86; correction, ibid 115 (1984), 493-494.

[9] A.G. Earnest and G. Nipp, "On the theta series of positive quaternary quadratic forms", *C.R. Math. Rep. Acad. Sci. Canada* XIII (1991), 33-38.

[10] G. Eisenstein, "Tabelle der reducirten positiven quadratischen Former nebst den Resultaten neuerer Forschungen", *J. reine angew. Math.* 41 (1851), 141-189.

[11] D.R. Estes and G. Nipp, "Factorization in quaternion orders", *J. Number Theory* 33 (1989), 224-236.

[12] W. Furter, *Tabelle der inaequivalenten, reduzierten quarternaeren quadratischen Formen mit Diskriminante kleiner oder gleich 128*, Diss. Uni. Zurich, 1973.

[13] K. Germann, "Tabellen reduzierter, positiver quaternaerer quadratischer Formen", *Comment. Math. Helv.* 38 (1963), 56-83.

[14] J.S. Hsia, "Regular positive ternary quadratic forms", *Mathematika 28* (1981), 231-238.

[15] J.S. Hsia and D.C. Hung, "Theta series of ternary and quaternary quadratic forms", *Invent. math.* 73 (1983), 151-156.

[16] B.W. Jones, "A table of Eisenstein-reduced positive ternary quadratic forms of determinant < 200", *National Research Council Bulletin* 97 (1935), 1-51.

[17] M. Kneser, "Darstellungsmasse indefiniter quadratischer Formen", *Math. Zeitschr.* 77 (1961), 188-194.

[18] J. Kramer, "On the linear independence of certain theta-series", *Math. Ann.* 281 (1988), 219-228.

[19] J.L. Lagrange, *Oeuvres* III, 693-758.

[20] H. Minkowski, *Gesammelte Abhandlungen*, Chelsea, New York, 1967.

[21] G. Nipp, "The spinor genus of quaternion orders", *Trans. Amer. Math. Soc.* 211 (1975), 299-309.

[22] O. T. O'Meara, *Introduction to quadratic forms*, Die Grundlehren der math. Wissenschaften, Band 117, Academic Press, New York; Springer-Verlag, Berlin, 1963.

[23] G. Pall, "On generalized quaternions", *Trans. Amer. Math. Soc.* 59 (1946), 280-332.

[24] G. Pall, "The weight of a genus of positive n-ary quadratic forms", *Proc. Sympos. Pure Math. vol. 8* (Amer. Math. Soc., Providence, R.I., 1965), 95-105.

[25] A. Schiemann, "Ein Beispiel positiv definiter quadratischer Formen der Dimension 4 mit gleichen Darstellungszahlen", *Arch. Math.* 54 (1990), 372-375.

[26] L.A. Seeber, *Untersuchungen uber die Eigenschaften der positiven ternaeren quadratischen Formen*, Freiburg, 1831.

[27] C. L. Siegel, "Ueber die analytische Theorie der quadratischen Formen", *Ann. of Math.* 36 (1935), 527-606.

[28] S.B. Townes, "Table of reduced positive quaternary quadratic forms", *Ann. of Math.* 41 (1940), 57-58.

[29] B. L. van der Waerden, "Die Reduktionstheorie der positiven quadratischen Formen", *Acta Math.* 96(1956), 265-309.

[30] B. L. van der Waerden and H. Gross, *Studien zur Theorie der quadratischen Formen*, Birkhaeuser Verlag, Basel and Stuttgart, 1968.

[31] G.L. Watson, *Integral quadratic forms*, Cambridge University Press, Cambridge, 1960.

[32] G. L. Watson, "The 2-adic density of a quadratic form", *Mathematika* 23 (1976), 94-106.

Tables of Reduced Positive Integral Quaternary Quadratic Forms with Discriminants Through 500 and of Discriminants 729 and 1729

```
D=    4
GENUS#: F11 F22 F33 F44 F12 F13 F23 F14 F24 F34: HASSE SYM   LEVEL AUTOS    MASS
    1:   1   1   1   1   0   0   0   1   1   1:  1              2    1152   1/1152

D=    5
GENUS#: F11 F22 F33 F44 F12 F13 F23 F14 F24 F34: HASSE SYM   LEVEL AUTOS    MASS
    1:   1   1   1   1   1   0   0   1   0   1:  -1-1           5     240   1/240

D=    8
GENUS#: F11 F22 F33 F44 F12 F13 F23 F14 F24 F34: HASSE SYM   LEVEL AUTOS    MASS
    1:   1   1   1   1   0   0   0   1   1   0:  1              8      96   1/96

D=    9
GENUS#: F11 F22 F33 F44 F12 F13 F23 F14 F24 F34: HASSE SYM   LEVEL AUTOS    MASS
    1:   1   1   1   1   1   0   0   0   0   1:  -1-1           3     288   1/288

D=   12
GENUS#: F11 F22 F33 F44 F12 F13 F23 F14 F24 F34: HASSE SYM   LEVEL AUTOS    MASS
    1:   1   1   1   2   1   1   0   1   0   0:  1 1           12      96   1/96
    2:   1   1   1   1   0   0   0   1   0   0:  -1-1          12      96   1/96

D=   13
GENUS#: F11 F22 F33 F44 F12 F13 F23 F14 F24 F34: HASSE SYM   LEVEL AUTOS    MASS
    1:   1   1   1   2   1   1   0   0   1   0:  -1-1          13      48   1/48

D=   16
GENUS#: F11 F22 F33 F44 F12 F13 F23 F14 F24 F34: HASSE SYM   LEVEL AUTOS    MASS
    1:   1   1   1   1   0   0   0   0   0   0:  1              4     384   1/384
    2:   1   1   1   2   1   1   0   0   0   0:  1              8      96   1/96

D=   17
GENUS#: F11 F22 F33 F44 F12 F13 F23 F14 F24 F34: HASSE SYM   LEVEL AUTOS    MASS
    1:   1   1   1   2   1   0   0   1   0   1:  -1-1          17      24   1/24

D=   20
GENUS#: F11 F22 F33 F44 F12 F13 F23 F14 F24 F34: HASSE SYM   LEVEL AUTOS    MASS
    1:   1   1   1   3   1   1   0   1   0   0:  -1-1          10      96   1/96
    2:   1   1   1   2   0   0   0   1   1   1:  1 1           10      96   1/96
    3:   1   1   1   2   1   0   0   1   0   0:  -1-1          20      24   1/24

D=   21
GENUS#: F11 F22 F33 F44 F12 F13 F23 F14 F24 F34: HASSE SYM   LEVEL AUTOS    MASS
    1:   1   1   1   2   1   0   0   0   0   1:  -1-1 1        21      48   1/48
    2:   1   1   1   3   1   1   0   0   1   0:  -1 1-1        21      48   1/48

D=   24
GENUS#: F11 F22 F33 F44 F12 F13 F23 F14 F24 F34: HASSE SYM   LEVEL AUTOS    MASS
    1:   1   1   1   2   1   0   0   0   0   0:  1 1           24      48   1/32
    1:   1   1   1   3   1   1   0   0   0   0:  1 1           24      96
    2:   1   1   1   2   0   0   0   1   1   0:  -1-1          24      32   1/32

D=   25
GENUS#: F11 F22 F33 F44 F12 F13 F23 F14 F24 F34: HASSE SYM   LEVEL AUTOS    MASS
    1:   1   1   2   2   1   1   0   1   1   2:  -1-1           5      72   1/72

D=   28
GENUS#: F11 F22 F33 F44 F12 F13 F23 F14 F24 F34: HASSE SYM   LEVEL AUTOS    MASS
    1:   1   1   1   2   0   0   0   1   0   0:  -1-1          28      32   1/24
    1:   1   1   1   4   1   1   0   1   0   0:  -1-1          28      96
    2:   1   1   2   2   1   1   0   0   1   1:  1 1           28      24   1/24

D=   29
GENUS#: F11 F22 F33 F44 F12 F13 F23 F14 F24 F34: HASSE SYM   LEVEL AUTOS    MASS
    1:   1   1   1   3   1   0   0   1   0   1:  -1-1          29      24   1/16
    1:   1   1   1   4   1   1   0   0   1   0:  -1-1          29      48

D=   32
GENUS#: F11 F22 F33 F44 F12 F13 F23 F14 F24 F34: HASSE SYM   LEVEL AUTOS    MASS
    1:   1   1   1   2   0   0   0   0   0   0:  1              8      96   1/96
    2:   1   1   1   3   1   0   0   1   0   0:  1             32      24   1/24
    3:   1   1   1   4   1   1   0   0   0   0:  1             16      96   1/96
    4:   1   1   2   2   0   0   0   1   1   2:  1             16      32   1/32
```

										HASSE SYM	LEVEL	AUTOS	MASS
5:	1 1 2 2 1 1 0 1 0 0:	1	32	24	1/24

D= 33

GENUS#:	F11 F22 F33 F44 F12 F13 F23 F14 F24 F34:	HASSE SYM	LEVEL	AUTOS	MASS
1:	1 1 1 3 1 0 0 0 0 1:	-1 1-1	33	48	1/16
1:	1 1 2 2 1 1 0 0 1 0:	-1 1-1	33	24	
2:	1 1 2 2 0 1 1 1 0 2:	-1-1 1	33	16	1/16

D= 36

GENUS#:	F11 F22 F33 F44 F12 F13 F23 F14 F24 F34:	HASSE SYM	LEVEL	AUTOS	MASS
1:	1 1 1 5 1 1 0 1 0 0:	1 1	18	96	1/96
2:	1 1 1 3 0 0 0 1 1 1:	1 1	18	96	1/96
3:	1 1 1 3 1 0 0 0 0 0:	-1-1	12	48	1/48
4:	1 1 2 2 1 0 0 0 0 2:	1 1	6	144	1/144
5:	1 1 2 2 0 1 1 1 1 1:	-1-1	6	64	1/64

D= 37

GENUS#:	F11 F22 F33 F44 F12 F13 F23 F14 F24 F34:	HASSE SYM	LEVEL	AUTOS	MASS
1:	1 1 1 5 1 1 0 0 1 0:	-1-1	37	48	5/48
1:	1 1 2 2 1 0 0 1 0 1:	-1-1	37	12	

D= 40

GENUS#:	F11 F22 F33 F44 F12 F13 F23 F14 F24 F34:	HASSE SYM	LEVEL	AUTOS	MASS
1:	1 1 1 5 1 1 0 0 0 0:	-1-1	40	96	7/96
1:	1 1 2 2 0 0 0 0 1 2:	-1-1	40	16	
2:	1 1 1 3 0 0 0 1 1 0:	1 1	40	32	7/96
2:	1 1 2 2 1 0 0 1 0 0:	1 1	40	24	

D= 41

GENUS#:	F11 F22 F33 F44 F12 F13 F23 F14 F24 F34:	HASSE SYM	LEVEL	AUTOS	MASS
1:	1 1 1 4 1 0 0 1 0 1:	-1-1	41	24	1/6
1:	1 1 2 2 0 1 1 1 0 0:	-1-1	41	8	

D= 44

GENUS#:	F11 F22 F33 F44 F12 F13 F23 F14 F24 F34:	HASSE SYM	LEVEL	AUTOS	MASS
1:	1 1 1 3 0 0 0 1 0 0:	-1-1	44	32	7/96
1:	1 1 1 4 1 0 0 1 0 0:	-1-1	44	24	
2:	1 1 1 6 1 1 0 1 0 0:	1 1	44	96	7/96
2:	1 1 2 2 0 1 1 0 0 1:	1 1	44	16	

D= 45

GENUS#:	F11 F22 F33 F44 F12 F13 F23 F14 F24 F34:	HASSE SYM	LEVEL	AUTOS	MASS
1:	1 1 1 4 1 0 0 0 0 1:	-1 1-1	15	48	1/48
2:	1 1 1 6 1 1 0 0 1 0:	-1 1-1	45	48	1/16
2:	1 1 2 3 1 1 0 1 1 2:	-1 1-1	45	24	
3:	1 1 2 2 1 0 0 0 0 1:	-1-1 1	15	48	1/48
4:	1 1 2 2 0 1 0 0 1 1:	-1 1-1	45	16	1/16

D= 48

GENUS#:	F11 F22 F33 F44 F12 F13 F23 F14 F24 F34:	HASSE SYM	LEVEL	AUTOS	MASS
1:	1 1 1 3 0 0 0 0 0 0:	-1-1	12	96	1/96
2:	1 1 1 4 1 0 0 0 0 0:	-1-1	48	48	1/12
2:	1 1 2 2 0 1 0 1 0 0:	-1-1	48	16	
3:	1 1 1 6 1 1 0 0 0 0:	1 1	24	96	1/96
4:	1 1 2 3 1 1 0 0 1 1:	1 1	48	12	1/12
5:	1 1 2 2 0 1 1 0 0 0:	1 1	24	32	1/32
6:	1 1 2 2 0 0 0 0 0 2:	1 1	12	96	1/96
7:	1 1 2 2 1 0 0 0 0 0:	-1-1	24	96	1/96
8:	1 2 2 2 0 0 0 1 2 2:	-1-1	24	32	1/32

D= 49

GENUS#:	F11 F22 F33 F44 F12 F13 F23 F14 F24 F34:	HASSE SYM	LEVEL	AUTOS	MASS
1:	1 1 2 2 0 1 0 0 1 0:	-1-1	7	32	1/32

D= 52

GENUS#:	F11 F22 F33 F44 F12 F13 F23 F14 F24 F34:	HASSE SYM	LEVEL	AUTOS	MASS
1:	1 1 1 4 0 0 0 1 1 1:	1 1	26	96	5/96
1:	1 2 2 2 0 0 2 1 2 0:	1 1	26	24	
2:	1 1 1 7 1 1 0 1 0 0:	-1-1	26	96	5/96
2:	1 1 2 3 1 0 0 1 0 2:	-1-1	26	24	
3:	1 1 2 2 0 1 0 0 0 1:	-1-1	52	8	5/24
3:	1 1 2 3 1 1 0 1 0 0:	-1-1	52	12	

```
D=  53
GENUS#: F11 F22 F33 F44 F12 F13 F23 F14 F24 F34: HASSE SYM   LEVEL AUTOS    MASS
    1:   1   1   1   5   1   0   0   1   0   1: -1-1            53    24    7/48
    1:   1   1   1   7   1   1   0   0   1   0: -1-1            53    48
    1:   1   1   2   3   1   1   0   0   1   0: -1-1            53    12

D=  56
GENUS#: F11 F22 F33 F44 F12 F13 F23 F14 F24 F34: HASSE SYM   LEVEL AUTOS    MASS
    1:   1   1   1   5   1   0   0   1   0   0:  1 1            56    24    5/48
    1:   1   1   2   3   0   1   1   1   1   0:  1 1            56    16
    2:   1   1   1   4   0   0   0   1   1   0: -1-1            56    32    5/48
    2:   1   1   1   7   1   1   0   0   0   0: -1-1            56    96
    2:   1   1   2   2   0   0   0   1   1   0: -1-1            56    16

D=  57
GENUS#: F11 F22 F33 F44 F12 F13 F23 F14 F24 F34: HASSE SYM   LEVEL AUTOS    MASS
    1:   1   1   1   5   1   0   0   0   0   1: -1-1 1          57    48    7/48
    1:   1   2   2   2   0   0   1   1   2   2: -1-1 1          57     8
    2:   1   1   2   3   0   1   1   1   0   2: -1 1-1          57    16    7/48
    2:   1   1   2   3   1   1   0   0   0   1: -1 1-1          57    12

D=  60
GENUS#: F11 F22 F33 F44 F12 F13 F23 F14 F24 F34: HASSE SYM   LEVEL AUTOS    MASS
    1:   1   1   1   5   1   0   0   0   0   0: -1 1-1          60    48    1/16
    1:   1   1   1   8   1   1   0   1   0   0: -1 1-1          60    96
    1:   1   1   2   2   0   0   0   0   0   1: -1 1-1          60    32
    2:   1   1   1   4   0   0   0   1   0   0: -1-1 1          60    32    1/16
    2:   1   1   2   3   0   1   1   1   1   1: -1-1 1          60    32
    3:   1   1   2   3   1   0   0   0   0   2:  1 1 1          60    48    1/16
    3:   1   1   2   3   1   1   0   0   0   0:  1 1 1          60    24
    4:   1   2   2   2   1   0   2   0   1   2:  1-1-1          60    16    1/16

D=  61
GENUS#: F11 F22 F33 F44 F12 F13 F23 F14 F24 F34: HASSE SYM   LEVEL AUTOS    MASS
    1:   1   1   1   8   1   1   0   0   1   0: -1-1            61    48   11/48
    1:   1   1   2   3   0   1   0   0   1   2: -1-1            61     8
    1:   1   1   2   3   1   0   0   1   1   1: -1-1            61    12

D=  64
GENUS#: F11 F22 F33 F44 F12 F13 F23 F14 F24 F34: HASSE SYM   LEVEL AUTOS    MASS
    1:   1   1   1   4   0   0   0   0   0   0:  1              16    96    1/96
    2:   1   1   1   8   1   1   0   0   0   0:  1              32    96    1/24
    2:   1   1   2   3   0   0   0   1   1   2:  1              32    32
    3:   1   1   2   2   0   0   0   0   0   0:  1               8    64    1/64
    4:   1   1   3   3   1   1   0   1   1   3:  1               8    72    1/72
    5:   1   1   2   3   1   0   0   1   0   0:  1              32    24    1/24
    6:   1   2   2   2   0   0   2   0   2   0:  1              16    96    1/96

D=  65
GENUS#: F11 F22 F33 F44 F12 F13 F23 F14 F24 F34: HASSE SYM   LEVEL AUTOS    MASS
    1:   1   1   1   6   1   0   0   1   0   1: -1-1 1          65    24    1/6
    1:   1   2   2   2   1   0   1   1   1   2: -1-1 1          65     8
    2:   1   1   2   3   0   1   1   0   1   0: -1 1-1          65     8    1/6
    2:   1   1   2   4   1   1   0   1   1   2: -1 1-1          65    24

D=  68
GENUS#: F11 F22 F33 F44 F12 F13 F23 F14 F24 F34: HASSE SYM   LEVEL AUTOS    MASS
    1:   1   1   1   5   0   0   0   1   1   1:  1 1            34    96    1/48
    1:   1   1   1   9   1   1   0   1   0   0:  1 1            34    96
    2:   1   1   1   6   1   0   0   1   0   0: -1-1            68    24    1/4
    2:   1   1   2   3   0   1   0   0   0   2: -1-1            68     8
    2:   1   1   2   4   1   1   0   0   1   1: -1-1            68    12
    3:   1   1   2   3   0   1   1   0   0   1: -1-1            34    16    3/16
    3:   1   2   2   2   1   1   0   0   0   2: -1-1            34     8

D=  69
GENUS#: F11 F22 F33 F44 F12 F13 F23 F14 F24 F34: HASSE SYM   LEVEL AUTOS    MASS
    1:   1   1   1   6   1   0   0   0   0   1: -1 1-1          69    48    1/8
    1:   1   1   1   9   1   1   0   0   1   0: -1 1-1          69    48
    1:   1   1   2   3   1   0   0   0   0   1: -1 1-1          69    24
    1:   1   1   3   3   1   1   0   0   1   2: -1 1-1          69    24
    2:   1   1   2   3   0   1   0   1   1   0: -1-1 1          69     8    1/8
```

```
D=  72
GENUS#: F11 F22 F33 F44 F12 F13 F23 F14 F24 F34: HASSE SYM  LEVEL AUTOS   MASS
    1:   1   1   1   6   1   0   0   0   0   0:  1 1          24    48    5/96
    1:   1   1   2   3   0   1   1   0   0   0:  1 1          24    32
    2:   1   1   1   9   1   1   0   0   0   0:  1 1          72    96    5/32
    2:   1   1   2   3   0   0   0   1   0   2:  1 1          72    16
    2:   1   1   2   4   1   1   0   1   0   0:  1 1          72    12
    3:   1   1   1   5   0   0   0   1   1   0:  1 1          72    32    5/32
    3:   1   2   2   2   1   0   2   0   1   0:  1 1          72     8
    4:   1   1   2   3   1   0   0   0   0   0: -1-1          24    48    5/96
    4:   1   2   2   2   0   0   2   0   1   2: -1-1          24    32

D=  73
GENUS#: F11 F22 F33 F44 F12 F13 F23 F14 F24 F34: HASSE SYM  LEVEL AUTOS   MASS
    1:   1   1   2   3   0   1   0   0   1   0: -1-1          73     8    11/24
    1:   1   1   2   4   1   1   0   0   1   0: -1-1          73    12
    1:   1   2   2   2   1   0   2   0   0   1: -1-1          73     4

D=  76
GENUS#: F11 F22 F33 F44 F12 F13 F23 F14 F24 F34: HASSE SYM  LEVEL AUTOS   MASS
    1:   1   1   1  10   1   1   0   1   0   0:  1 1          76    96    19/96
    1:   1   1   2   3   0   0   0   1   1   1:  1 1          76    16
    1:   1   2   2   2   1   0   0   0   1   2:  1 1          76     8
    2:   1   1   1   5   0   0   0   0   0   1: -1-1          76    32    19/96
    2:   1   1   2   3   0   1   0   1   0   0: -1-1          76     8
    2:   1   1   2   4   1   0   0   1   0   2: -1-1          76    24

D=  77
GENUS#: F11 F22 F33 F44 F12 F13 F23 F14 F24 F34: HASSE SYM  LEVEL AUTOS   MASS
    1:   1   1   1  10   1   1   0   0   1   0: -1-1 1        77    48    1/8
    1:   1   1   2   3   0   1   0   0   1   0: -1-1 1        77    16
    1:   1   1   3   3   1   1   0   1   1   2: -1-1 1        77    24
    2:   1   1   1   7   1   0   0   1   0   1: -1 1-1        77    24    1/8
    2:   1   1   2   4   1   1   0   0   0   1: -1 1-1        77    12

D=  80
GENUS#: F11 F22 F33 F44 F12 F13 F23 F14 F24 F34: HASSE SYM  LEVEL AUTOS   MASS
    1:   1   1   1  10   1   1   0   0   0   0: -1-1          40    96    1/24
    1:   1   1   2   3   0   0   0   1   1   0: -1-1          40    32
    2:   1   1   1   5   0   0   0   0   0   0:  1 1          20    96    1/96
    3:   1   1   1   7   1   0   0   1   0   0: -1-1          80    24    1/6
    3:   1   1   2   3   0   1   0   0   0   1: -1-1          80     8
    4:   1   1   3   3   1   1   0   0   1   1: -1-1          20    24    1/24
    5:   1   1   2   3   0   0   0   0   0   2: -1-1          20    32    1/32
    6:   1   1   2   4   1   1   0   0   0   0: -1-1          80    24    1/6
    6:   1   2   2   2   1   0   1   1   0   1: -1-1          80     8
    7:   1   1   2   4   0   1   1   1   1   0:  1 1          40    16    1/8
    7:   1   2   2   2   0   0   0   1   2   0:  1 1          40    16
    8:   1   2   2   2   1   1   0   1   0   0: -1-1          20    24    1/24

D=  81
GENUS#: F11 F22 F33 F44 F12 F13 F23 F14 F24 F34: HASSE SYM  LEVEL AUTOS   MASS
    1:   1   1   1   7   1   0   0   0   0   1: -1-1          27    48    1/16
    1:   1   2   2   2   1   1   0   1   1   0: -1-1          27    24
    2:   1   1   2   4   0   1   1   1   0   2: -1-1          27    16    1/16
    3:   1   1   3   3   1   0   0   0   0   3: -1-1           9   144    1/144
    4:   2   2   2   2   1   1   1   2   2  -1: -1-1           9   144    1/144

D=  84
GENUS#: F11 F22 F33 F44 F12 F13 F23 F14 F24 F34: HASSE SYM  LEVEL AUTOS   MASS
    1:   1   1   1  11   1   1   0   1   0   0: -1 1-1        42    96    5/96
    1:   1   1   3   3   1   1   0   1   0   0: -1 1-1        42    24
    2:   1   1   1   6   0   0   0   1   1   1: -1-1 1        42    96    5/96
    2:   2   2   2   2   1   0   0   2   2   2:  1-1-1        42    24
    3:   1   1   1   7   1   0   0   0   0   0: -1-1 1        84    48    5/24
    3:   1   1   2   3   0   1   0   0   0   0: -1-1 1        84    16
    3:   1   2   2   2   1   1   0   0  -1   1: -1-1 1        84     8
    4:   1   1   2   3   0   0   0   0   1   1: -1 1-1        84     8    5/24
    4:   1   1   3   3   1   1   0   0   0   2: -1-1 1        84    12
    5:   1   1   2   4   0   1   1   1   1   1:  1 1 1        42    32    5/96
    5:   1   2   2   2   0   0   2   1   0   0:  1 1 1        42    48
    6:   1   1   2   4   1   0   0   0   0   2: -1-1 1        42    48    5/96
```

```
        6:   1   1   3   3   0   0   0   1   1   3: -1-1 1      42    32

D=  85
GENUS#: F11 F22 F33 F44 F12 F13 F23 F14 F24 F34: HASSE SYM   LEVEL AUTOS   MASS
    1:   1   1   1  11   1   1   0   0   1   0: -1-1 1        85    48    3/16
    1:   1   1   3   3   1   1   0   0   1   0: -1-1 1        85    24
    1:   1   2   2   2   1   0   1   0  -1   1: -1-1 1        85     8
    2:   1   1   2   4   1   0   0   1   0   1: -1 1-1        85    12    3/16
    2:   1   1   2   5   1   1   0   1   1   2: -1 1-1        85    24
    2:   1   1   3   3   0   1   0   0   1   3: -1 1-1        85    16

D=  88
GENUS#: F11 F22 F33 F44 F12 F13 F23 F14 F24 F34: HASSE SYM   LEVEL AUTOS   MASS
    1:   1   1   1  11   1   1   0   0   0   0:  1 1          88    96   23/96
    1:   1   1   2   3   0   0   0   1   0   0:  1 1          88    16
    1:   1   1   2   4   1   0   0   1   0   0:  1 1          88    24
    1:   1   2   2   3   1   0   2   0   2   0:  1 1          88     8
    2:   1   1   1   6   0   0   0   0   1   1: -1-1          88    32   23/96
    2:   1   1   2   5   1   1   0   0   1   1: -1-1          88    12
    2:   1   2   2   2   0   0   2   0   1   1: -1-1          88     8

D=  89
GENUS#: F11 F22 F33 F44 F12 F13 F23 F14 F24 F34: HASSE SYM   LEVEL AUTOS   MASS
    1:   1   1   1   8   1   0   0   1   0   1: -1-1          89    24   13/24
    1:   1   1   2   4   0   1   0   0   1   2: -1-1          89     8
    1:   1   1   2   4   0   1   1   1   0   0: -1-1          89     8
    1:   1   2   2   2   1   1   0   0   0   1: -1-1          89     4

D=  92
GENUS#: F11 F22 F33 F44 F12 F13 F23 F14 F24 F34: HASSE SYM   LEVEL AUTOS   MASS
    1:   1   1   1  12   1   1   0   1   0   0: -1-1          92    96    5/24
    1:   1   1   1   6   0   0   0   1   0   0: -1-1          92    32
    1:   1   1   1   8   1   0   0   1   0   0: -1-1          92    24
    1:   1   1   2   3   0   0   0   0   0   1: -1-1          92    16
    1:   1   1   2   4   0   1   1   0   0   1: -1-1          92    16
    2:   1   1   2   5   1   1   0   1   0   0:  1 1          92    12    5/24
    2:   1   2   2   3   1   0   2   0   0   2:  1 1          92     8

D=  93
GENUS#: F11 F22 F33 F44 F12 F13 F23 F14 F24 F34: HASSE SYM   LEVEL AUTOS   MASS
    1:   1   1   1  12   1   1   0   0   1   0: -1 1-1        93    48    3/16
    1:   1   1   2   5   1   1   0   0   1   0: -1 1-1        93    12
    1:   1   1   3   3   1   0   0   1   0   1: -1 1-1        93    12
    2:   1   1   1   8   1   0   0   0   0   1: -1-1 1        93    48    3/16
    2:   1   1   2   4   1   0   0   0   0   1: -1-1 1        93    24
    2:   1   2   2   2   1   0   1   0   1   1: -1-1 1        93     8

D=  96
GENUS#: F11 F22 F33 F44 F12 F13 F23 F14 F24 F34: HASSE SYM   LEVEL AUTOS   MASS
    1:   1   1   1  12   1   1   0   0   0   0:  1 1          48    96    1/32
    1:   1   1   2   4   1   0   0   0   0   0:  1 1          48    48
    2:   1   1   1   6   0   0   0   0   0   0: -1-1          24    96    1/32
    2:   1   2   2   2   0   0   0   0   2   0: -1-1          24    48
    3:   1   1   1   8   1   0   0   0   0   0:  1 1          96    48    1/8
    3:   1   1   3   3   0   0   0   0   1   3:  1 1          96    16
    3:   1   1   3   4   1   1   0   1   1   3:  1 1          96    24
    4:   1   1   2   3   0   0   0   0   0   0:  1 1          24    32    1/32
    5:   1   1   3   3   1   0   0   1   0   0:  1 1          96    24    1/8
    5:   1   1   3   3   1   0   0   0   0   2:  1 1          96    48
    5:   1   2   2   3   1   1   0   1   2   2:  1 1          96    16
    6:   1   1   2   4   0   0   0   1   1   2:  1 1          48    32    3/32
    6:   1   2   2   2   1   0   0   1   0   0:  1 1          48    16
    7:   1   1   2   4   0   1   1   0   0   0: -1-1          48    32    3/32
    7:   2   2   2   2   2   1  -1   0   0   2: -1-1          48    16
    8:   1   1   2   4   0   1   0   0   0   2: -1-1          96     8    1/8
    9:   1   1   3   3   0   1   1   1   1   0: -1-1          48    32    1/32
   10:   1   2   2   2   0   0   0   1   1   1: -1-1          96     8    1/8

D=  97
GENUS#: F11 F22 F33 F44 F12 F13 F23 F14 F24 F34: HASSE SYM   LEVEL AUTOS   MASS
    1:   1   1   2   4   0   1   0   1   1   0: -1-1          97     8   17/24
    1:   1   1   2   5   1   1   0   0   0   1: -1-1          97    12
```

GENUS#	F11	F22	F33	F44	F12	F13	F23	F14	F24	F34:	HASSE SYM	LEVEL	AUTOS
1:	1	2	2	2	1	0	1	0	0	1:	−1−1	97	4
1:	1	2	2	3	1	0	2	0	2	1:	−1−1	97	4

D= 100

GENUS#	F11	F22	F33	F44	F12	F13	F23	F14	F24	F34:	HASSE SYM	LEVEL	AUTOS	MASS
1:	1	1	1	13	0	1	0	1	0	0:	1 1	50	96	5/96
1:	1	2	2	3	0	0	2	1	2	0:	1 1	50	24	
2:	1	1	1	7	0	0	0	1	1	1:	1 1	50	96	5/96
2:	1	1	2	5	0	0	1	0	0	2:	1 1	50	24	
3:	1	1	3	3	0	1	1	1	1	1:	1 1	10	64	1/64
4:	1	1	2	5	1	1	0	0	0	0:	−1−1	20	24	1/12
4:	1	2	2	2	0	0	1	0	−1	1:	−1−1	20	24	
5:	1	2	2	3	1	0	2	0	1	2:	−1−1	10	16	1/16

D= 101

GENUS#	F11	F22	F33	F44	F12	F13	F23	F14	F24	F34:	HASSE SYM	LEVEL	AUTOS	MASS
1:	1	1	1	13	1	1	0	0	1	0:	−1−1	101	48	19/48
1:	1	1	1	9	1	0	0	1	0	1:	−1−1	101	24	
1:	1	1	2	4	0	1	0	0	1	1:	−1−1	101	8	
1:	1	1	3	3	0	1	1	1	0	2:	−1−1	101	8	
1:	1	1	3	4	1	1	0	0	1	2:	−1−1	101	12	

D= 104

GENUS#	F11	F22	F33	F44	F12	F13	F23	F14	F24	F34:	HASSE SYM	LEVEL	AUTOS	MASS
1:	1	1	1	13	1	1	0	0	0	0:	−1−1	104	96	25/96
1:	1	1	2	4	0	0	0	0	1	2:	−1−1	104	16	
1:	1	1	2	4	0	1	0	1	0	0:	−1−1	104	8	
1:	1	1	2	5	0	1	1	1	1	0:	−1−1	104	16	
2:	1	1	1	7	0	0	0	1	1	1:	1 1	104	32	25/96
2:	1	1	1	9	1	0	0	1	0	1:	1 1	104	24	
2:	1	1	3	3	0	1	1	0	0	2:	1 1	104	16	
2:	1	2	2	2	0	0	0	1	0	1:	1 1	104	8	

D= 105

GENUS#	F11	F22	F33	F44	F12	F13	F23	F14	F24	F34:	HASSE SYM	LEVEL	AUTOS	MASS
1:	1	1	1	9	1	0	0	0	0	1:	−1 1−1 1	105	48	3/16
1:	1	1	2	6	1	1	0	1	1	2:	−1 1−1 1	105	24	
1:	1	2	2	3	1	0	0	0	2	1:	−1 1−1 1	105	8	
2:	1	1	2	4	0	1	0	0	1	0:	−1−1−1−1	105	16	3/16
2:	2	2	2	2	2	1	0	0	1	2:	−1−1−1−1	105	8	
3:	1	1	2	5	0	1	1	1	0	2:	−1−1 1 1	105	16	3/16
3:	1	2	2	3	1	0	2	1	1	1:	−1−1 1 1	105	8	
4:	1	1	3	3	1	0	0	0	0	1:	−1 1 1−1	105	48	3/16
4:	1	1	3	3	0	1	0	0	1	2:	−1 1 1−1	105	16	
4:	1	1	3	4	1	0	0	1	0	3:	−1 1 1−1	105	24	
4:	1	2	2	2	1	0	0	0	0	1:	−1 1 1−1	105	16	

D= 108

GENUS#	F11	F22	F33	F44	F12	F13	F23	F14	F24	F34:	HASSE SYM	LEVEL	AUTOS	MASS
1:	1	1	1	14	1	1	0	1	0	0:	1 1	108	96	9/32
1:	1	1	2	4	0	0	0	1	1	1:	1 1	108	16	
1:	1	1	2	6	1	1	0	0	1	1:	1 1	108	12	
1:	1	2	2	3	1	0	0	0	2	2:	1 1	108	8	
2:	1	1	1	9	1	0	0	0	0	0:	−1−1	36	48	1/32
2:	1	1	3	3	0	0	0	0	0	3:	−1−1	36	96	
3:	1	1	2	5	1	0	0	0	0	2:	−1−1	36	48	1/16
3:	1	2	2	2	0	0	1	0	1	1:	−1−1	36	24	
4:	1	1	1	7	0	0	0	0	1	0:	−1−1	108	32	9/32
4:	1	1	2	4	0	1	0	0	0	1:	−1−1	108	8	
4:	1	2	2	3	1	0	2	1	0	0:	−1−1	108	8	
5:	1	1	2	5	0	1	1	1	1	1:	1 1	36	32	1/32
6:	1	1	3	3	1	0	0	0	0	0:	1 1	12	96	1/96
7:	1	2	2	3	0	0	2	−1	1	−1:	1 1	36	16	1/16
8:	2	2	2	2	2	1	−1	−1	1	−1:	−1−1	12	96	1/96

D= 109

GENUS#	F11	F22	F33	F44	F12	F13	F23	F14	F24	F34:	HASSE SYM	LEVEL	AUTOS	MASS
1:	1	1	1	14	1	1	0	0	1	0:	−1−1	109	48	9/16
1:	1	1	2	5	1	0	0	1	0	1:	−1−1	109	12	
1:	1	1	3	3	0	1	1	1	0	0:	−1−1	109	8	
1:	1	1	3	4	1	1	0	1	1	2:	−1−1	109	12	
1:	1	2	2	3	1	0	1	1	0	2:	−1−1	109	4	

D= 112

GENUS#	F11	F22	F33	F44	F12	F13	F23	F14	F24	F34	HASSE SYM	LEVEL	AUTOS	MASS
1:	1	1	1	14	1	1	0	0	0	0:	-1-1	56	96	1/24
1:	1	2	2	3	0	0	0	1	2	2:	-1-1	56	32	
2:	1	1	1	7	0	0	0	0	0	0:	-1-1	28	96	1/24
2:	1	1	2	4	0	0	0	0	0	2:	-1-1	28	32	
3:	1	1	2	5	1	0	0	1	0	0:	1 1	56	24	1/24
4:	1	1	2	6	1	1	0	1	0	0:	1 1	112	12	1/3
4:	1	2	2	3	1	0	1	0	2	2:	1 1	112	4	
5:	1	1	2	4	0	1	0	0	0	0:	-1-1	112	16	1/3
5:	1	1	3	3	0	1	0	1	0	2:	-1-1	112	16	
5:	1	1	3	4	1	1	0	0	1	1:	-1-1	112	12	
5:	1	2	2	2	0	0	1	0	1	0:	-1-1	112	8	
6:	1	1	2	4	0	0	0	1	1	0:	-1-1	56	32	1/8
6:	1	2	2	2	0	0	0	1	0	0:	-1-1	56	32	
6:	2	2	2	2	2	1	0	2	0	0:	-1-1	56	16	
7:	1	2	2	3	0	0	2	0	2	0:	1 1	28	24	1/24
8:	1	2	2	3	1	0	2	0	1	0:	1 1	56	8	1/8

D= 113

GENUS#	F11	F22	F33	F44	F12	F13	F23	F14	F24	F34	HASSE SYM	LEVEL	AUTOS	MASS
1:	1	1	1	10	1	0	0	1	0	1:	-1-1	113	24	3/4
1:	1	1	2	5	0	1	1	0	1	0:	-1-1	113	8	
1:	1	1	2	6	1	1	0	0	1	0:	-1-1	113	12	
1:	1	2	2	3	1	1	0	1	2	1:	-1-1	113	4	
1:	1	2	2	3	1	0	2	0	1	1:	-1-1	113	4	

D= 116

GENUS#	F11	F22	F33	F44	F12	F13	F23	F14	F24	F34	HASSE SYM	LEVEL	AUTOS	MASS
1:	1	1	1	10	1	0	0	1	0	0:	-1-1	116	24	5/8
1:	1	1	2	4	0	0	0	0	1	1:	-1-1	116	8	
1:	1	1	3	3	0	1	0	0	0	2:	-1-1	116	8	
1:	1	1	3	4	1	1	0	1	0	0:	-1-1	116	12	
1:	1	2	2	3	1	1	0	0	2	0:	-1-1	116	4	
2:	1	1	1	15	1	1	0	1	0	0:	-1-1	58	96	5/32
2:	1	1	3	3	0	0	0	1	1	1:	-1-1	58	16	
2:	1	1	3	4	1	1	0	0	0	2:	-1-1	58	12	
3:	1	1	1	8	0	0	0	1	1	1:	1 1	58	96	5/32
3:	1	1	2	5	0	1	1	0	0	1:	1 1	58	16	
3:	2	2	2	2	2	1	0	0	2	0:	1 1	58	12	

D= 117

GENUS#	F11	F22	F33	F44	F12	F13	F23	F14	F24	F34	HASSE SYM	LEVEL	AUTOS	MASS
1:	1	1	1	10	1	0	0	0	0	1:	-1-1 1	39	48	1/24
1:	1	1	3	4	1	0	0	0	0	3:	-1-1 1	39	48	
2:	1	1	1	15	1	1	0	0	1	0:	-1 1-1	117	48	1/4
2:	1	1	2	6	1	1	0	0	0	1:	-1 1-1	117	12	
2:	1	1	3	3	0	1	0	0	1	1:	-1 1-1	117	16	
2:	1	1	3	4	1	1	0	0	1	0:	-1 1-1	117	12	
3:	1	1	2	5	1	0	0	0	0	1:	-1 1-1	39	24	1/6
3:	1	2	2	3	0	0	1	1	2	2:	-1 1-1	39	8	
4:	1	1	2	5	0	1	0	0	1	2:	-1 1-1	117	8	1/4
4:	1	2	2	3	1	0	1	1	1	2:	-1 1-1	117	8	

D= 120

GENUS#	F11	F22	F33	F44	F12	F13	F23	F14	F24	F34	HASSE SYM	LEVEL	AUTOS	MASS
1:	1	1	1	10	1	0	0	0	0	0:	1-1-1	120	48	17/96
1:	1	1	3	3	0	1	1	0	0	0:	1-1-1	120	32	
1:	2	2	2	2	1	1	1	2	2	0:	1-1-1	120	8	
2:	1	1	1	15	1	1	0	0	0	0:	-1 1-1	120	96	17/96
2:	1	2	2	4	0	0	0	0	1	0:	-1 1-1	120	16	
2:	1	1	2	6	1	1	0	0	0	0:	-1 1-1	120	24	
2:	1	2	2	3	0	0	2	0	2	1:	-1 1-1	120	16	
3:	1	1	1	8	0	0	0	0	1	1:	-1-1 1	120	32	17/96
3:	1	1	2	5	1	0	0	0	0	0:	-1-1 1	120	48	
3:	1	2	2	2	0	0	0	1	0	0:	-1-1 1	120	16	
3:	1	2	2	3	1	0	2	0	0	0:	-1-1 1	120	16	
4:	1	1	2	5	0	1	1	0	0	0:	1 1 1	120	32	17/96
4:	1	1	3	3	0	1	0	1	0	0:	1 1 1	120	16	
4:	1	1	3	4	1	0	0	1	1	2:	1 1 1	120	12	

D= 121

GENUS#	F11	F22	F33	F44	F12	F13	F23	F14	F24	F34:	HASSE	SYM	LEVEL	AUTOS	MASS
1:	1	1	3	3	0	1	0	0	1	0:	-1-1		11	32	25/288
1:	1	1	4	4	1	1	0	1	1	4:	-1-1		11	72	
1:	2	2	2	2	2	1	0	1	1	2:	-1-1		11	24	

D= 124

GENUS#	F11	F22	F33	F44	F12	F13	F23	F14	F24	F34:	HASSE	SYM	LEVEL	AUTOS	MASS
1:	1	1	1	16	1	1	0	1	0	0:	-1-1		124	96	5/12
1:	1	1	1	8	0	0	0	0	1	0:	-1-1		124	32	
1:	1	1	2	4	0	0	0	0	0	1:	-1-1		124	16	
1:	1	1	2	5	0	1	0	0	0	2:	-1-1		124	8	
1:	1	1	3	4	0	1	1	0	0	3:	-1-1		124	16	
1:	1	2	2	3	1	0	0	1	0	2:	-1-1		124	8	
2:	1	1	2	6	1	0	0	1	0	2:	1 1		124	24	5/12
2:	1	2	2	3	1	0	1	0	2	0:	1 1		124	4	
2:	1	2	2	3	0	0	2	1	1	1:	1 1		124	8	

D= 125

GENUS#	F11	F22	F33	F44	F12	F13	F23	F14	F24	F34:	HASSE	SYM	LEVEL	AUTOS	MASS
1:	1	1	1	11	1	0	0	1	0	1:	-1-1		125	24	25/48
1:	1	1	1	16	1	1	0	0	1	0:	-1-1		125	48	
1:	1	1	2	5	0	1	0	1	1	0:	-1-1		125	8	
1:	1	1	3	4	1	1	0	0	0	1:	-1-1		125	12	
1:	1	2	2	3	1	0	1	1	2	0:	-1-1		125	4	
2:	1	1	2	7	1	1	0	1	1	2:	-1-1		25	24	1/24
3:	2	2	2	2	1	1	-1	-1	1	1:	-1-1		5	240	1/240
4:	1	1	3	4	0	1	1	1	0	3:	-1-1		25	16	1/16

D= 128

GENUS#	F11	F22	F33	F44	F12	F13	F23	F14	F24	F34:	HASSE	SYM	LEVEL	AUTOS	MASS
1:	1	1	1	11	1	0	0	1	0	0:	1		128	24	1/3
1:	1	1	3	3	0	0	0	0	1	1:	1		128	8	
1:	1	1	4	4	1	1	0	0	1	3:	1		128	24	
1:	1	2	2	3	1	1	0	1	0	0:	1		128	8	
2:	1	1	1	16	1	1	0	0	0	0:	1		64	96	1/24
2:	1	1	2	5	0	0	0	1	1	2:	1		64	32	
3:	1	1	1	8	0	0	0	0	0	0:	1		32	96	1/96
4:	1	1	2	4	0	0	0	0	0	0:	1		16	32	1/32
5:	1	1	2	6	0	1	1	1	0	1:	1		64	16	1/8
5:	2	2	2	2	2	1	0	0	0	2:	1		64	16	
6:	1	1	2	7	1	1	0	0	1	1:	1		128	12	1/3
6:	1	2	2	3	1	0	1	0	0	2:	1		128	4	
7:	1	1	3	4	1	1	0	0	0	0:	1		32	24	1/24
8:	1	1	3	5	1	1	0	1	1	3:	1		64	24	1/24
9:	1	2	2	2	0	0	0	0	0	0:	1		8	96	1/96
10:	1	2	2	4	1	0	2	0	2	0:	1		64	8	1/8
11:	1	1	3	3	0	0	0	0	0	2:	1		32	32	1/32
12:	1	2	2	3	1	1	0	0	1	1:	1		32	8	1/8
13:	1	2	2	3	0	0	0	0	2	2:	1		32	32	1/32
14:	2	2	2	3	2	2	0	2	0	0:	1		32	96	1/96

D= 129

GENUS#	F11	F22	F33	F44	F12	F13	F23	F14	F24	F34:	HASSE	SYM	LEVEL	AUTOS	MASS
1:	1	1	1	11	1	0	0	0	0	1:	-1-1	1	129	48	25/48
1:	1	1	2	5	0	1	0	0	1	1:	-1-1	1	129	8	
1:	1	2	2	3	1	1	0	1	1	1:	-1-1	1	129	8	
1:	2	2	2	2	1	1	1	1	2	2:	-1-1	1	129	4	
2:	1	1	2	6	0	1	1	1	0	2:	-1 1-1		129	16	25/48
2:	1	1	3	4	1	0	0	0	1	3:	-1 1-1		129	8	
2:	1	1	3	4	1	0	0	1	0	1:	-1 1-1		129	12	
2:	1	2	2	3	1	0	0	1	2	1:	-1 1-1		129	4	

D= 132

GENUS#	F11	F22	F33	F44	F12	F13	F23	F14	F24	F34:	HASSE	SYM	LEVEL	AUTOS	MASS
1:	1	1	1	11	1	0	0	0	0	0:	-1 1-1		132	48	3/8
1:	1	1	2	7	1	1	0	1	0	0:	-1 1-1		132	12	
1:	1	1	3	3	0	0	0	0	1	0:	-1 1-1		132	16	
1:	1	1	3	4	1	0	0	0	0	2:	-1 1-1		132	24	
1:	1	1	3	4	1	0	0	1	0	0:	-1 1-1		132	24	
1:	1	2	2	3	0	0	1	0	2	2:	-1 1-1		132	8	
2:	1	1	1	17	1	1	0	1	0	0:	1 1 1		66	96	1/32

	1	1	2	6	1	0	0	0	0	2:	1 1 1	66	48	
2:	1	1	2	6	1	0	0	0	0	2:	1 1 1	66	48	
3:	1	1	1	9	0	0	0	1	1	1:	1-1-1	66	96	1/32
3:	1	2	2	3	0	0	2	1	0	0:	1-1-1	66	48	
4:	1	1	2	5	0	1	0	1	0	0:	-1-1 1	132	8	3/8
4:	1	2	2	3	1	0	1	1	0	1:	-1-1 1	132	4	
5:	1	1	2	6	0	1	1	1	1	1:	-1-1 1	66	32	9/32
5:	1	2	2	4	1	0	2	0	0	2:	-1-1 1	66	8	
5:	2	2	2	2	2	1	-1	0	0	1:	-1-1 1	66	8	
6:	1	1	3	4	0	0	0	1	1	3:	-1 1-1	66	32	9/32
6:	1	2	2	3	1	1	0	0	-1	1:	-1 1-1	66	8	
6:	1	2	2	3	1	0	0	0	1	2:	-1 1-1	66	8	

D= 133

GENUS#:	F11	F22	F33	F44	F12	F13	F23	F14	F24	F34:	HASSE SYM	LEVEL	AUTOS	MASS
1:	1	1	1	17	1	1	0	0	1	0:	-1-1 1	133	48	17/48
1:	1	1	3	5	1	1	0	0	1	2:	-1-1 1	133	12	
1:	1	2	2	3	0	0	1	1	0	2:	-1-1 1	133	4	
2:	1	1	2	5	0	1	0	0	1	0:	-1 1-1	133	16	17/48
2:	1	1	2	6	1	0	0	1	0	1:	-1 1-1	133	12	
2:	1	1	2	7	1	1	0	0	1	0:	-1 1-1	133	12	
2:	1	2	3	3	1	0	2	1	1	3:	-1 1-1	133	8	

D= 136

GENUS#:	F11	F22	F33	F44	F12	F13	F23	F14	F24	F34:	HASSE SYM	LEVEL	AUTOS	MASS
1:	1	1	1	17	1	1	0	0	0	0:	1 1	136	96	23/48
1:	1	1	1	9	0	0	0	0	1	1:	1 1	136	32	
1:	1	1	2	5	0	0	0	1	0	2:	1 1	136	16	
1:	1	2	2	3	0	0	0	1	1	2:	1 1	136	8	
1:	1	2	2	3	1	0	1	1	1	1:	1 1	136	4	
2:	1	1	2	5	0	1	0	0	0	1:	-1-1	136	8	23/48
2:	1	1	2	6	1	0	0	1	0	0:	-1-1	136	24	
2:	1	1	3	4	0	1	1	1	1	0:	-1-1	136	16	
2:	1	2	2	3	0	0	2	0	1	0:	-1-1	136	8	
2:	1	2	2	3	1	0	0	0	2	0:	-1-1	136	8	

D= 137

GENUS#:	F11	F22	F33	F44	F12	F13	F23	F14	F24	F34:	HASSE SYM	LEVEL	AUTOS	MASS
1:	1	1	1	12	1	0	0	1	0	1:	-1-1	137	24	1/1
1:	1	1	2	6	0	1	1	1	0	0:	-1-1	137	8	
1:	1	1	2	7	1	1	0	0	0	1:	-1-1	137	12	
1:	1	2	2	3	1	0	1	0	-1	1:	-1-1	137	4	
1:	1	2	2	3	1	1	0	0	0	1:	-1-1	137	4	
1:	1	2	2	4	1	0	2	0	2	1:	-1-1	137	4	

D= 140

GENUS#:	F11	F22	F33	F44	F12	F13	F23	F14	F24	F34:	HASSE SYM	LEVEL	AUTOS	MASS
1:	1	1	1	12	1	0	0	1	0	0:	-1-1 1	140	24	19/96
1:	1	1	2	5	0	1	0	0	0	0:	-1-1 1	140	16	
1:	1	1	3	3	0	0	0	0	0	1:	-1-1 1	140	32	
1:	1	2	2	4	1	0	2	0	1	2:	-1-1 1	140	16	
2:	1	1	1	18	1	1	0	1	0	0:	1-1-1	140	96	19/96
2:	1	1	2	5	0	0	0	1	1	1:	1-1-1	140	16	
2:	2	2	2	2	1	1	-1	0	0	2:	1-1-1	140	8	
3:	1	1	2	6	0	1	1	0	0	1:	1 1 1	140	16	19/96
3:	1	1	2	7	1	1	0	0	0	0:	1 1 1	140	24	
3:	1	1	3	4	0	1	1	1	1	1:	1 1 1	140	32	
3:	1	2	2	3	1	0	0	0	0	2:	1 1 1	140	16	
4:	1	1	1	9	0	0	0	0	1	0:	-1 1-1	140	32	19/96
4:	1	1	3	4	0	1	0	0	0	3:	-1 1-1	140	8	
4:	1	1	4	4	1	1	0	1	1	3:	-1 1-1	140	24	

D= 141

GENUS#:	F11	F22	F33	F44	F12	F13	F23	F14	F24	F34:	HASSE SYM	LEVEL	AUTOS	MASS
1:	1	1	1	12	1	0	0	0	0	1:	-1 1-1	141	48	3/8
1:	1	1	1	18	1	1	0	0	1	0:	-1 1-1	141	48	
1:	1	1	2	6	1	0	0	0	0	1:	-1 1-1	141	24	
1:	1	1	3	4	1	0	0	0	0	1:	-1 1-1	141	24	
1:	1	1	3	5	1	0	0	1	0	3:	-1 1-1	141	24	
1:	1	1	3	5	1	1	0	1	1	2:	-1 1-1	141	12	
1:	1	2	3	3	1	0	2	0	0	3:	-1 1-1	141	8	
2:	1	1	3	4	0	1	1	0	1	2:	-1-1 1	141	8	3/8
2:	1	2	2	3	1	0	1	1	0	0:	-1-1 1	141	4	

D= 144

GENUS#	F11	F22	F33	F44	F12	F13	F23	F14	F24	F34:	HASSE SYM	LEVEL	AUTOS	MASS
1:	1	1	1	12	1	0	0	0	0	0:	-1-1	48	48	1/12
1:	1	1	3	4	0	0	0	0	1	3:	-1-1	48	16	
2:	1	1	1	18	1	1	0	0	0	0:	1 1	72	96	1/8
2:	1	1	2	5	0	0	0	1	1	0:	1 1	72	32	
2:	1	1	3	5	1	1	0	0	1	1:	1 1	72	12	
3:	1	1	1	9	0	0	0	0	0	0:	1 1	36	96	1/32
3:	2	2	2	3	2	2	0	0	2	0:	1 1	36	48	
4:	1	1	3	3	0	0	0	0	0	0:	-1-1	12	64	1/64
5:	1	1	4	4	1	0	0	0	0	4:	-1-1	12	144	1/144
6:	1	2	2	4	1	1	0	1	2	2:	-1-1	12	16	1/16
7:	1	1	2	5	0	0	0	0	0	2:	1 1	36	32	1/32
8:	1	1	2	6	0	1	1	0	0	0:	-1-1	24	32	1/16
8:	2	2	2	2	2	1	-1	0	0	0:	-1-1	24	32	
9:	1	1	2	6	1	0	0	0	0	0:	1 1	24	48	1/12
9:	1	2	3	3	0	1	2	1	2	0:	1 1	24	16	
10:	1	1	3	4	1	0	0	0	0	0:	-1-1	48	48	1/12
10:	1	2	2	3	1	1	0	0	0	0:	-1-1	48	16	
11:	1	1	3	4	0	1	1	0	0	2:	1 1	72	16	1/8
11:	1	2	2	3	0	0	0	1	2	1:	1 1	72	16	
12:	1	2	2	3	0	0	2	0	0	0:	1 1	12	48	1/48

D= 145

GENUS#	F11	F22	F33	F44	F12	F13	F23	F14	F24	F34:	HASSE SYM	LEVEL	AUTOS	MASS
1:	1	1	2	6	0	1	0	0	1	2:	-1-1 1	145	8	2/3
1:	1	1	2	8	1	1	0	1	1	2:	-1-1 1	145	24	
1:	1	2	2	3	1	0	1	0	1	1:	-1-1 1	145	4	
1:	2	2	2	2	2	1	0	-1	-1	1:	-1-1 1	145	4	
2:	1	1	3	4	0	1	0	1	1	2:	-1 1-1	145	8	2/3
2:	1	1	4	4	1	1	0	0	1	2:	-1 1-1	145	24	
2:	1	2	2	3	1	0	1	0	1	1:	-1 1-1	145	4	
2:	1	2	2	3	0	0	1	1	-1	1:	-1 1-1	145	8	
2:	1	2	2	4	1	0	2	1	1	1:	-1 1-1	145	8	

D= 148

GENUS#	F11	F22	F33	F44	F12	F13	F23	F14	F24	F34:	HASSE SYM	LEVEL	AUTOS	MASS
1:	1	1	1	10	0	0	0	1	1	1:	1 1	74	96	25/96
1:	1	2	2	4	0	0	2	1	2	0:	1 1	74	24	
1:	1	2	2	4	1	0	2	1	0	0:	1 1	74	8	
1:	2	2	2	3	2	2	0	2	0	1:	1 1	74	12	
2:	1	1	1	19	1	1	0	1	0	0:	-1-1	74	96	25/96
2:	1	1	2	7	1	0	0	1	0	2:	-1-1	74	24	
2:	1	1	3	5	1	1	0	1	0	0:	-1-1	74	12	
2:	1	2	3	3	0	0	2	1	0	3:	-1-1	74	8	
3:	1	1	2	5	0	0	0	1	0	1:	-1-1	148	8	25/24
3:	1	1	2	8	1	1	0	0	1	1:	-1-1	148	12	
3:	1	1	3	5	1	1	0	0	0	2:	-1-1	148	12	
3:	1	2	2	3	1	0	1	0	1	0:	-1-1	148	4	
3:	1	2	2	3	0	0	1	0	0	2:	-1-1	148	4	
3:	1	2	3	3	1	0	2	1	0	2:	-1-1	148	4	

D= 149

GENUS#	F11	F22	F33	F44	F12	F13	F23	F14	F24	F34:	HASSE SYM	LEVEL	AUTOS	MASS
1:	1	1	1	13	1	0	0	1	0	1:	-1-1	149	24	35/48
1:	1	1	1	19	1	1	0	0	1	0:	-1-1	149	48	
1:	1	1	3	4	0	1	1	1	0	0:	-1-1	149	8	
1:	1	1	3	4	0	1	0	0	1	2:	-1-1	149	8	
1:	1	1	3	5	1	1	0	0	1	0:	-1-1	149	12	
1:	1	1	4	4	1	1	0	0	0	3:	-1-1	149	12	
1:	1	2	2	3	1	0	1	0	0	1:	-1-1	149	4	

D= 152

GENUS#	F11	F22	F33	F44	F12	F13	F23	F14	F24	F34:	HASSE SYM	LEVEL	AUTOS	MASS
1:	1	1	1	13	1	0	0	1	0	0:	1 1	152	24	41/96
1:	1	1	1	19	1	1	0	0	0	0:	1 1	152	96	
1:	1	1	2	5	0	0	0	0	1	0:	1 1	152	16	
1:	1	1	3	4	0	0	0	1	1	2:	1 1	152	16	
1:	1	2	2	3	1	0	0	1	0	0:	1 1	152	8	
1:	1	2	2	4	1	0	2	0	1	0:	1 1	152	8	
2:	1	1	1	10	0	0	0	1	1	0:	-1-1	152	32	41/96
2:	1	1	2	6	0	1	0	0	0	2:	-1-1	152	8	

GENUS#	F11	F22	F33	F44	F12	F13	F23	F14	F24	F34:	HASSE	SYM	LEVEL	AUTOS	MASS
2:	1	1	2	7	0	1	1	1	1	0:	-1-1		152	16	
2:	1	1	2	8	1	1	0	1	0	0:	-1-1		152	12	
2:	1	2	2	3	0	0	0	0	2	1:	-1-1		152	8	

D= 153

GENUS#	F11	F22	F33	F44	F12	F13	F23	F14	F24	F34:	HASSE	SYM	LEVEL	AUTOS	MASS
1:	1	1	1	13	1	0	0	0	0	1:	-1	1-1	51	48	5/24
1:	1	1	2	7	0	1	1	1	0	2:	-1	1-1	51	16	
1:	1	2	2	4	1	1	0	0	2	1:	-1	1-1	51	8	
2:	1	1	2	6	0	1	0	1	1	0:	-1	1-1	153	8	5/8
2:	1	2	2	4	1	0	2	0	0	1:	-1	1-1	153	4	
2:	2	2	2	2	1	1	0	1	1	2:	-1	1-1	153	4	
3:	1	1	2	8	1	1	0	0	1	0:	-1	1-1	153	12	5/8
3:	1	1	3	4	0	1	0	1	1	0:	-1	1-1	153	8	
3:	1	1	4	4	1	1	0	1	1	2:	-1	1-1	153	24	
3:	1	2	2	3	1	0	0	0	1	-1:	-1	1-1	153	4	
3:	1	2	3	3	1	0	2	0	1	3:	-1	1-1	153	8	
4:	1	1	3	5	1	0	0	0	0	3:	-1-1	1	51	48	5/24
4:	1	2	2	3	0	0	1	1	1	1:	-1-1	1	51	8	
4:	2	2	2	3	2	1	-1	1	2	1:	-1-1	1	51	16	

D= 156

GENUS#	F11	F22	F33	F44	F12	F13	F23	F14	F24	F34:	HASSE	SYM	LEVEL	AUTOS	MASS
1:	1	1	1	10	0	0	0	0	0	1:	-1-1	1	156	32	13/48
1:	1	1	1	13	1	0	0	0	0	0:	-1-1	1	156	48	
1:	1	1	3	4	0	1	1	0	0	1:	-1-1	1	156	16	
1:	1	1	3	4	0	0	0	0	0	3:	-1-1	1	156	32	
1:	2	2	2	2	1	1	1	2	0	0:	-1-1	1	156	8	
2:	1	1	1	20	1	1	0	1	0	0:	-1	1-1	156	96	13/48
2:	1	1	2	5	0	0	0	0	0	1:	-1	1-1	156	16	
2:	1	1	2	7	0	1	1	1	1	1:	-1	1-1	156	32	
2:	1	1	3	4	0	1	0	1	0	2:	-1	1-1	156	8	
2:	1	1	4	4	1	1	0	0	1	1:	-1	1-1	156	24	
3:	1	1	2	7	1	0	0	0	0	2:	1-1-1		156	48	13/48
3:	1	2	2	3	1	0	1	0	0	0:	1-1-1		156	8	
3:	2	2	2	3	2	1	-1	2	0	0:	1-1-1		156	8	
4:	1	1	3	5	1	0	0	1	1	2:	1	1 1	156	12	13/48
4:	1	2	2	4	0	0	2	1	2	1:	1	1 1	156	16	
4:	1	2	3	3	1	0	2	0	2	0:	1	1 1	156	8	

D= 157

GENUS#	F11	F22	F33	F44	F12	F13	F23	F14	F24	F34:	HASSE	SYM	LEVEL	AUTOS	MASS
1:	1	1	1	20	1	1	0	0	1	0:	-1-1		157	48	43/48
1:	1	1	2	6	0	1	0	0	1	1:	-1-1		157	8	
1:	1	1	2	7	1	0	0	1	1	1:	-1-1		157	12	
1:	1	1	2	8	1	1	0	0	0	1:	-1-1		157	12	
1:	1	1	3	5	1	1	0	0	0	1:	-1-1		157	12	
1:	1	2	2	3	0	0	1	1	1	0:	-1-1		157	4	
1:	1	2	3	3	0	0	1	1	2	3:	-1-1		157	4	

D= 160

GENUS#	F11	F22	F33	F44	F12	F13	F23	F14	F24	F34:	HASSE	SYM	LEVEL	AUTOS	MASS
1:	1	1	1	10	0	0	0	0	0	0:	1	1	40	96	7/96
1:	1	2	2	3	0	0	0	0	0	2:	1	1	40	16	
2:	1	1	1	20	1	1	0	0	0	0:	-1-1		80	96	7/96
2:	1	2	3	3	0	1	2	1	2	1:	-1-1		80	16	
3:	1	1	2	5	0	0	0	0	0	0:	-1-1		40	32	7/96
3:	1	2	2	4	0	0	2	0	0	0:	-1-1		40	24	
4:	1	1	2	7	1	0	0	1	0	0:	1	1	80	24	7/96
4:	1	1	3	4	0	1	1	0	0	0:	1	1	80	32	
5:	1	1	2	8	1	1	0	0	0	0:	1	1	160	24	7/24
5:	1	2	2	3	0	0	0	1	1	-1:	1	1	160	8	
5:	1	2	3	3	0	0	2	1	-1	2:	1	1	160	8	
6:	1	1	2	6	0	1	0	1	0	0:	-1-1		160	8	7/24
6:	1	1	3	5	1	1	0	0	0	0:	-1-1		160	24	
6:	1	2	2	3	0	0	1	0	1	-1:	-1-1		160	8	
7:	1	1	3	6	1	1	0	1	1	3:	-1-1		160	24	7/24
7:	1	2	3	3	1	0	2	1	2	0:	-1-1		160	8	
7:	2	2	2	2	1	1	-1	1	-1	0:	-1-1		160	8	
8:	1	1	2	6	0	0	0	1	1	2:	-1-1		80	32	7/32
8:	1	2	2	4	1	0	2	0	0	0:	-1-1		80	16	
8:	2	2	2	2	1	1	0	2	0	0:	-1-1		80	8	

	F11	F22	F33	F44	F12	F13	F23	F14	F24	F34:	HASSE	SYM	LEVEL	AUTOS	MASS
9:	1	1	3	4	0	1	0	0	0	2:	1 1		160	8	7/24
9:	1	1	4	4	1	1	0	1	0	0:	1 1		160	24	
9:	1	2	3	3	1	0	1	1	1	3:	1 1		160	8	
10:	1	1	4	4	0	0	1	1	1	4:	1 1		80	32	7/32
10:	1	2	2	3	1	0	0	0	1	0:	1 1		80	8	
10:	2	2	2	3	2	2	0	1	-1	1:	1 1		80	16	

D= 161

GENUS#:	F11	F22	F33	F44	F12	F13	F23	F14	F24	F34:	HASSE	SYM	LEVEL	AUTOS	MASS
1:	1	1	1	14	1	0	0	1	0	1:	-1 1-1		161	24	2/3
1:	1	1	2	7	0	1	1	1	0	0:	-1 1-1		161	8	
1:	1	2	2	4	1	0	1	1	0	2:	-1 1-1		161	4	
1:	1	2	2	4	1	1	0	1	1	2:	-1 1-1		161	4	
2:	1	1	2	6	0	1	0	0	1	0:	-1-1 1		161	16	2/3
2:	1	1	3	4	0	1	0	0	1	1:	-1-1 1		161	8	
2:	1	1	4	4	0	1	0	0	1	4:	-1-1 1		161	16	
2:	1	1	4	4	1	1	0	0	1	0:	-1-1 1		161	24	
2:	1	2	2	3	1	0	0	0	0	1:	-1-1 1		161	8	
2:	2	2	2	2	2	1	0	0	1	0:	-1-1 1		161	4	

D= 164

GENUS#:	F11	F22	F33	F44	F12	F13	F23	F14	F24	F34:	HASSE	SYM	LEVEL	AUTOS	MASS
1:	1	1	1	11	0	0	0	1	1	1:	1 1		82	96	1/12
1:	1	1	1	21	1	1	0	1	0	0:	1 1		82	96	
1:	1	1	3	5	0	1	1	0	0	3:	1 1		82	16	
2:	1	1	1	14	0	0	0	1	0	0:	-1-1		164	24	1/1
2:	1	1	2	6	0	1	0	0	0	1:	-1-1		164	8	
2:	1	1	3	4	0	1	0	1	0	0:	-1-1		164	8	
2:	1	1	3	4	0	0	0	0	1	2:	-1-1		164	8	
2:	1	1	4	4	1	1	0	0	0	2:	-1-1		164	12	
2:	1	2	2	4	1	0	1	0	2	2:	-1-1		164	4	
2:	1	2	3	3	1	1	0	1	2	2:	-1-1		164	4	
3:	1	1	2	7	0	1	1	0	0	1:	-1-1		82	16	3/4
3:	1	1	3	4	0	0	0	1	1	1:	-1-1		82	16	
3:	1	2	2	4	1	1	0	0	0	2:	-1-1		82	4	
3:	1	2	2	4	1	0	0	0	2	2:	-1-1		82	8	
3:	2	2	2	2	1	0	0	1	0	2:	-1-1		82	4	

D= 165

GENUS#:	F11	F22	F33	F44	F12	F13	F23	F14	F24	F34:	HASSE	SYM	LEVEL	AUTOS	MASS
1:	1	1	1	14	1	0	0	0	0	1:	-1-1-1-1		165	48	11/48
1:	1	1	4	4	1	0	0	0	0	3:	-1-1-1-1		165	48	
1:	1	2	2	3	0	0	1	1	0	0:	-1-1-1-1		165	16	
1:	2	2	2	2	1	1	-1	-1	-1	1:	-1-1-1-1		165	8	
2:	1	1	1	21	1	1	0	0	1	0:	-1 1-1 1		165	48	11/48
2:	1	1	3	6	1	0	0	0	1	2:	-1 1-1 1		165	12	
2:	1	2	3	3	1	0	2	0	2	1:	-1 1-1 1		165	8	
3:	1	1	2	7	1	0	0	0	0	1:	-1-1 1 1		165	24	11/48
3:	1	1	3	5	0	1	1	1	0	3:	-1-1 1 1		165	16	
3:	1	2	3	3	1	1	0	0	0	3:	-1-1 1 1		165	8	
4:	1	1	2	9	1	1	0	1	1	2:	-1 1 1-1		165	24	11/48
4:	1	1	3	4	0	1	0	0	1	0:	-1 1 1-1		165	16	
4:	1	1	3	5	1	0	0	1	1	1:	-1 1 1-1		165	12	
4:	1	1	4	5	1	1	0	1	1	4:	-1 1 1-1		165	24	

D= 168

GENUS#:	F11	F22	F33	F44	F12	F13	F23	F14	F24	F34:	HASSE	SYM	LEVEL	AUTOS	MASS
1:	1	1	1	14	1	0	0	0	0	0:	1 1 1		168	48	9/32
1:	1	1	2	7	1	0	0	0	0	0:	1 1 1		168	48	
1:	1	1	2	9	1	1	0	0	1	1:	1 1 1		168	12	
1:	1	1	3	4	0	0	0	1	1	0:	1 1 1		168	32	
1:	1	2	2	3	1	0	0	0	0	0:	1 1 1		168	16	
1:	1	2	3	3	0	0	0	0	2	3:	1 1 1		168	16	
2:	1	1	1	21	1	1	0	0	0	0:	-1 1-1		168	96	9/32
2:	1	1	2	6	0	0	0	0	1	2:	-1 1-1		168	16	
2:	1	1	3	5	1	0	0	0	0	2:	-1 1-1		168	24	
2:	1	1	3	5	1	0	0	1	0	0:	-1 1-1		168	24	
2:	1	2	2	3	0	0	1	0	1	1:	-1 1-1		168	8	
3:	1	1	2	6	0	1	0	0	0	0:	-1-1 1		168	16	9/32
3:	1	1	2	7	0	1	1	0	0	0:	-1-1 1		168	32	
3:	1	2	2	4	0	0	2	0	-1	1:	-1-1 1		168	16	
3:	1	2	2	5	1	0	2	0	2	0:	-1-1 1		168	8	

4:	1	1		11	0	0	0	1	1	0:	1-1-1	168	32	9/32
4:	1	2	-	3	0	0	0	1	1	0:	1-1-1	168	8	
4:	2	2	2	3	2	1	-1	2	0	1:	1-1-1	168	8	

D= 169

GENUS#:	F11	F22	F33	F44	F12	F13	F23	F14	F24	F34:	HASSE SYM	LEVEL	AUTOS	MASS
1:	1	2	2	4	1	0	1	1	1	2:	-1-1	13	8	1/8

D= 172

GENUS#:	F11	F22	F33	F44	F12	F13	F23	F14	F24	F34:	HASSE SYM	LEVEL	AUTOS	MASS
1:	1	1	1	22	1	1	0	1	0	0:	1 1	172	96	21/32
1:	1	1	2	6	0	0	0	1	1	1:	1 1	172	16	
1:	1	1	2	9	1	1	0	1	0	0:	1 1	172	12	
1:	1	2	2	4	0	0	2	1	1	1:	1 1	172	8	
1:	1	2	3	3	0	1	2	0	2	0:	1 1	172	8	
1:	2	2	2	3	2	1	0	1	2	2:	1 1	172	4	
2:	1	1	1	11	0	0	0	1	0	0:	-1-1	172	32	21/32
2:	1	1	2	8	1	0	0	1	0	2:	-1-1	172	24	
2:	1	1	3	4	0	1	0	0	0	1:	-1-1	172	8	
2:	1	1	4	5	1	1	0	0	1	3:	-1-1	172	12	
2:	1	2	2	3	0	0	1	0	1	0:	-1-1	172	4	
2:	1	2	2	5	1	0	2	0	0	2:	-1-1	172	8	

D= 173

GENUS#:	F11	F22	F33	F44	F12	F13	F23	F14	F24	F34:	HASSE SYM	LEVEL	AUTOS	MASS
1:	1	1	1	15	1	0	0	1	0	1:	-1-1	173	24	13/16
1:	1	1	1	22	1	1	0	0	1	0:	-1-1	173	48	
1:	1	1	2	7	0	1	0	0	1	2:	-1-1	173	8	
1:	1	1	2	9	1	1	0	0	1	0:	-1-1	173	12	
1:	1	1	3	5	0	1	0	0	1	3:	-1-1	173	8	
1:	1	1	3	6	1	1	0	1	1	2:	-1-1	173	12	
1:	1	1	4	4	1	0	0	1	0	1:	-1-1	173	12	
1:	1	2	3	3	1	1	0	0	2	1:	-1-1	173	4	

D= 176

GENUS#:	F11	F22	F33	F44	F12	F13	F23	F14	F24	F34:	HASSE SYM	LEVEL	AUTOS	MASS
1:	1	1	1	11	0	0	0	0	0	0:	-1-1	44	96	7/96
1:	1	1	3	4	0	0	0	0	0	2:	-1-1	44	16	
2:	1	1	1	15	1	0	0	1	0	0:	-1-1	176	24	7/12
2:	1	1	3	4	0	1	0	0	0	0:	-1-1	176	16	
2:	1	1	3	4	0	0	0	1	0	1:	-1-1	176	8	
2:	1	1	4	4	0	0	0	0	1	4:	-1-1	176	16	
2:	1	1	4	4	1	1	0	0	0	0:	-1-1	176	24	
2:	1	2	2	3	0	0	0	0	1	-1:	-1-1	176	8	
2:	1	2	2	4	1	1	0	0	1	1:	-1-1	176	8	
3:	1	1	1	22	1	1	0	0	0	0:	1 1	88	96	7/96
3:	1	1	3	5	0	1	1	1	1	0:	1 1	88	16	
4:	1	1	2	6	0	0	0	1	1	0:	1 1	88	32	7/32
4:	1	1	2	8	0	1	1	1	1	0:	1 1	88	16	
4:	2	2	2	2	1	0	0	0	2	0:	1 1	88	8	
5:	1	1	3	6	1	1	0	0	1	1:	1 1	176	12	7/12
5:	1	2	2	4	1	0	1	0	2	0:	1 1	176	4	
5:	1	2	3	3	1	0	2	0	0	2:	1 1	176	4	
6:	1	1	4	5	1	0	0	1	0	4:	-1-1	88	24	7/96
6:	1	2	2	3	0	0	0	1	0	0:	-1-1	88	32	
7:	1	1	2	6	0	0	0	0	0	2:	1 1	44	32	7/96
7:	2	2	2	3	2	0	0	2	0	2:	1 1	44	24	
8:	1	2	2	4	1	0	0	1	0	0:	-1-1	88	8	7/32
8:	1	2	2	4	0	0	0	1	2	2:	-1-1	88	32	
8:	2	2	2	3	2	2	0	0	-1	1:	-1-1	88	16	

D= 177

GENUS#:	F11	F22	F33	F44	F12	F13	F23	F14	F24	F34:	HASSE SYM	LEVEL	AUTOS	MASS
1:	1	1	1	15	1	0	0	0	0	1:	-1 1-1	177	48	13/16
1:	1	1	2	9	1	1	0	0	0	1:	-1 1-1	177	12	
1:	1	1	3	5	1	0	0	0	0	1:	-1 1-1	177	24	
1:	1	1	3	6	1	0	0	1	0	3:	-1 1-1	177	24	
1:	1	2	2	4	0	0	1	1	2	2:	-1 1-1	177	8	
1:	1	2	2	4	1	1	0	1	1	0:	-1 1-1	177	8	
1:	1	2	3	3	1	0	2	0	-1	1:	-1 1-1	177	4	
1:	1	2	3	3	1	0	0	0	1	3:	-1 1-1	177	8	
2:	1	1	2	8	0	1	1	1	0	2:	-1-1 1	177	16	13/16

	F11	F22	F33	F44	F12	F13	F23	F14	F24	F34:	HASSE SYM	LEVEL	AUTOS	MASS
2:	1	2	2	4	1	0	1	1	2	0:	-1-1 1	177	4	
2:	1	2	2	5	1	0	2	1	0	1:	-1-1 1	177	4	
2:	2	2	2	3	2	1	-1	0	1	1:	-1-1 1	177	4	

D= 180

GENUS#:	F11	F22	F33	F44	F12	F13	F23	F14	F24	F34:	HASSE SYM	LEVEL	AUTOS	MASS
1:	1	1	1	12	0	0	0	1	1	1:	1 1 1	90	96	5/32
1:	1	2	2	5	1	0	2	0	1	2:	1 1 1	90	16	
1:	2	2	2	3	2	2	0	1	1	0:	1 1 1	90	12	
2:	1	1	1	15	1	0	0	0	0	0:	-1 1-1	60	48	5/24
2:	1	1	3	4	0	0	0	1	0	0:	-1 1-1	60	16	
2:	1	2	2	4	1	1	0	0	-1	1:	-1 1-1	60	8	
3:	1	1	1	23	1	1	0	1	0	0:	-1 1-1	90	96	5/32
3:	1	1	3	6	1	1	0	0	0	2:	-1 1-1	90	12	
3:	1	2	3	3	0	0	2	1	2	1:	-1 1-1	90	16	
4:	1	1	2	7	0	1	0	0	0	2:	-1 1-1	180	8	5/8
4:	1	2	2	4	1	0	1	0	0	2:	-1 1-1	180	4	
4:	2	2	2	2	1	1	1	-1	1	0:	-1 1-1	180	4	
5:	1	1	2	6	0	0	0	0	1	1:	-1 1-1	180	8	5/8
5:	1	1	2	9	1	1	0	0	0	0:	-1 1-1	180	24	
5:	1	1	3	6	1	0	1	0	0	0:	-1 1-1	180	12	
5:	1	2	3	3	1	0	1	0	2	2:	-1 1-1	180	4	
5:	1	2	3	3	0	0	2	0	1	-2:	-1 1-1	180	8	
6:	1	1	3	5	1	0	0	0	0	0:	-1-1 1	60	48	5/24
6:	1	2	2	3	0	0	1	0	0	0:	-1-1 1	60	16	
6:	1	2	3	0	1	1	0	0	0	3:	-1-1 1	60	8	
7:	1	2	2	4	0	0	2	1	0	0:	1-1-1	30	48	5/96
7:	2	2	2	3	2	1	-1	-1	1	-1:	1-1-1	30	32	
8:	1	1	2	8	1	0	0	0	0	2:	-1-1 1	30	48	5/96
8:	1	3	3	3	0	0	0	1	3	3:	-1-1 1	30	32	
9:	1	1	2	8	0	1	1	1	1	1:	1 1 1	30	32	5/96
9:	2	2	2	2	1	0	0	0	0	2:	1 1 1	30	48	
10:	1	1	3	5	0	1	1	1	1	1:	-1 1-1	90	32	5/32
10:	1	2	3	3	0	1	0	1	2	2:	-1 1-1	90	8	
11:	1	1	3	5	0	0	0	1	1	3:	-1 1-1	30	32	5/96
11:	1	1	4	4	1	0	0	0	0	2:	-1 1-1	30	48	
12:	1	1	4	4	0	1	1	1	1	3:	1 1 1	90	32	5/32
12:	1	2	2	4	1	0	0	1	0	2:	1 1 1	90	8	

D= 181

GENUS#:	F11	F22	F33	F44	F12	F13	F23	F14	F24	F34:	HASSE SYM	LEVEL	AUTOS	MASS
1:	1	1	1	23	1	1	0	0	1	0:	-1-1	181	48	19/16
1:	1	1	2	7	0	1	0	1	1	0:	-1-1	181	8	
1:	1	1	2	8	1	0	0	1	0	1:	-1-1	181	12	
1:	1	1	3	5	0	1	1	1	0	2:	-1-1	181	8	
1:	1	1	3	6	1	1	0	0	1	0:	-1-1	181	12	
1:	1	2	3	3	0	1	1	1	2	0:	-1-1	181	4	
1:	1	2	3	3	0	1	0	1	1	3:	-1-1	181	4	
1:	1	2	3	3	1	1	0	1	2	1:	-1-1	181	4	

D= 184

GENUS#:	F11	F22	F33	F44	F12	F13	F23	F14	F24	F34:	HASSE SYM	LEVEL	AUTOS	MASS
1:	1	1	1	12	0	0	0	1	1	0:	-1-1	184	32	37/48
1:	1	1	1	23	1	1	0	0	0	0:	-1-1	184	96	
1:	1	1	2	6	0	0	0	1	0	0:	-1-1	184	16	
1:	1	1	2	8	1	0	0	1	0	0:	-1-1	184	24	
1:	1	2	2	3	0	0	0	0	1	0:	-1-1	184	8	
1:	1	2	2	4	0	0	2	0	1	0:	-1-1	184	8	
1:	1	2	2	4	1	0	1	1	0	1:	-1-1	184	4	
1:	1	2	3	3	0	1	2	0	0	2:	-1-1	184	8	
2:	1	1	3	5	0	1	1	0	0	2:	1 1	184	16	37/48
2:	1	1	3	5	0	1	0	0	0	3:	1 1	184	8	
2:	1	1	4	5	1	1	0	1	1	3:	1 1	184	12	
2:	1	2	3	3	1	1	0	0	2	0:	1 1	184	4	
2:	2	2	2	3	2	1	0	0	2	1:	1 1	184	4	

D= 185

GENUS#:	F11	F22	F33	F44	F12	F13	F23	F14	F24	F34:	HASSE SYM	LEVEL	AUTOS	MASS
1:	1	1	1	16	1	0	0	1	0	1:	-1-1 1	185	24	19/24
1:	1	1	2	7	0	1	0	0	1	1:	-1-1 1	185	8	
1:	1	2	2	4	1	1	0	0	0	1:	-1-1 1	185	4	
1:	1	2	2	5	1	0	2	1	1	1:	-1-1 1	185	8	

1:	2	2	2	2	1	1	0	1	0	-1:	-1-1 1	185	4	
2:	1	1	2	10	1	1	0	1	1	2:	-1 1-1	185	24	19/24
2:	1	1	2	8	0	1	1	0	1	0:	-1 1-1	185	8	
2:	1	1	4	4	0	1	1	0	1	3:	-1 1-1	185	8	
2:	1	2	2	4	1	0	0	0	2	1:	-1 1-1	185	4	
2:	1	2	3	3	1	1	0	0	1	2:	-1 1-1	185	4	

D= 188

GENUS#:	F11	F22	F33	F44	F12	F13	F23	F14	F24	F34:	HASSE	SYM	LEVEL	AUTOS	MASS
1:	1	1	1	12	0	0	0	1	0	0:	-1-1		188	32	7/12
1:	1	1	1	16	1	0	0	1	0	0:	-1-1		188	24	
1:	1	1	1	24	1	1	0	1	0	0:	-1-1		188	96	
1:	1	1	2	6	0	0	0	0	0	1:	-1-1		188	16	
1:	1	1	2	7	0	1	0	1	0	0:	-1-1		188	8	
1:	1	1	2	8	0	1	1	0	0	1:	-1-1		188	16	
1:	1	1	3	4	0	0	0	0	0	1:	-1-1		188	16	
1:	1	1	4	4	0	0	0	1	1	3:	-1-1		188	16	
1:	1	2	2	4	1	0	0	0	1	2:	-1-1		188	8	
2:	1	1	2	10	1	1	0	0	1	1:	1 1		188	12	7/12
2:	1	2	2	4	1	0	1	1	1	1:	1 1		188	4	
2:	1	2	2	5	1	0	2	1	0	0:	1 1		188	8	
2:	1	2	3	3	0	0	2	1	0	2:	1 1		188	8	

D= 189

GENUS#:	F11	F22	F33	F44	F12	F13	F23	F14	F24	F34:	HASSE	SYM	LEVEL	AUTOS	MASS
1:	1	1	1	16	1	0	0	0	0	1:	-1-1 1		63	48	1/16
1:	1	1	4	4	1	0	0	0	0	1:	-1-1 1		63	48	
1:	1	2	3	3	1	0	0	0	0	3:	-1-1 1		63	48	
2:	1	1	1	24	1	1	0	0	1	0:	-1 1-1		189	48	9/16
2:	1	1	3	5	0	1	0	1	1	2:	-1 1-1		189	8	
2:	1	1	3	6	1	1	0	0	0	1:	-1 1-1		189	12	
2:	1	1	4	5	1	1	0	0	1	2:	-1 1-1		189	12	
2:	1	2	3	3	0	0	1	1	2	2:	-1 1-1		189	4	
3:	1	1	2	7	0	1	0	0	1	0:	-1-1 1		189	16	9/16
3:	1	1	3	5	0	1	1	1	0	0:	-1-1 1		189	8	
3:	1	2	2	4	1	0	1	0	1	-1:	-1-1 1		189	4	
3:	1	2	3	3	1	1	0	1	1	2:	-1-1 1		189	8	
4:	1	1	2	8	1	0	0	0	0	1:	-1-1 1		63	24	1/8
4:	1	3	3	3	0	0	3	1	3	0:	-1-1 1		63	24	
4:	2	2	2	3	1	1	1	2	2	-1:	-1-1 1		63	24	
5:	2	2	2	2	1	1	1	1	1	1:	-1-1 1		21	48	1/48
6:	1	1	3	6	1	0	0	0	0	3:	-1 1-1		21	48	1/48
7:	1	1	4	4	0	1	0	0	1	3:	-1 1-1		63	16	1/16
8:	1	2	3	3	0	1	1	1	2	1:	-1 1-1		63	8	1/8

D= 192

GENUS#:	F11	F22	F33	F44	F12	F13	F23	F14	F24	F34:	HASSE	SYM	LEVEL	AUTOS	MASS
1:	1	1	1	12	0	0	0	0	0	0:	-1-1		48	96	1/48
1:	2	2	2	3	2	2	0	0	0	0:	-1-1		48	96	
2:	1	1	1	16	1	0	0	0	0	0:	-1-1		192	48	1/3
2:	1	1	3	5	0	0	0	1	0	3:	-1-1		192	16	
2:	1	2	3	3	1	1	0	1	0	2:	-1-1		192	8	
2:	2	2	2	2	1	1	1	1	1	0:	-1-1		192	8	
3:	1	1	1	24	1	1	0	0	0	0:	1 1		96	96	1/12
3:	1	1	2	7	0	0	0	1	1	2:	1 1		96	32	
3:	1	1	3	6	1	1	0	0	0	0:	1 1		96	24	
4:	1	1	2	10	1	1	0	1	0	0:	1 1		192	12	1/3
4:	1	2	3	3	0	0	2	1	1	2:	1 1		192	4	
5:	1	1	2	6	0	0	0	0	0	0:	1 1		24	32	1/32
6:	1	1	3	4	0	0	0	0	0	0:	-1-1		48	32	1/16
6:	1	2	2	4	0	0	0	0	2	2:	-1-1		48	32	
7:	1	1	3	7	1	1	0	1	1	3:	1 1		48	24	1/24
8:	1	1	2	7	0	1	0	0	0	1:	-1-1		192	8	1/3
8:	1	1	4	5	1	0	0	0	0	4:	-1-1		192	48	
8:	1	2	2	4	0	0	1	0	2	2:	-1-1		192	8	
8:	1	2	2	5	1	1	0	1	2	2:	-1-1		192	16	
9:	1	2	2	3	0	0	0	0	0	0:	-1-1		24	32	1/32
10:	1	1	2	8	0	1	1	0	0	0:	1 1		96	32	1/4
10:	1	1	4	4	0	1	1	1	1	0:	1 1		96	32	
10:	1	2	2	4	1	0	0	0	2	0:	1 1		96	8	
10:	2	2	2	3	2	0	0	-1	1	-2:	1 1		96	16	
11:	1	1	2	8	1	0	0	0	0	0:	-1-1		96	48	1/12

```
11:  1  2  3  3  0  1  0  0  0  3: -1-1      96    16
12:  1  1  3  6  1  0  0  1  1  2:  1 1     192    12        1/3
12:  1  2  3  3  1  0  2  0  1  0:  1 1     192     4
13:  1  1  4  4  0  0  0  0  0  4: -1-1      12    96       1/96
14:  1  1  4  4  1  0  0  0  0  0: -1-1      48    96       1/24
14:  1  3  3  3  1  1 -1  1  3  2: -1-1      48    32
15:  1  2  2  4  1  1  0  0  0  0: -1-1      48    16        1/8
15:  2  2  2  2  1  1  0  0  1  1: -1-1      48    16
16:  1  2  2  4  0  0  2  0  0  0:  1 1      48    48       1/48
17:  1  2  2  5  1  0  2  0  1  0: -1-1      96     8        1/4
17:  2  2  2  3  2  1 -1  1  1  1: -1-1      96     8
18:  1  2  3  3  1  0  1  0 -1  2:  1 1      48     8        1/8
19:  1  2  3  3  0  0  2  0  2  0:  1 1      48    16       1/16
20:  2  2  2  3  0  0  0  2  2  2:  1 1      12    96       1/96
```

D= 193

GENUS#:	F11	F22	F33	F44	F12	F13	F23	F14	F24	F34:	HASSE	SYM	LEVEL	AUTOS	MASS
1:	1	1	2	10	1	1	0	0	1	0:	-1-1		193	12	49/24
1:	1	1	3	5	0	1	0	0	1	2:	-1-1		193	8	
1:	1	1	4	5	1	1	0	0	0	3:	-1-1		193	12	
1:	1	2	2	4	1	0	1	1	0	0:	-1-1		193	4	
1:	1	2	2	4	0	0	1	1	0	0:	-1-1		193	4	
1:	1	2	2	5	1	0	2	0	0	1:	-1-1		193	4	
1:	1	2	3	3	1	1	0	0	1	-2:	-1-1		193	4	
1:	1	2	3	3	1	0	1	1	2	0:	-1-1		193	4	
1:	2	2	2	3	2	1	0	2	1	2:	-1-1		193	2	

D= 196

GENUS#:	F11	F22	F33	F44	F12	F13	F23	F14	F24	F34:	HASSE	SYM	LEVEL	AUTOS	MASS
1:	1	1	1	13	0	0	0	1	1	1:	1 1		98	96	7/48
1:	1	1	1	25	1	1	0	1	0	0:	1 1		98	96	
1:	1	2	3	3	0	1	0	1	2	0:	1 1		98	8	
2:	1	1	2	7	0	1	0	0	0	0:	-1-1		28	16	3/16
2:	1	2	3	3	0	1	1	-1	-1	1:	-1-1		28	8	
3:	1	1	2	9	1	0	0	1	0	2:	1 1		98	24	7/48
3:	1	1	3	5	0	1	1	0	0	1:	1 1		98	16	
3:	1	2	2	5	0	0	2	1	2	0:	1 1		98	24	
4:	1	1	4	4	0	1	1	1	1	1:	-1-1		14	64	9/64
4:	1	2	2	4	1	0	0	0	0	2:	-1-1		14	16	
4:	2	2	2	2	1	1	0	0	1	-1:	-1-1		14	16	
5:	1	1	5	5	1	1	0	1	1	5:	1 1		14	72	1/36
5:	2	2	3	3	2	2	0	2	2	3:	1 1		14	72	

D= 197

GENUS#:	F11	F22	F33	F44	F12	F13	F23	F14	F24	F34:	HASSE	SYM	LEVEL	AUTOS	MASS
1:	1	1	1	17	1	0	0	1	0	1:	-1-1		197	24	49/48
1:	1	1	1	25	1	1	0	0	1	0:	-1-1		197	48	
1:	1	1	2	10	1	1	0	0	0	1:	-1-1		197	12	
1:	1	1	3	5	0	1	0	1	1	0:	-1-1		197	8	
1:	1	1	3	7	1	1	0	0	1	2:	-1-1		197	12	
1:	1	1	4	5	1	1	0	1	0	2:	-1-1		197	12	
1:	1	1	4	5	1	0	0	1	0	3:	-1-1		197	12	
1:	1	2	2	4	1	0	1	0	1	1:	-1-1		197	4	
1:	1	2	3	3	1	0	0	0	2	1:	-1-1		197	4	

D= 200

GENUS#:	F11	F22	F33	F44	F12	F13	F23	F14	F24	F34:	HASSE	SYM	LEVEL	AUTOS	MASS
1:	1	1	1	17	1	0	0	1	0	0:	1 1		200	24	65/96
1:	1	1	1	25	1	1	0	0	0	0:	1 1		200	96	
1:	1	1	2	7	0	0	0	1	0	2:	1 1		200	16	
1:	1	1	3	5	0	0	0	1	1	2:	1 1		200	16	
1:	1	2	2	4	1	0	1	0	1	0:	1 1		200	4	
1:	2	2	2	3	2	1	0	2	0	0:	1 1		200	4	
2:	1	1	2	10	1	1	0	0	0	0:	1 1		40	24	13/96
2:	1	2	2	5	1	0	2	0	0	0:	1 1		40	16	
2:	1	2	3	3	0	0	2	0	2	1:	1 1		40	32	
3:	1	1	1	13	0	0	0	1	1	0:	1 1		200	32	65/96
3:	1	1	2	9	0	1	1	1	1	0:	1 1		200	16	
3:	1	1	3	5	0	1	0	1	0	2:	1 1		200	8	
3:	1	1	4	5	1	1	0	0	1	1:	1 1		200	12	
3:	1	2	2	4	0	0	0	1	2	1:	1 1		200	8	
3:	1	2	3	3	1	0	0	1	0	2:	1 1		200	4	

	F11	F22	F33	F44	F12	F13	F23	F14	F24	F34:	HASSE SYM	LEVEL	AUTOS	MASS
4:	1	1	3	5	0	1	1	0	0	0:	-1-1	40	32	13/96
4:	1	1	4	4	0	1	0	1	0	3:	-1-1	40	16	
4:	2	2	2	2	1	1	-1	0	0	0:	-1-1	40	24	

D= 201

GENUS#:	F11	F22	F33	F44	F12	F13	F23	F14	F24	F34:	HASSE SYM	LEVEL	AUTOS	MASS
1:	1	1	1	17	1	0	0	0	0	1:	-1-1 1	201	48	49/48
1:	1	1	2	8	0	1	0	0	1	2:	-1-1 1	201	8	
1:	1	2	2	4	1	0	1	0	0	1:	-1-1 1	201	4	
1:	1	2	2	5	1	1	0	0	2	1:	-1-1 1	201	8	
1:	2	2	2	2	1	1	1	0	1	0:	-1-1 1	201	4	
1:	2	2	2	3	2	1	-1	1	0	1:	-1-1 1	201	4	
2:	1	1	2	9	0	1	1	1	0	2:	-1 1-1	201	16	49/48
2:	1	1	3	6	1	0	0	1	1	1:	-1 1-1	201	12	
2:	1	1	4	4	0	1	1	1	0	2:	-1 1-1	201	8	
2:	1	2	2	4	1	0	0	1	0	1:	-1 1-1	201	4	
2:	1	2	3	3	0	1	1	0	-1	2:	-1 1-1	201	4	
2:	1	2	3	4	1	0	2	1	1	3:	-1 1-1	201	4	

D= 204

GENUS#:	F11	F22	F33	F44	F12	F13	F23	F14	F24	F34:	HASSE SYM	LEVEL	AUTOS	MASS
1:	1	1	1	17	1	0	0	0	0	0:	-1 1-1	204	48	13/32
1:	1	1	2	9	1	0	0	0	0	2:	-1 1-1	204	48	
1:	1	1	3	5	0	0	0	0	0	3:	-1 1-1	204	32	
1:	1	1	3	5	0	1	0	0	0	2:	-1 1-1	204	8	
1:	1	1	4	5	1	1	0	1	0	0:	-1 1-1	204	12	
1:	1	2	3	3	0	0	0	0	1	3:	-1 1-1	204	8	
2:	1	1	1	26	1	1	0	1	0	0:	1 1 1	204	96	13/32
2:	1	1	2	7	0	0	0	1	1	1:	1 1 1	204	16	
2:	1	1	3	6	1	0	0	0	0	2:	1 1 1	204	24	
2:	1	1	3	6	1	0	0	1	0	0:	1 1 1	204	24	
2:	1	2	3	3	1	0	0	0	2	0:	1 1 1	204	8	
2:	2	2	2	3	1	0	0	2	2	2:	1 1 1	204	8	
3:	1	1	1	13	0	0	0	0	0	1:	-1-1 1	204	32	13/32
3:	1	1	4	4	0	1	0	0	0	3:	-1-1 1	204	8	
3:	1	2	3	3	1	1	0	1	0	0:	-1-1 1	204	8	
3:	2	2	2	3	2	1	-1	1	1	0:	-1-1 1	204	8	
4:	1	1	2	9	0	1	1	1	1	1:	1-1-1	204	32	13/32
4:	1	1	3	6	0	1	1	0	0	3:	1-1-1	204	16	
4:	1	2	2	5	0	0	2	1	1	2:	1-1-1	204	16	
4:	2	2	2	3	2	1	0	0	2	0:	1-1-1	204	4	

D= 205

GENUS#:	F11	F22	F33	F44	F12	F13	F23	F14	F24	F34:	HASSE SYM	LEVEL	AUTOS	MASS
1:	1	1	1	26	1	1	0	0	1	0:	-1-1 1	205	48	17/24
1:	1	1	2	11	1	1	0	1	1	2:	-1-1 1	205	24	
1:	1	1	3	6	0	1	1	1	0	3:	-1-1 1	205	16	
1:	1	1	3	7	1	1	0	1	1	2:	-1-1 1	205	12	
1:	1	2	3	3	0	1	2	0	1	1:	-1-1 1	205	4	
1:	2	2	2	3	1	1	-1	0	2	1:	-1-1 1	205	4	
2:	1	1	2	9	1	0	0	1	0	1:	-1 1-1	205	12	17/24
2:	1	1	3	5	0	1	0	0	1	1:	-1 1-1	205	8	
2:	1	1	4	5	1	1	0	0	1	0:	-1 1-1	205	12	
2:	1	1	5	5	1	1	0	0	1	4:	-1 1-1	205	24	
2:	1	2	2	4	0	0	1	1	1	-1:	-1 1-1	205	8	
2:	1	2	3	3	1	1	0	1	1	0:	-1 1-1	205	8	
2:	1	3	3	3	0	0	1	1	3	3:	-1 1-1	205	8	

D= 208

GENUS#:	F11	F22	F33	F44	F12	F13	F23	F14	F24	F34:	HASSE SYM	LEVEL	AUTOS	MASS
1:	1	1	1	13	0	0	0	0	0	0:	1 1	52	96	5/96
1:	1	2	2	5	0	0	2	0	2	0:	1 1	52	24	
2:	1	1	1	26	1	1	0	0	0	0:	-1-1	104	96	5/24
2:	1	1	2	7	0	0	0	1	1	0:	-1-1	104	32	
2:	1	1	2	9	1	0	0	1	0	0:	-1-1	104	24	
2:	1	2	3	3	0	0	0	1	2	1:	-1-1	104	8	
3:	1	1	2	11	1	1	0	0	1	1:	-1-1	208	12	5/6
3:	1	2	2	4	1	0	1	0	0	0:	-1-1	208	8	
3:	1	2	3	3	0	1	1	0	2	0:	-1-1	208	4	
3:	1	2	3	3	0	1	1	1	1	2:	-1-1	208	8	
3:	2	2	2	3	1	1	0	0	2	2:	-1-1	208	4	
4:	1	1	2	8	0	1	0	0	0	2:	-1-1	208	8	5/6

	F11	F22	F33	F44	F12	F13	F23	F14	F24	F34	HASSE SYM	LEVEL	AUTOS	MASS
4:	1	1	3	5	0	1	0	1	0	0:	-1-1	208	8	
4:	1	1	4	5	1	1	0	0	0	2:	-1-1	208	12	
4:	1	2	2	4	0	0	1	0	0	2:	-1-1	208	4	
4:	1	2	3	3	1	0	1	1	0	1:	-1-1	208	4	
5:	1	1	3	7	1	1	0	0	1	1:	-1-1	52	12	5/24
5:	1	3	3	3	1	0	1	1	3	3:	-1-1	52	8	
6:	1	1	4	4	0	1	1	0	0	2:	1 1	104	16	5/8
6:	1	2	2	4	1	0	0	1	0	0:	1 1	104	8	
6:	1	2	2	4	0	0	0	1	2	0:	1 1	104	16	
6:	1	2	2	6	1	0	2	0	2	0:	1 1	104	8	
6:	2	2	2	3	2	1	0	1	2	1:	1 1	104	4	
7:	1	2	3	3	1	0	1	0	1	2:	-1-1	52	8	5/24
7:	2	2	2	2	1	1	0	1	0	0:	-1-1	52	12	
8:	1	1	2	7	0	0	0	0	0	2:	-1-1	52	32	5/32
8:	1	2	3	3	0	0	2	0	0	2:	-1-1	52	8	

D= 209

GENUS#:	F11	F22	F33	F44	F12	F13	F23	F14	F24	F34	HASSE SYM	LEVEL	AUTOS	MASS
1:	1	1	1	18	1	0	0	1	0	1:	-1-1 1	209	24	47/48
1:	1	1	4	4	0	1	0	1	1	0:	-1-1 1	209	8	
1:	1	1	4	4	0	1	0	0	1	2:	-1-1 1	209	16	
1:	1	2	2	4	1	0	0	0	1	-1:	-1-1 1	209	4	
1:	1	2	2	5	1	1	0	1	2	1:	-1-1 1	209	4	
1:	2	2	2	3	2	1	0	1	1	2:	-1-1 1	209	4	
2:	1	1	2	8	0	1	0	1	1	0:	-1 1-1	209	8	47/48
2:	1	1	2	9	0	1	1	1	0	0:	-1 1-1	209	8	
2:	1	1	3	5	0	1	0	0	1	0:	-1 1-1	209	16	
2:	1	1	4	6	1	1	0	1	1	4:	-1 1-1	209	24	
2:	1	2	3	4	1	1	0	1	2	3:	-1 1-1	209	8	
2:	1	2	3	4	1	0	2	0	0	3:	-1 1-1	209	4	
2:	2	2	2	2	1	1	0	0	1	0:	-1 1-1	209	4	

D= 212

GENUS#:	F11	F22	F33	F44	F12	F13	F23	F14	F24	F34	HASSE SYM	LEVEL	AUTOS	MASS
1:	1	1	1	14	0	0	0	1	1	1:	1 1	106	96	35/96
1:	1	1	2	9	0	1	1	0	0	1:	1 1	106	16	
1:	1	2	2	6	1	0	2	0	0	2:	1 1	106	8	
1:	2	2	2	3	2	0	0	2	2	1:	1 1	106	12	
1:	2	2	2	4	2	2	0	1	2	1:	1 1	106	12	
2:	1	1	1	18	1	0	0	1	0	0:	-1-1	212	24	35/24
2:	1	1	2	11	1	1	0	1	0	0:	-1-1	212	12	
2:	1	1	2	7	0	0	0	1	0	1:	-1-1	212	8	
2:	1	1	3	5	0	0	0	0	1	2:	-1-1	212	8	
2:	1	1	3	7	1	1	0	0	0	2:	-1-1	212	12	
2:	1	2	2	5	1	1	0	0	2	0:	-1-1	212	4	
2:	1	2	3	3	1	1	0	0	-1	1:	-1-1	212	4	
2:	1	2	3	3	0	0	2	0	1	-1:	-1-1	212	4	
2:	1	2	3	3	1	0	0	0	1	-2:	-1-1	212	4	
3:	1	1	1	27	1	1	0	1	0	0:	-1-1	106	96	35/96
3:	1	1	3	5	0	0	0	1	1	1:	-1-1	106	16	
3:	1	1	3	7	1	1	0	1	0	0:	-1-1	106	12	
3:	1	1	4	5	1	0	0	1	1	2:	-1-1	106	12	
3:	1	2	3	3	0	1	0	0	2	1:	-1-1	106	8	

D= 213

GENUS#:	F11	F22	F33	F44	F12	F13	F23	F14	F24	F34	HASSE SYM	LEVEL	AUTOS	MASS
1:	1	1	1	18	1	0	0	0	0	1:	-1 1-1	213	48	5/8
1:	1	1	1	27	1	1	0	0	1	0:	-1 1-1	213	48	
1:	1	1	2	11	1	0	0	0	1	0:	-1 1-1	213	12	
1:	1	1	2	9	1	0	0	0	0	1:	-1 1-1	213	24	
1:	1	1	3	6	1	0	0	0	0	1:	-1 1-1	213	24	
1:	1	1	3	7	1	0	0	1	0	3:	-1 1-1	213	24	
1:	1	1	3	7	1	1	0	0	1	0:	-1 1-1	213	12	
1:	1	1	4	5	1	0	0	0	0	3:	-1 1-1	213	24	
1:	1	2	2	4	0	0	1	1	1	1:	-1 1-1	213	8	
1:	1	3	3	3	1	0	3	0	1	3:	-1 1-1	213	8	
2:	1	1	2	8	0	1	0	0	1	1:	-1-1 1	213	8	5/8
2:	1	2	2	5	1	0	1	1	0	2:	-1-1 1	213	4	
2:	1	2	3	3	0	1	1	1	0	2:	-1-1 1	213	4	

D= 216

GENUS#:	F11	F22	F33	F44	F12	F13	F23	F14	F24	F34	HASSE SYM	LEVEL	AUTOS	MASS

GENUS#	F11	F22	F33	F44	F12	F13	F23	F14	F24	F34	HASSE SYM	LEVEL	AUTOS	MASS
1:	1	1	1	18	1	0	0	0	0	0:	1 1	72	48	3/32
1:	1	1	3	5	0	0	0	1	1	0:	1 1	72	32	
1:	2	2	2	2	1	1	1	0	0	0:	1 1	72	24	
2:	1	1	1	27	1	1	0	0	0	0:	1 1	216	96	27/32
2:	1	1	2	7	0	0	0	0	1	0:	1 1	216	16	
2:	1	1	3	5	0	1	0	0	0	1:	1 1	216	8	
2:	1	1	4	5	0	1	1	0	0	4:	1 1	216	16	
2:	1	1	4	6	1	1	0	0	1	3:	1 1	216	12	
2:	1	2	2	4	1	0	0	0	1	0:	1 1	216	8	
2:	1	2	3	4	0	1	2	1	2	0:	1 1	216	8	
2:	1	2	3	4	1	0	2	1	0	2:	1 1	216	4	
3:	1	1	1	14	0	0	0	1	1	0:	-1-1	216	32	27/32
3:	1	1	2	8	0	1	0	1	0	0:	-1-1	216	8	
3:	1	1	3	6	0	1	1	1	1	0:	-1-1	216	16	
3:	1	2	2	4	0	0	0	0	1	2:	-1-1	216	8	
3:	1	2	2	5	1	0	1	0	2	2:	-1-1	216	4	
3:	2	2	2	3	2	1	0	0	0	2:	-1-1	216	4	
4:	1	1	2	9	0	1	1	0	0	0:	-1-1	72	32	3/32
4:	1	1	4	4	0	1	0	1	0	2:	-1-1	72	16	
5:	1	1	2	9	1	0	0	0	0	0:	1 1	72	48	3/16
5:	1	2	3	3	0	0	0	0	0	3:	1 1	72	48	
5:	1	2	3	3	0	0	0	1	2	0:	1 1	72	16	
5:	2	2	2	3	1	1	1	2	2	2:	1 1	72	12	
6:	1	1	3	6	1	0	0	0	0	0:	1 1	24	48	1/32
6:	1	3	3	3	0	0	0	0	3	3:	1 1	24	96	
7:	1	2	2	5	0	0	2	0	2	1:	-1-1	72	16	3/16
7:	1	2	3	3	0	1	1	1	1	0:	-1-1	72	8	
8:	2	2	2	3	2	1	-1	0	0	0:	-1-1	24	32	1/32

D= 217

GENUS#	F11	F22	F33	F44	F12	F13	F23	F14	F24	F34	HASSE SYM	LEVEL	AUTOS	MASS
1:	1	1	2	11	1	1	0	0	0	1:	-1 1-1	217	12	29/24
1:	1	1	2	8	0	1	0	0	1	0:	-1 1-1	217	16	
1:	1	1	4	5	0	1	1	1	0	4:	-1 1-1	217	16	
1:	1	2	2	4	1	0	0	0	0	1:	-1 1-1	217	8	
1:	1	2	2	6	1	0	2	0	2	1:	-1 1-1	217	4	
1:	1	2	3	3	1	0	1	0	1	-1:	-1 1-1	217	8	
1:	1	2	3	3	1	0	1	1	0	0:	-1 1-1	217	4	
1:	2	2	2	3	1	1	0	-1	1	1:	-1 1-1	217	4	
2:	1	1	3	6	0	0	0	0	1	3:	-1-1 1	217	8	29/24
2:	1	1	4	5	1	1	0	0	0	1:	-1-1 1	217	12	
2:	1	2	2	4	0	0	1	1	0	1:	-1-1 1	217	4	
2:	1	2	3	3	0	0	1	1	1	2:	-1-1 1	217	4	
2:	2	2	2	3	2	1	0	-1	1	0:	-1-1 1	217	2	

D= 220

GENUS#	F11	F22	F33	F44	F12	F13	F23	F14	F24	F34	HASSE SYM	LEVEL	AUTOS	MASS
1:	1	1	1	28	1	1	0	1	0	0:	-1-1 1	220	96	23/48
1:	1	1	2	7	0	0	0	0	0	1:	-1-1 1	220	16	
1:	1	1	2	8	0	1	0	0	0	1:	-1-1 1	220	8	
1:	1	1	3	6	0	1	1	1	1	1:	-1-1 1	220	32	
1:	1	2	3	4	0	1	2	1	0	3:	-1-1 1	220	8	
1:	2	2	2	3	2	1	0	1	2	0:	-1-1 1	220	8	
2:	1	1	2	10	1	0	0	1	0	2:	1 1 1	220	24	23/48
2:	1	2	2	6	1	0	2	0	1	2:	1 1 1	220	16	
2:	1	2	3	3	0	0	0	1	1	2:	1 1 1	220	4	
2:	1	2	3	3	0	1	1	1	1	1:	1 1 1	220	8	
3:	1	1	2	11	1	1	0	0	0	0:	1-1-1	220	24	23/48
3:	1	2	2	5	0	0	2	1	1	0:	1-1-1	220	8	
3:	1	2	3	3	0	0	2	1	0	0:	1-1-1	220	16	
3:	2	2	2	3	1	1	-1	2	2	0:	1-1-1	220	4	
4:	1	1	1	14	0	0	0	1	0	0:	-1 1-1	220	32	23/48
4:	1	1	3	5	0	1	0	0	0	0:	-1 1-1	220	16	
4:	1	1	4	4	0	1	1	0	0	1:	-1 1-1	220	16	
4:	1	1	4	4	0	0	0	0	0	3:	-1 1-1	220	32	
4:	1	1	4	5	1	1	0	0	0	0:	-1 1-1	220	24	
4:	1	2	2	4	0	0	1	0	1	-1:	-1 1-1	220	8	
4:	1	2	2	5	1	0	0	0	2	2:	-1 1-1	220	8	

D= 221

GENUS#	F11	F22	F33	F44	F12	F13	23	F14	F24	F34	HASSE SYM	LEVEL	AUTOS	MASS
1:	1	1	1	19	1	0	0	1	0	1:	-1 1-1	221	24	2/3

```
1:   1   1   3   6   0   1   1   0   1   2: -1 1-1    221     8
1:   1   1   4   5   1   0   0   1   0   1: -1 1-1    221    12
1:   1   1   5   5   1   1   0   1   1   4: -1 1-1    221    24
1:   1   2   3   3   0   1   0   0  -1   2: -1 1-1    221     4
1:   1   2   3   4   1   0   2   0   1   3: -1 1-1    221     8
2:   1   1   1  28   1   1   0   0   1   0: -1-1 1    221    48   2/3
2:   1   1   3   7   1   1   0   0   0   1: -1-1 1    221    12
2:   1   1   4   4   0   1   0   0   1   1: -1-1 1    221    16
2:   1   1   4   5   0   1   0   0   1   4: -1-1 1    221     8
2:   1   2   2   5   1   0   1   1   1   2: -1-1 1    221     8
2:   1   2   3   3   1   1   0   0   0   1: -1-1 1    221     4
```

D= 224

GENUS#:	F11	F22	F33	F44	F12	F13	F23	F14	F24	F34:	HASSE SYM	LEVEL	AUTOS	MASS
1:	1	1	1	14	0	0	0	0	0	0:	-1-1	56	96	5/48
1:	1	1	2	7	0	0	0	0	0	0:	-1-1	56	32	
1:	1	2	2	4	0	0	0	0	0	2:	-1-1	56	16	
2:	1	1	1	19	1	0	0	1	0	0:	1 1	224	24	5/12
2:	1	1	3	5	0	0	0	1	0	1:	1 1	224	8	
2:	1	2	3	4	1	1	0	0	2	2:	1 1	224	4	
3:	1	1	1	28	1	1	0	0	0	0:	-1-1	112	96	5/48
3:	1	1	4	5	0	0	0	1	1	4:	-1-1	112	32	
3:	1	2	3	3	0	1	0	1	0	2:	-1-1	112	16	
4:	1	1	2	10	0	1	1	1	1	0:	1 1	112	16	5/16
4:	1	2	2	5	1	1	0	0	1	1:	1 1	112	8	
4:	2	2	2	3	2	0	0	0	1	2:	1 1	112	8	
5:	1	1	3	6	0	1	1	0	0	2:	1 1	112	16	5/48
5:	1	1	4	5	1	0	0	1	0	0:	1 1	112	24	
6:	1	1	3	7	1	1	0	0	0	0:	-1-1	224	24	5/12
6:	1	2	2	4	0	0	0	-1	1	-1:	-1-1	224	8	
6:	1	2	3	3	1	0	0	0	0	2:	-1-1	224	16	
6:	1	3	3	3	0	0	2	-1	2	-2:	-1-1	224	16	
6:	2	2	2	3	1	1	0	1	2	2:	-1-1	224	8	
7:	1	1	2	8	0	1	0	0	0	0:	-1-1	224	16	5/12
7:	1	1	3	8	1	1	0	1	1	3:	-1-1	224	24	
7:	1	1	4	4	0	0	0	0	1	2:	-1-1	224	8	
7:	1	1	4	4	0	1	0	1	0	0:	-1-1	224	16	
7:	1	2	3	3	0	0	1	0	-1	2:	-1-1	224	8	
8:	1	1	4	6	1	0	0	1	0	4:	1 1	224	24	5/12
8:	1	2	2	5	1	1	0	1	0	0:	1 1	224	8	
8:	1	2	3	4	1	0	2	0	2	0:	1 1	224	4	
9:	1	1	2	8	0	0	0	1	1	2:	-1-1	112	32	5/16
9:	1	1	4	4	0	1	1	0	0	0:	-1-1	112	32	
9:	1	2	2	4	1	0	0	0	0	0:	-1-1	112	16	
9:	2	2	2	2	1	1	0	0	0	0:	-1-1	112	8	
9:	2	2	2	4	2	2	0	0	1	1:	-1-1	112	16	
10:	1	1	3	5	0	0	0	0	0	2:	1 1	56	16	5/48
10:	2	2	2	3	2	0	0	2	0	0:	1 1	56	24	

D= 225

GENUS#:	F11	F22	F33	F44	F12	F13	F23	F14	F24	F34:	HASSE SYM	LEVEL	AUTOS	MASS
1:	1	1	1	19	1	0	0	0	0	1:	-1-1 1	75	48	5/24
1:	1	2	2	5	1	1	0	1	1	0:	-1-1 1	75	8	
1:	2	2	2	4	2	1	-1	1	2	1:	-1-1 1	75	16	
2:	1	1	2	10	0	1	1	1	0	2:	-1-1 1	75	16	5/24
2:	1	1	3	7	1	0	0	0	0	3:	-1-1 1	75	48	
2:	1	2	3	4	1	0	0	0	2	3:	-1-1 1	75	8	
3:	1	1	2	12	1	1	0	1	1	2:	-1 1-1	45	24	1/6
3:	1	2	3	3	1	0	1	0	1	1:	-1 1-1	45	8	
4:	1	1	4	4	0	1	0	0	1	0:	-1 1-1	15	32	1/16
4:	2	2	2	2	1	0	0	0	0	1:	-1 1-1	15	32	
5:	1	1	5	5	1	0	0	0	0	5:	-1 1-1	15	144	1/9
5:	1	2	2	4	0	0	1	1	0	2:	-1 1-1	15	16	
5:	2	2	3	3	2	1	-1	1	2	2:	-1 1-1	15	24	
6:	1	2	2	6	1	0	2	1	1	1:	-1 1-1	45	8	1/6
6:	2	2	2	3	1	1	-1	-1	1	1:	-1 1-1	45	24	

D= 228

GENUS#:	F11	F22	F33	F44	F12	F13	F23	F14	F24	F34:	HASSE SYM	LEVEL	AUTOS	MASS
1:	1	1	1	15	0	0	0	1	1	1:	1-1-1	114	96	7/96
1:	1	1	2	10	1	0	0	0	0	2:	1-1-1	114	48	
1:	2	2	3	3	2	2	0	0	2	1:	1-1-1	114	24	

GENUS#	F11	F22	F33	F44	F12	F13	F23	F14	F24	F34:	HASSE	SYM	LEVEL	AUTOS	MASS
2:	1	1	1	19	1	0	0	0	0	0:	-1 -1	1	228	48	7/8
2:	1	1	3	5	0	0	0	0	1	0:	-1 -1	1	228	16	
2:	1	1	4	5	1	0	0	0	0	2:	-1 -1	1	228	24	
2:	1	2	2	4	0	0	1	0	1	1:	-1 -1	1	228	8	
2:	1	2	2	5	1	0	1	0	2	0:	-1 -1	1	228	4	
2:	1	2	3	3	1	1	0	0	0	0:	-1 -1	1	228	8	
2:	2	2	2	3	1	1	1	2	2	0:	-1 -1	1	228	4	
3:	1	1	1	29	1	1	0	1	0	0:	1 1	1	114	96	7/96
3:	1	1	5	5	1	1	0	0	1	3:	1 1	1	114	24	
3:	1	2	2	5	0	0	2	1	0	0:	1 1	1	114	48	
4:	1	1	2	12	1	1	0	0	1	1:	-1 1	-1	228	12	7/8
4:	1	1	3	6	0	1	0	0	0	3:	-1 1	-1	228	8	
4:	1	1	3	7	1	0	0	1	0	2:	-1 1	-1	228	12	
4:	1	1	4	6	1	1	0	1	1	3:	-1 1	-1	228	12	
4:	1	2	3	3	0	0	1	0	2	0:	-1 1	-1	228	4	
4:	1	2	3	4	1	0	2	1	2	0:	-1 1	-1	228	4	
5:	1	1	3	6	0	0	0	1	1	3:	-1 -1	1	114	32	21/32
5:	1	2	2	5	1	1	0	0	1	-1:	-1 -1	1	114	8	
5:	1	2	2	6	1	0	2	1	0	0:	-1 -1	1	114	8	
5:	2	2	2	3	2	1	0	1	0	-1:	-1 -1	1	114	4	
5:	2	2	2	4	2	1	-1	2	0	0:	-1 -1	1	114	8	
6:	1	1	2	10	0	1	1	1	1	1:	-1 1	-1	114	32	21/32
6:	1	2	3	3	1	0	1	0	1	0:	-1 1	-1	114	8	
6:	1	2	3	4	1	0	1	1	0	3:	-1 1	-1	114	4	
6:	1	2	3	4	0	0	2	1	0	3:	-1 1	-1	114	8	
6:	2	2	2	3	2	0	0	2	1	1:	-1 1	-1	114	8	

D= 229

GENUS#	F11	F22	F33	F44	F12	F13	F23	F14	F24	F34:	HASSE	SYM	LEVEL	AUTOS	MASS
1:	1	1	1	29	1	1	0	0	1	0:	-1 -1		229	48	27/16
1:	1	1	2	10	1	0	0	1	0	1:	-1 -1		229	12	
1:	1	1	2	9	0	1	0	0	1	2:	-1 -1		229	8	
1:	1	1	3	6	0	1	1	0	1	0:	-1 -1		229	8	
1:	1	1	3	8	1	1	0	0	1	2:	-1 -1		229	12	
1:	1	2	2	5	1	0	1	1	2	0:	-1 -1		229	4	
1:	1	2	3	3	0	1	1	1	0	0:	-1 -1		229	4	
1:	1	2	3	3	0	0	1	-1	-1	1:	-1 -1		229	4	
1:	1	2	3	4	0	1	2	0	1	3:	-1 -1		229	4	
1:	1	3	3	3	1	0	3	1	1	2:	-1 -1		229	4	

D= 232

GENUS#	F11	F22	F33	F44	F12	F13	F23	F14	F24	F34:	HASSE	SYM	LEVEL	AUTOS	MASS
1:	1	1	1	29	1	1	0	0	0	0:	-1 -1		232	96	33/32
1:	1	1	2	8	0	0	0	0	1	2:	-1 -1		232	16	
1:	1	1	5	5	1	0	0	1	0	4:	-1 -1		232	12	
1:	1	2	2	4	0	0	1	0	1	0:	-1 -1		232	4	
1:	1	2	2	5	0	0	2	0	1	1:	-1 -1		232	8	
1:	1	2	2	6	1	0	2	0	1	0:	-1 -1		232	8	
1:	1	2	3	4	0	1	2	1	2	1:	-1 -1		232	8	
1:	1	2	3	4	1	0	2	1	1	2:	-1 -1		232	4	
2:	1	1	1	15	0	0	0	1	1	0:	1 1		232	32	33/32
2:	1	1	2	10	1	0	0	1	0	0:	1 1		232	24	
2:	1	1	2	12	1	1	0	1	0	0:	1 1		232	12	
2:	1	2	2	4	0	0	0	1	1	0:	1 1		232	8	
2:	1	2	2	5	1	0	1	0	0	2:	1 1		232	4	
2:	1	2	3	3	0	0	0	0	2	1:	1 1		232	8	
2:	1	2	3	3	0	1	0	0	0	2:	1 1		232	8	
2:	2	2	3	3	2	2	0	1	2	2:	1 1		232	4	

D= 233

GENUS#	F11	F22	F33	F44	F12	F13	F23	F14	F24	F34:	HASSE	SYM	LEVEL	AUTOS	MASS
1:	1	1	1	20	1	0	0	1	0	1:	-1 -1		233	24	53/24
1:	1	1	2	10	0	1	1	1	0	0:	-1 -1		233	8	
1:	1	1	2	12	1	1	0	0	1	0:	-1 -1		233	12	
1:	1	1	3	6	0	1	0	1	1	2:	-1 -1		233	8	
1:	1	1	4	6	1	1	0	0	1	2:	-1 -1		233	12	
1:	1	2	2	5	1	1	0	0	1	0:	-1 -1		233	4	
1:	1	2	2	6	1	0	2	0	1	1:	-1 -1		233	4	
1:	1	2	3	3	1	0	1	0	0	1:	-1 -1		233	4	
1:	1	2	3	4	1	0	2	0	2	1:	-1 -1		233	4	
1:	1	2	3	4	1	0	1	0	2	3:	-1 -1		233	4	
1:	2	2	2	3	2	1	0	1	1	-1:	-1 -1		233	2	

```
D= 236
GENUS#:  F11 F22 F33 F44 F12 F13 F23 F14 F24 F34:  HASSE SYM   LEVEL  AUTOS   MASS
    1:    1   1   1  15   0   0   0   1   0   0:  -1-1        236     32    85/96
    1:    1   1   1  20   1   0   0   1   0   0:  -1-1        236     24
    1:    1   1   2   9   0   1   0   0   0   2:  -1-1        236      8
    1:    1   1   3   5   0   0   0   0   0   1:  -1-1        236     16
    1:    1   1   4   4   0   0   0   1   0   1:  -1-1        236      8
    1:    1   1   4   5   0   1   0   0   0   4:  -1-1        236      8
    1:    1   2   3   4   0   1   0   0   2   3:  -1-1        236      8
    1:    1   2   3   4   1   1   0   1   1   3:  -1-1        236      4
    2:    1   1   1  30   1   1   0   1   0   0:   1 1        236     96    85/96
    2:    1   1   2  10   0   1   1   0   0   1:   1 1        236     16
    2:    1   1   2   8   0   0   0   1   1   1:   1 1        236     16
    2:    1   1   3   6   0   1   1   0   0   1:   1 1        236     16
    2:    1   1   4   5   0   1   1   1   1   3:   1 1        236     16
    2:    1   2   2   5   1   0   1   1   0   1:   1 1        236      4
    2:    1   2   2   5   1   0   0   1   0   2:   1 1        236      8
    2:    2   2   2   3   1   0   0   0   2   2:   1 1        236      4

D= 237
GENUS#:  F11 F22 F33 F44 F12 F13 F23 F14 F24 F34:  HASSE SYM   LEVEL  AUTOS   MASS
    1:    1   1   1  20   1   0   0   0   0   1:  -1-1 1      237     48    35/48
    1:    1   1   2  10   1   0   0   0   0   1:  -1-1 1      237     24
    1:    1   1   2   9   0   1   0   1   1   0:  -1-1 1      237      8
    1:    1   1   4   5   1   0   0   0   0   1:  -1-1 1      237     24
    1:    1   2   2   5   0   0   1   1   2   2:  -1-1 1      237      8
    1:    1   3   3   3   1   1  -1   0   0   3:  -1-1 1      237      8
    1:    2   2   2   3   1   1   1   1   2   2:  -1-1 1      237      4
    2:    1   1   1  30   1   1   0   0   1   0:  -1 1-1      237     48    35/48
    2:    1   1   2  12   1   1   0   0   0   1:  -1 1-1      237     12
    2:    1   1   3   6   0   1   0   0   1   2:  -1 1-1      237      8
    2:    1   1   3   7   1   0   0   1   1   1:  -1 1-1      237     12
    2:    1   1   3   8   1   1   0   1   1   2:  -1 1-1      237     12
    2:    1   1   4   6   1   1   0   0   0   3:  -1 1-1      237     12
    2:    1   2   3   3   0   1   1   0   1   1:  -1 1-1      237      4

D= 240
GENUS#:  F11 F22 F33 F44 F12 F13 F23 F14 F24 F34:  HASSE SYM   LEVEL  AUTOS   MASS
    1:    1   1   1  15   0   0   0   0   0   0:  -1-1 1       60     96    1/16
    1:    1   1   4   4   0   0   0   0   0   2:  -1-1 1       60     32
    1:    2   2   2   3   2   0   0   0   0   2:  -1-1 1       60     48
    2:    1   1   1  20   1   0   0   0   0   0:  -1 1-1      240     48    1/2
    2:    1   1   3   6   0   0   0   1   0   3:  -1 1-1      240     16
    2:    1   1   3   8   1   1   0   0   1   1:  -1 1-1      240     12
    2:    1   1   4   5   1   0   0   0   0   0:  -1 1-1      240     48
    2:    1   2   2   4   0   0   1   0   0   0:  -1 1-1      240     16
    2:    1   2   2   5   1   1   0   0   0   0:  -1 1-1      240     16
    2:    1   2   3   3   0   0   1   0   1   2:  -1 1-1      240      8
    2:    1   3   3   3   0   0   1   0   3   3:  -1 1-1      240     16
    3:    1   1   1  30   1   1   0   0   0   0:  -1 1-1      120     96    1/16
    3:    1   1   4   6   1   0   0   0   0   4:  -1 1-1      120     48
    3:    2   2   3   3   0   0   0   2   2   3:  -1 1-1      120     32
    4:    1   1   2  12   1   1   0   0   0   0:   1 1 1      240     24    1/2
    4:    1   1   3   7   1   0   0   0   0   2:   1 1 1      240     24
    4:    1   1   3   7   1   0   0   1   0   0:   1 1 1      240     24
    4:    1   2   3   3   1   0   1   0   0   0:   1 1 1      240      8
    4:    1   2   3   4   0   0   2   1   1   3:   1 1 1      240      8
    4:    1   3   3   3   0   0   0   1   2   3:   1 1 1      240      8
    5:    1   1   2   8   0   0   0   0   0   2:  -1 1-1       60     32    1/16
    5:    1   1   3   5   0   0   0   0   0   0:  -1 1-1       60     32
    6:    1   1   2  10   1   0   0   0   0   0:   1 1 1      120     48    1/16
    6:    1   1   5   5   1   1   0   1   1   3:   1 1 1      120     24
    7:    1   1   2   8   0   0   0   1   1   0:  -1 1-1      120     32    3/16
    7:    1   2   2   6   1   1   0   1   2   2:  -1 1-1      120     16
    7:    2   2   2   2   1   0   0   0   0   0:  -1 1-1      120     32
    7:    2   2   2   3   0   0   0   2   2   1:  -1 1-1      120     16
    8:    1   1   2  10   0   0   1   0   0   0:  -1-1 1      120     32    3/16
    8:    1   2   2   4   0   0   0   1   0   0:  -1-1 1      120     32
    8:    2   2   2   3   2   0   0  -1   :   0:  -1-1 1      120     16
    8:    2   2   2   4   2   2   0   1   0   0:  -1-1 1      120     16
    9:    1   1   3   6   0   1   1   0   0   0:  -1-1 1      120     32    1/16
    9:    1   2   2   5   0   0   0   1   2   2:  -1-1 1      120     32
```

GENUS#	F11	F22	F33	F44	F12	F13	F23	F14	F24	F34:	HASSE SYM	LEVEL	AUTOS	MASS
10:	1	1	4	4	0	0	0	0	1	0:	-1-1 1	240	16	1/2
10:	1	1	4	5	0	0	0	1	0	4:	-1-1 1	240	16	
10:	1	2	2	4	0	0	0	0	-1	1:	-1-1 1	240	8	
10:	1	2	3	4	1	1	0	1	2	2:	-1-1 1	240	4	
11:	1	2	2	5	1	0	1	1	1	1:	1-1-1	240	4	1/2
11:	2	2	2	3	1	1	-1	0	2	0:	1-1-1	240	4	
12:	1	2	2	5	0	0	2	0	0	0:	1-1-1	60	48	1/16
12:	2	2	3	3	2	2	0	2	0	0:	1-1-1	60	24	
13:	1	2	2	6	1	0	2	0	0	0:	1-1-1	120	16	3/16
13:	2	2	2	4	2	1	-1	2	0	1:	1-1-1	120	8	
14:	1	2	3	3	0	1	0	1	0	0:	1-1-1	120	16	1/16
15:	1	2	3	3	0	0	0	0	2	0:	1 1 1	60	16	1/16
16:	1	2	3	4	0	0	0	1	2	3:	1 1 1	120	16	3/16
16:	1	2	3	4	1	0	1	1	1	3:	1 1 1	120	8	

D= 241

GENUS#	F11	F22	F33	F44	F12	F13	F23	F14	F24	F34:	HASSE SYM	LEVEL	AUTOS	MASS
1:	1	1	2	9	0	1	0	0	1	1:	-1-1	241	8	71/24
1:	1	1	3	6	0	1	0	1	1	0:	-1-1	241	8	
1:	1	1	4	5	0	1	1	1	0	3:	-1-1	241	8	
1:	1	1	4	6	1	1	0	1	1	2:	-1-1	241	12	
1:	1	2	2	5	1	0	1	0	-1	1:	-1-1	241	4	
1:	1	2	2	5	1	0	0	1	2	1:	-1-1	241	4	
1:	1	2	3	3	0	0	1	1	1	0:	-1-1	241	4	
1:	1	2	3	4	1	0	2	1	0	1:	-1-1	241	4	
1:	1	2	3	4	0	1	1	1	2	3:	-1-1	241	4	
1:	1	2	3	4	1	0	0	0	3	3:	-1-1	241	4	
1:	2	2	2	3	2	1	0	0	1	-1:	-1-1	241	2	
1:	2	2	2	3	1	1	0	2	1	2:	-1-1	241	2	

D= 244

GENUS#	F11	F22	F33	F44	F12	F13	F23	F14	F24	F34:	HASSE SYM	LEVEL	AUTOS	MASS
1:	1	1	1	16	0	0	0	1	1	1:	1 1	122	96	55/96
1:	1	1	4	5	0	1	1	0	0	3:	1 1	122	16	
1:	1	2	2	5	1	0	0	0	1	2:	1 1	122	8	
1:	1	2	2	6	0	0	2	1	2	0:	1 1	122	24	
1:	2	2	2	3	2	1	0	1	0	1:	1 1	122	4	
1:	2	2	2	4	2	2	0	1	1	0:	1 1	122	12	
2:	1	1	1	31	1	1	0	1	0	0:	-1-1	122	96	55/96
2:	1	1	2	11	1	0	0	1	0	2:	-1-1	122	24	
2:	1	1	3	7	0	1	1	0	0	3:	-1-1	122	16	
2:	1	1	3	8	1	1	0	0	0	2:	-1-1	122	12	
2:	1	2	3	4	0	1	2	0	2	0:	-1-1	122	8	
2:	1	3	3	3	1	1	-1	1	0	0:	1 1	122	4	
3:	1	1	2	8	0	0	0	0	1	1:	-1-1	244	8	55/24
3:	1	1	2	9	0	1	0	1	0	0:	-1-1	244	8	
3:	1	1	3	6	0	1	0	1	0	2:	-1-1	244	8	
3:	1	1	3	8	1	1	0	1	0	0:	-1-1	244	12	
3:	1	1	4	6	1	1	0	0	1	1:	-1-1	244	12	
3:	1	2	3	3	0	0	0	1	1	1:	-1-1	244	4	
3:	1	2	3	3	0	0	0	0	1	2:	-1-1	244	4	
3:	1	2	3	4	1	0	2	0	0	2:	-1-1	244	4	
3:	1	2	3	4	0	1	1	0	2	3:	-1-1	244	4	
3:	1	3	3	3	1	0	3	0	0	2:	-1-1	244	4	
3:	2	2	2	3	1	1	0	2	2	0:	-1-1	244	2	

D= 245

GENUS#	F11	F22	F33	F44	F12	F13	F23	F14	F24	F34:	HASSE SYM	LEVEL	AUTOS	MASS
1:	1	1	1	21	1	0	0	1	0	1:	-1-1 1	245	24	35/48
1:	1	1	2	13	1	1	0	1	1	2:	-1-1 1	245	24	
1:	1	1	3	7	0	1	1	1	0	3:	-1-1 1	245	16	
1:	1	1	4	6	1	0	0	1	0	3:	-1-1 1	245	12	
1:	1	2	2	5	1	0	1	1	0	0:	-1-1 1	245	4	
1:	1	2	3	4	1	0	2	0	1	-1:	-1-1 1	245	4	
2:	1	1	1	31	1	1	0	0	1	0:	-1-1 1	245	48	35/48
2:	1	1	3	8	1	1	0	0	1	0:	-1-1 1	245	12	
2:	1	1	4	5	0	1	0	1	1	3:	-1-1 1	245	8	
2:	1	2	3	3	0	1	0	0	1	1:	-1-1 1	245	4	
2:	2	2	2	3	1	1	-1	1	2	0:	-1-1 1	245	4	
3:	1	1	2	9	0	1	0	0	1	1:	-1-1 1	35	16	5/48
3:	1	3	3	3	1	1	-1	1	2	2:	-1-1 1	35	24	
4:	1	1	5	5	1	1	0	0	1	2:	-1 1-1	35	24	5/48

GENUS#:	F11	F22	F33	F44	F12	F13	F23	F14	F24	F34:	HASSE SYM	LEVEL	AUTOS	MASS
4:	1	2	3	3	1	0	0	0	0	1:	-1 1-1	35	16	

D= 248

GENUS#:	F11	F22	F33	F44	F12	F13	F23	F14	F24	F34:	HASSE SYM	LEVEL	AUTOS	MASS
1:	1	1	1	21	1	0	0	1	0	0:	1 1	248	24	7/8
1:	1	1	2	11	0	1	1	1	1	0:	1 1	248	16	
1:	1	1	3	6	0	0	0	1	1	2:	1 1	248	16	
1:	1	1	3	6	0	1	0	0	0	2:	1 1	248	8	
1:	1	1	4	6	1	1	0	1	0	0:	1 1	248	12	
1:	1	2	3	4	1	0	2	0	1	2:	1 1	248	4	
1:	2	2	2	3	2	1	0	1	0	0:	1 1	248	4	
2:	1	1	1	16	0	0	0	1	1	0:	-1-1	248	32	7/8
2:	1	1	1	31	1	1	0	0	0	0:	-1-1	248	96	
2:	1	1	2	13	1	1	0	0	1	1:	-1-1	248	12	
2:	1	1	2	8	0	0	0	1	0	0:	-1-1	248	16	
2:	1	1	2	9	0	1	0	0	0	1:	-1-1	248	8	
2:	1	1	4	5	0	1	1	1	1	0:	-1-1	248	16	
2:	1	2	2	4	0	0	0	0	1	0:	-1-1	248	8	
2:	1	2	2	5	1	0	0	0	2	0:	-1-1	248	8	
2:	1	2	2	7	1	0	2	0	2	0:	-1-1	248	8	
2:	1	2	3	4	0	0	2	0	0	3:	-1-1	248	8	

D= 249

GENUS#:	F11	F22	F33	F44	F12	F13	F23	F14	F24	F34:	HASSE SYM	LEVEL	AUTOS	MASS
1:	1	1	1	21	1	0	0	0	0	1:	-1 1-1	249	48	23/16
1:	1	1	3	6	0	1	0	0	1	1:	-1 1-1	249	8	
1:	1	1	3	7	1	0	0	0	1	1:	-1 1-1	249	24	
1:	1	1	3	8	1	0	0	1	0	3:	-1 1-1	249	24	
1:	1	1	4	6	1	1	0	0	1	0:	-1 1-1	249	12	
1:	1	2	2	6	1	1	0	0	2	1:	-1 1-1	249	8	
1:	1	2	3	4	1	0	2	1	1	1:	-1 1-1	249	4	
1:	1	2	3	4	0	0	1	1	2	3:	-1 1-1	249	4	
1:	1	2	3	4	1	0	0	1	0	3:	-1 1-1	249	8	
1:	1	3	3	3	1	0	0	0	2	3:	-1 1-1	249	8	
1:	2	2	2	3	1	0	0	2	1	1:	-1 1-1	249	4	
2:	1	1	2	11	0	1	1	1	0	2:	-1-1 1	249	16	23/16
2:	1	1	4	5	0	1	0	0	1	3:	-1-1 1	249	8	
2:	1	2	2	5	1	0	0	1	0	1:	-1-1 1	249	4	
2:	1	2	3	4	1	1	0	0	2	1:	-1-1 1	249	4	
2:	2	2	2	3	2	1	0	0	1	0:	-1-1 1	249	2	
2:	2	2	2	4	2	1	-1	1	2	0:	-1-1 1	249	4	

D= 252

GENUS#:	F11	F22	F33	F44	F12	F13	F23	F14	F24	F34:	HASSE SYM	LEVEL	AUTOS	MASS
1:	1	1	1	21	1	0	0	0	0	0:	-1-1 1	84	48	1/12
1:	1	1	2	11	0	1	1	1	1	1:	-1-1 1	84	32	
1:	1	1	3	6	0	0	0	0	0	3:	-1-1 1	84	32	
2:	1	1	1	32	1	1	0	1	0	0:	-1 1-1	252	96	1/2
2:	1	1	2	8	0	0	0	0	0	1:	-1 1-1	252	16	
2:	1	1	3	6	0	1	0	1	0	0:	-1 1-1	252	8	
2:	1	1	4	5	0	1	1	1	1	1:	-1 1-1	252	32	
2:	1	1	4	6	1	1	0	0	0	2:	-1 1-1	252	12	
2:	1	2	2	5	1	0	0	0	0	2:	-1 1-1	252	16	
2:	1	2	3	3	0	0	1	0	1	-1:	-1 1-1	252	8	
3:	1	1	2	13	1	1	0	1	0	0:	1 1 1	252	12	1/2
3:	1	1	5	6	1	1	0	1	1	5:	1 1 1	252	24	
3:	1	2	3	4	1	0	2	1	0	0:	1 1 1	252	4	
3:	1	2	3	4	0	0	2	1	2	0:	1 1 1	252	8	
4:	1	1	1	16	0	0	0	0	0	1:	-1 1-1	252	32	1/2
4:	1	1	2	9	0	1	0	0	0	0:	-1 1-1	252	16	
4:	1	1	4	4	0	0	0	0	0	1:	-1 1-1	252	32	
4:	1	1	4	5	0	0	0	1	1	3:	-1 1-1	252	16	
4:	1	2	3	4	0	1	2	0	2	1:	-1 1-1	252	16	
4:	2	2	2	3	2	1	0	0	0	1:	-1 1-1	252	4	
5:	1	1	3	7	1	0	0	0	0	0:	1-1-1	84	48	1/12
5:	1	2	3	3	1	1	0	0	0	0:	1-1-1	84	32	
5:	2	2	2	4	2	1	-1	-1	1	-1:	1-1-1	84	32	
6:	1	1	5	5	1	0	0	0	0	4:	-1 1-1	84	48	1/3
6:	1	2	2	5	0	0	1	0	2	2:	-1 1-1	84	8	
6:	1	2	3	4	0	1	2	1	0	2:	-1 1-1	84	8	
6:	2	2	3	3	2	0	0	1	2	3:	-1 1-1	84	16	
7:	1	2	2	5	1	0	1	0	1	0:	1 1 1	252	4	1/2

```
7:   1   2   2   7   1   0   2   0   0   2: 1 1 1      252    8
7:   2   2   3   3   2   2   0   0   1   2: 1 1 1      252    8
8:   1   1   2  11   1   0   0   0   0   2: 1 1 1       84   48    1/3
8:   1   2   2   6   0   0   2   1   1   2: 1 1 1       84   16
8:   1   2   3   3   0   1   1   0   0   0: 1 1 1       84    8
8:   2   2   3   3   1   2  -1   1   1   3: 1 1 1       84    8
```

D= 253

GENUS#:	F11	F22	F33	F44	F12	F13	F23	F14	F24	F34:	HASSE	SYM	LEVEL	AUTOS	MASS
1:	1	1	1	32	1	0	0	0	1	0:	-1 1-1		253	48	15/16
1:	1	1	2	11	1	0	0	1	0	1:	-1 1-1		253	12	
1:	1	1	3	8	1	1	0	0	0	1:	-1 1-1		253	12	
1:	1	2	2	5	1	0	1	0	0	1:	-1 1-1		253	4	
1:	1	2	3	4	0	1	2	1	1	2:	-1 1-1		253	4	
1:	1	3	3	3	1	0	1	1	3	2:	-1 1-1		253	4	
2:	1	1	2	13	1	1	0	0	1	0:	-1-1 1		253	12	15/16
2:	1	1	3	6	0	1	0	0	1	0:	-1-1 1		253	16	
2:	1	1	4	7	1	1	0	1	1	4:	-1-1 1		253	24	
2:	1	1	5	5	1	1	0	0	0	3:	-1-1 1		253	12	
2:	1	1	5	5	1	1	0	1	1	2:	-1-1 1		253	24	
2:	1	2	2	5	0	0	1	1	2	0:	-1-1 1		253	4	
2:	1	2	3	3	0	1	0	0	1	0:	-1-1 1		253	8	
2:	2	2	3	3	1	2	0	1	2	3:	-1-1 1		253	4	

D= 256

GENUS#:	F11	F22	F33	F44	F12	F13	F23	F14	F24	F34:	HASSE	SYM	LEVEL	AUTOS	MASS
1:	1	1	1	16	0	0	0	0	0	0:	1		64	96	1/24
1:	1	2	2	5	0	0	0	0	2	2:	1		64	32	
2:	1	1	1	32	1	1	0	0	0	0:	1		128	96	1/6
2:	1	1	2	9	0	0	0	1	1	2:	1		128	32	
2:	1	2	3	4	0	1	2	0	0	2:	1		128	8	
3:	1	1	2	11	1	0	0	1	0	0:	1		128	24	1/6
3:	1	2	3	3	0	1	0	0	0	1:	1		128	8	
4:	1	1	2	8	0	0	0	0	0	0:	1		32	32	1/16
4:	2	2	2	3	0	0	2	2	0	0:	1		32	32	
5:	1	1	3	7	0	1	1	1	1	0:	1		128	16	1/6
5:	1	1	3	9	1	1	0	1	1	3:	1		128	24	
5:	2	2	3	3	0	2	0	2	3	3:	1		128	16	
6:	1	1	3	8	1	1	0	0	0	0:	1		32	24	1/6
6:	1	3	3	3	0	0	2	1	3	1:	1		32	8	
7:	1	1	4	4	0	0	0	0	0	0:	1		16	64	1/64
8:	1	1	4	5	0	0	0	0	0	4:	1		64	32	1/24
8:	2	2	2	5	2	2	0	2	0	0:	1		64	96	
9:	1	1	5	5	1	1	0	0	1	1:	1		128	24	1/6
9:	1	3	3	3	1	0	3	1	1	1:	1		128	8	
10:	1	2	2	4	0	0	0	0	0	0:	1		16	32	1/32
11:	1	2	2	6	0	0	2	0	2	0:	1		64	24	1/24
12:	1	2	3	3	0	0	0	0	0	2:	1		32	16	1/16
13:	1	3	3	3	0	0	2	0	2	-2:	1		8	96	1/96
14:	2	2	3	3	2	2	2	2	2	2:	1		16	64	1/64
15:	2	2	3	3	2	2	0	0	2	0:	1		64	24	1/24

D= 257

GENUS#:	F11	F22	F33	F44	F12	F13	F23	F14	F24	F34:	HASSE	SYM	LEVEL	AUTOS	MASS
1:	1	1	1	22	1	0	0	1	0	1:	-1-1		257	24	5/2
1:	1	1	2	10	0	1	0	0	1	2:	-1-1		257	8	
1:	1	1	2	11	0	1	1	1	0	0:	-1-1		257	8	
1:	1	1	2	13	1	1	0	0	0	1:	-1-1		257	12	
1:	1	1	4	5	0	0	1	1	0	2:	-1-1		257	8	
1:	1	2	2	5	1	0	0	1	1	1:	-1-1		257	4	
1:	1	2	2	6	1	1	0	1	2	1:	-1-1		257	4	
1:	1	2	2	7	1	0	2	0	2	1:	-1-1		257	4	
1:	1	2	3	4	1	0	1	0	0	3:	-1-1		257	4	
1:	1	2	3	4	1	1	0	1	2	1:	-1-1		257	4	
1:	1	2	3	4	1	0	1	1	2	2:	-1-1		257	4	
1:	2	2	2	3	1	1	0	0	1	2:	-1-1		257	2	

D= 260

GENUS#:	F11	F22	F33	F44	F12	F13	F23	F14	F24	F34:	HASSE	SYM	LEVEL	AUTOS	MASS
1:	1	1	1	17	0	0	0	1	1	1:	1 1 1		130	96	1/12
1:	1	1	3	7	0	1	1	1	1	1:	1 1 1		130	32	
1:	1	1	5	5	1	1	0	1	0	0:	1 1 1		130	24	

GENUS#:	F11	F22	F33	F44	F12	F13	F23	F14	F24	F34:	HASSE SYM	LEVEL	AUTOS	MASS
2:	1	1	1	22	1	0	0	1	0	0:	-1-1 1	260	24	1/1
2:	1	1	3	6	0	0	0	0	1	2:	-1-1 1	260	8	
2:	1	1	4	6	1	0	0	1	1	2:	-1-1 1	260	12	
2:	1	2	2	5	1	0	1	0	0	0:	-1-1 1	260	8	
2:	1	2	3	3	0	0	1	0	1	1:	-1-1 1	260	8	
2:	1	2	3	4	1	0	2	0	1	0:	-1-1 1	260	4	
2:	2	2	2	3	1	1	-1	-1	0	1:	-1-1 1	260	4	
3:	1	1	1	33	1	1	0	1	0	0:	1-1-1	130	96	1/12
3:	1	1	5	5	0	0	0	1	1	5:	1-1-1	130	32	
3:	2	2	3	3	2	2	0	2	0	1:	1-1-1	130	24	
4:	1	1	2	11	0	1	1	0	0	1:	-1 1-1	130	16	3/4
4:	1	2	3	4	0	0	2	1	2	1:	-1 1-1	130	16	
4:	1	2	3	4	1	0	1	0	2	2:	-1 1-1	130	4	
4:	1	2	4	4	1	0	2	1	1	4:	-1 1-1	130	8	
4:	2	2	2	3	2	0	0	1	1	1:	-1 1-1	130	4	
5:	1	1	2	13	1	1	0	0	0	0:	-1 1-1	260	24	1/1
5:	1	1	3	6	0	1	0	0	0	1:	-1 1-1	260	8	
5:	1	1	4	5	0	1	0	1	0	3:	-1 1-1	260	8	
5:	1	1	4	7	1	1	0	0	1	3:	-1 1-1	260	12	
5:	1	2	3	4	1	0	0	1	2	2:	-1 1-1	260	4	
5:	1	2	3	4	0	0	2	0	1	-2:	-1 1-1	260	8	
5:	1	2	3	4	1	1	0	0	2	0:	-1 1-1	260	4	
6:	1	1	3	6	0	0	0	1	1	1:	-1-1 1	130	16	3/4
6:	1	2	2	6	1	1	0	0	2	0:	-1-1 1	130	4	
6:	1	2	2	7	1	0	2	0	1	2:	-1-1 1	130	16	
6:	2	2	2	3	1	0	0	-1	1	-2:	-1-1 1	130	8	
6:	2	2	2	4	2	1	0	1	2	2:	-1-1 1	130	4	

D= 261

GENUS#:	F11	F22	F33	F44	F12	F13	F23	F14	F24	F34:	HASSE SYM	LEVEL	AUTOS	MASS
1:	1	1	1	22	1	0	0	0	0	1:	-1 1-1	87	48	5/16
1:	1	1	4	5	0	1	0	1	1	2:	-1 1-1	87	8	
1:	1	1	4	6	1	0	0	0	0	3:	-1 1-1	87	24	
1:	1	2	3	4	1	0	0	0	1	3:	-1 1-1	87	8	
2:	1	1	1	33	1	1	0	0	1	0:	-1 1-1	261	48	15/16
2:	1	1	3	7	0	1	0	0	1	3:	-1 1-1	261	8	
2:	1	1	3	9	1	1	0	0	1	2:	-1 1-1	261	12	
2:	1	1	4	6	1	1	0	0	0	1:	-1 1-1	261	12	
2:	1	1	5	5	1	1	0	0	1	0:	-1 1-1	261	24	
2:	1	1	5	6	1	1	0	0	1	4:	-1 1-1	261	12	
2:	1	2	3	4	0	1	2	1	1	1:	-1 1-1	261	8	
2:	1	2	3	4	1	0	2	0	1	1:	-1 1-1	261	4	
2:	1	3	3	3	0	0	3	1	2	1:	-1 1-1	261	8	
3:	1	1	2	11	1	0	0	0	0	1:	-1-1 1	87	24	5/16
3:	1	1	3	8	1	0	0	0	0	3:	-1-1 1	87	48	
3:	1	3	3	3	1	1	0	0	3	0:	-1-1 1	87	8	
3:	2	2	3	3	1	0	0	2	2	3:	-1-1 1	87	8	
4:	1	1	3	7	0	1	1	0	1	2:	-1 1-1	261	8	15/16
4:	1	1	5	5	0	1	0	0	1	5:	-1 1-1	261	16	
4:	1	2	3	4	0	1	2	0	1	2:	-1 1-1	261	4	
4:	1	2	3	4	1	1	0	0	1	2:	-1 1-1	261	4	
4:	2	2	2	3	1	1	1	2	1	0:	-1 1-1	261	4	

D= 264

GENUS#:	F11	F22	F33	F44	F12	F13	F23	F14	F24	F34:	HASSE SYM	LEVEL	AUTOS	MASS
1:	1	1	1	17	0	0	0	1	1	0:	1-1-1	264	32	7/12
1:	1	1	1	22	1	0	0	0	0	0:	1-1-1	264	48	
1:	1	1	3	6	0	0	0	1	1	0:	1-1-1	264	32	
1:	1	2	2	5	0	0	0	1	2	1:	1-1-1	264	8	
1:	2	2	2	3	1	1	1	0	2	0:	1-1-1	264	4	
1:	2	2	2	3	2	1	0	0	0	0:	1-1-1	264	8	
2:	1	1	1	33	1	1	0	0	0	0:	1 1 1	264	96	7/12
2:	1	1	2	11	0	1	1	0	0	0:	1 1 1	264	32	
2:	1	1	2	9	0	0	0	1	0	2:	1 1 1	264	16	
2:	1	1	3	6	0	1	0	0	0	0:	1 1 1	264	16	
2:	1	1	4	6	1	1	0	0	0	0:	1 1 1	264	24	
2:	1	2	3	4	0	1	1	0	0	3:	1 1 1	264	4	
2:	1	3	3	3	1	1	0	0	2	2:	1 1 1	264	8	
3:	1	1	2	10	0	1	0	0	0	2:	-1-1 1	264	8	7/12
3:	1	1	2	11	1	0	0	0	0	0:	-1-1 1	264	48	
3:	1	1	3	7	0	1	1	0	0	2:	-1-1 1	264	16	
3:	1	2	3	3	0	0	0	1	0	0:	-1-1 1	264	16	

GENUS#	F11	F22	F33	F44	F12	F13	F23	F14	F24	F34:	HASSE SYM	LEVEL	AUTOS	MASS
3:	1	2	3	4	0	0	0	0	2	3:	-1-1 1	264	16	
3:	2	2	2	3	1	0	0	2	2	0:	-1-1 1	264	8	
3:	2	2	2	4	2	1	-1	1	1	1:	-1-1 1	264	8	
4:	1	1	3	8	1	0	0	1	1	2:	-1 1-1	264	12	7/12
4:	1	1	4	5	0	1	1	0	0	2:	-1 1-1	264	16	
4:	1	1	4	5	0	1	0	0	0	3:	-1 1-1	264	8	
4:	1	2	2	5	1	0	0	1	0	0:	-1 1-1	264	8	
4:	1	2	2	6	0	0	2	0	1	-1:	-1 1-1	264	16	
4:	1	2	3	3	0	0	1	0	1	0:	-1 1-1	264	8	

D= 265

GENUS#	F11	F22	F33	F44	F12	F13	F23	F14	F24	F34:	HASSE SYM	LEVEL	AUTOS	MASS
1:	1	1	2	10	0	1	0	1	1	0:	-1-1 1	265	8	5/3
1:	1	1	5	6	1	0	0	1	0	5:	-1-1 1	265	24	
1:	1	2	2	5	0	0	1	1	-1	1:	-1-1 1	265	8	
1:	1	2	2	6	1	0	1	1	0	2:	-1-1 1	265	4	
1:	1	2	2	7	1	0	2	1	1	1:	-1-1 1	265	8	
1:	1	2	3	4	1	0	2	0	0	1:	-1-1 1	265	4	
1:	2	2	2	3	1	1	-1	-1	-1	1:	-1-1 1	265	4	
1:	2	2	2	3	1	1	0	1	2	1:	-1-1 1	265	2	
2:	1	1	2	14	1	1	0	1	1	2:	-1 1-1	265	24	5/3
2:	1	1	4	5	0	1	1	1	0	0:	-1 1-1	265	8	
2:	1	2	2	5	1	0	0	0	1	1:	-1 1-1	265	4	
2:	1	2	3	4	0	1	1	1	0	3:	-1 1-1	265	4	
2:	1	2	3	4	1	0	1	1	0	2:	-1 1-1	265	4	
2:	1	2	3	4	0	1	1	1	2	0:	-1 1-1	265	4	
2:	1	2	3	4	1	1	0	1	1	2:	-1 1-1	265	4	
2:	2	2	3	3	2	1	0	2	1	3:	-1 1-1	265	4	

D= 268

GENUS#	F11	F22	F33	F44	F12	F13	F23	F14	F24	F34:	HASSE SYM	LEVEL	AUTOS	MASS
1:	1	1	1	34	1	1	0	1	0	0:	1 1	268	96	41/32
1:	1	1	2	14	1	1	0	0	1	1:	1 1	268	12	
1:	1	1	2	9	0	0	0	1	1	1:	1 1	268	16	
1:	1	2	2	6	1	0	1	0	2	2:	1 1	268	4	
1:	1	2	2	6	0	0	2	1	0	1:	1 1	268	8	
1:	1	2	3	4	0	0	2	1	0	2:	1 1	268	8	
1:	1	2	3	4	0	1	2	1	0	0:	1 1	268	8	
1:	2	2	3	3	2	1	-1	0	2	0:	1 1	268	4	
1:	2	2	3	3	0	1	0	2	2	3:	1 1	268	4	
2:	1	1	1	17	0	0	0	0	1	0:	-1-1	268	32	41/32
2:	1	1	2	12	1	0	0	1	0	2:	-1-1	268	24	
2:	1	1	4	5	0	0	0	1	1	3:	-1-1	268	8	
2:	1	1	5	5	1	0	0	1	1	2:	-1-1	268	12	
2:	1	2	2	5	0	0	1	0	0	2:	-1-1	268	4	
2:	1	2	2	7	1	0	2	1	0	0:	-1-1	268	8	
2:	1	2	3	3	0	0	0	0	1	1:	-1-1	268	4	
2:	1	2	3	4	0	1	0	1	2	2:	-1-1	268	8	
2:	1	2	3	4	1	1	0	1	0	2:	-1-1	268	4	

D= 269

GENUS#	F11	F22	F33	F44	F12	F13	F23	F14	F24	F34:	HASSE SYM	LEVEL	AUTOS	MASS
1:	1	1	1	23	1	0	0	1	0	1:	-1-1	269	24	83/48
1:	1	1	1	34	1	1	0	0	1	0:	-1-1	269	48	
1:	1	1	2	10	0	1	0	0	1	1:	-1-1	269	8	
1:	1	1	3	7	0	1	1	1	0	0:	-1-1	269	8	
1:	1	1	3	9	1	1	0	1	1	2:	-1-1	269	12	
1:	1	1	4	5	0	1	0	0	1	2:	-1-1	269	8	
1:	1	1	4	5	0	1	0	1	1	0:	-1-1	269	8	
1:	1	1	4	6	1	0	0	1	0	1:	-1-1	269	12	
1:	1	2	3	4	0	1	0	1	1	3:	-1-1	269	4	
1:	1	2	3	4	1	1	0	0	-1	2:	-1-1	269	4	
1:	1	2	3	5	1	0	2	1	1	3:	-1-1	269	4	
1:	1	3	3	3	1	1	-1	0	2	1:	-1-1	269	4	

D= 272

GENUS#	F11	F22	F33	F44	F12	F13	F23	F14	F24	F34:	HASSE SYM	LEVEL	AUTOS	MASS
1:	1	1	1	17	0	0	0	0	0	0:	1 1	68	96	1/16
1:	1	1	2	9	0	0	0	0	0	2:	1 1	68	32	
1:	2	2	2	5	2	2	0	0	2	0:	1 1	68	48	
2:	1	1	1	23	1	0	0	1	0	0:	-1-1	272	24	1/1
2:	1	1	2	10	0	1	0	1	0	0:	-1-1	272	8	

	F11	F22	F33	F44	F12	F13	F23	F14	F24	F34	HASSE SYM	LEVEL	AUTOS	MASS
2:	1	1	3	6	0	0	0	1	0	1:	-1-1	272	8	
2:	1	1	3	7	0	1	0	0	0	3:	-1-1	272	8	
2:	1	1	4	7	1	1	0	1	1	3:	-1-1	272	12	
2:	1	2	3	4	0	0	1	0	2	3:	-1-1	272	4	
2:	1	2	3	4	1	0	1	0	1	-2:	-1-1	272	4	
3:	1	1	1	34	1	1	0	0	0	0:	1 1	136	96	1/4
3:	1	1	2	9	0	0	0	1	1	0:	1 1	136	32	
3:	1	1	3	9	1	1	0	0	1	1:	1 1	136	12	
3:	1	1	4	5	0	0	0	1	1	2:	1 1	136	16	
3:	1	2	2	5	0	0	0	1	0	2:	1 1	136	16	
4:	1	1	2	12	0	1	1	1	1	0:	-1-1	136	16	3/4
4:	1	1	4	6	0	1	1	0	0	4:	-1-1	136	16	
4:	1	2	2	5	1	0	0	0	1	0:	-1-1	136	8	
4:	1	2	2	7	1	0	2	0	1	0:	-1-1	136	8	
4:	2	2	2	3	2	0	0	1	1	0:	-1-1	136	8	
4:	2	2	2	4	2	1	0	0	2	1:	-1-1	136	4	
5:	1	1	2	14	1	1	0	1	0	0:	-1-1	272	12	1/1
5:	1	1	4	6	1	0	0	1	0	0:	-1-1	272	24	
5:	1	2	2	6	1	1	0	0	1	1:	-1-1	272	8	
5:	1	2	3	4	1	0	2	0	0	0:	-1-1	272	8	
5:	1	2	3	4	0	0	2	1	1	2:	-1-1	272	4	
5:	1	3	3	3	1	1	-1	0	2	-2:	-1-1	272	8	
5:	2	2	2	3	1	1	0	2	0	0:	-1-1	272	4	
6:	1	1	3	6	0	0	0	0	0	2:	-1-1	68	16	3/16
6:	1	2	3	4	0	0	2	0	2	0:	-1-1	68	8	
7:	1	1	4	7	1	0	0	1	0	4:	-1-1	68	24	1/12
7:	1	3	3	3	1	1	-1	-1	-2	1:	-1-1	68	24	
8:	1	2	2	6	1	1	0	1	0	0:	-1-1	68	8	3/4
8:	1	2	3	4	1	0	1	0	2	0:	-1-1	68	4	
8:	1	2	4	4	1	0	0	0	2	4:	-1-1	68	8	
8:	2	2	2	3	1	1	0	-1	1	1:	-1-1	68	4	

D= 273

GENUS#:	F11	F22	F33	F44	F12	F13	F23	F14	F24	F34:	HASSE SYM	LEVEL	AUTOS	MASS
1:	1	1	1	23	1	0	0	0	0	1:	-1-1 1 1	273	48	37/48
1:	1	2	2	6	1	1	0	1	0	0:	-1-1 1 1	273	8	
1:	1	2	2	6	1	0	1	1	1	2:	-1-1 1 1	273	8	
1:	1	2	2	7	1	0	2	0	0	1:	-1-1 1 1	273	4	
1:	2	2	2	3	1	1	1	1	1	-1:	-1-1 1 1	273	4	
2:	1	1	2	10	0	1	0	0	1	0:	-1-1-1-1	273	16	37/48
2:	1	1	5	5	1	0	0	0	0	3:	-1-1-1-1	273	48	
2:	1	2	2	5	0	0	1	1	1	1:	-1-1-1-1	273	8	
2:	1	2	3	4	1	0	0	0	0	3:	-1-1-1-1	273	16	
2:	2	2	2	3	1	1	0	2	1	0:	-1-1-1-1	273	4	
2:	2	2	2	4	2	1	-1	1	0	1:	-1-1-1-1	273	4	
3:	1	1	2	12	0	1	1	1	0	2:	-1 1 1-1	273	16	37/48
3:	1	1	2	14	1	1	0	0	1	0:	-1 1 1-1	273	12	
3:	1	2	3	4	1	0	1	1	1	2:	-1 1 1-1	273	4	
3:	1	2	3	4	1	0	1	1	2	0:	-1 1 1-1	273	4	
3:	1	2	3	4	0	1	1	1	2	1:	-1 1 1-1	273	8	
4:	1	1	3	8	1	0	0	1	0	1:	-1 1-1 1	273	12	37/48
4:	1	1	4	6	0	1	1	1	0	4:	-1 1-1 1	273	16	
4:	1	2	2	5	1	0	0	0	0	1:	-1 1-1 1	273	8	
4:	1	2	3	4	0	0	1	1	0	3:	-1 1-1 1	273	4	
4:	1	3	3	3	1	0	2	0	-2	1:	-1 1-1 1	273	8	
4:	2	2	3	3	2	1	-1	2	1	2:	-1 1-1 1	273	8	

D= 276

GENUS#:	F11	F22	F33	F44	F12	F13	F23	F14	F24	F34:	HASSE SYM	LEVEL	AUTOS	MASS
1:	1	1	1	18	0	0	0	1	1	1:	1-1-1	138	96	5/16
1:	1	1	2	12	0	1	1	1	1	1:	1-1-1	138	32	
1:	1	2	2	6	0	0	2	1	0	0:	1-1-1	138	48	
1:	2	2	2	3	2	0	0	0	0	1:	1-1-1	138	24	
1:	2	2	2	4	2	1	-1	1	1	0:	1-1-1	138	8	
1:	2	2	2	5	2	2	0	2	0	1:	1-1-1	138	12	
2:	1	1	1	23	1	0	0	0	0	0:	-1 1-1	276	48	5/4
2:	1	1	2	9	0	0	0	0	0	1:	-1 1-1	276	8	
2:	1	1	3	6	0	0	0	0	1	0:	-1 1-1	276	16	
2:	1	1	3	8	1	0	0	1	0	0:	-1 1-1	276	24	
2:	1	1	3	8	1	0	0	0	0	2:	-1 1-1	276	24	
2:	1	1	3	9	1	1	0	0	0	2:	-1 1-1	276	12	
2:	1	2	2	6	1	1	0	0	-1	1:	-1 1-1	276	8	

GENUS#	F11	F22	F33	F44	F12	F13	F23	F14	F24	F34	HASSE SYM	LEVEL	AUTOS	MASS
2:	1	2	3	3	0	0	0	0	1	0:	-1 1-1	276	8	
2:	1	2	3	4	0	0	0	1	1	3:	-1 1-1	276	8	
2:	1	3	3	3	0	0	3	0	0	2:	-1 1-1	276	8	
2:	1	3	3	3	1	0	1	0	2	-2:	-1 1-1	276	4	
2:	1	3	3	4	1	0	3	0	3	0:	-1 1-1	276	8	
3:	1	1	1	35	1	1	0	1	0	0:	-1 1-1	138	96	5/16
3:	1	1	2	12	1	0	0	0	0	2:	-1 1-1	138	48	
3:	1	1	3	7	0	0	0	1	1	3:	-1 1-1	138	32	
3:	1	1	3	9	1	1	0	1	0	0:	-1 1-1	138	12	
3:	1	1	4	6	1	0	0	0	0	2:	-1 1-1	138	24	
3:	1	3	3	3	0	0	0	1	3	1:	-1 1-1	138	8	
4:	1	1	2	10	0	1	0	0	0	1:	-1-1 1	276	8	5/4
4:	1	1	4	5	0	1	0	1	0	2:	-1-1 1	276	8	
4:	1	2	3	4	1	1	0	0	0	2:	-1-1 1	276	4	
4:	1	2	3	4	0	1	1	0	2	2:	-1-1 1	276	4	
4:	2	2	2	3	1	1	0	0	2	0:	-1-1 1	276	2	
5:	1	1	3	7	0	1	1	0	0	1:	-1-1 1	138	16	5/16
5:	1	2	3	4	0	1	0	0	2	2:	-1-1 1	138	8	
5:	1	3	3	3	1	1	-1	1	2	1:	-1-1 1	138	8	
6:	1	1	4	5	0	1	1	0	0	1:	1 1 1	138	16	5/16
6:	1	2	2	6	1	0	0	0	2	2:	1 1 1	138	8	
6:	2	2	2	3	1	0	0	1	1	2:	1 1 1	138	8	

D= 277

GENUS#	F11	F22	F33	F44	F12	F13	F23	F14	F24	F34	HASSE SYM	LEVEL	AUTOS	MASS
1:	1	1	1	35	1	1	0	0	1	0:	-1-1	277	48	103/48
1:	1	1	2	12	1	0	0	1	0	1:	-1-1	277	12	
1:	1	1	2	14	1	1	0	0	0	1:	-1-1	277	12	
1:	1	1	3	7	0	1	0	1	1	2:	-1-1	277	8	
1:	1	1	3	9	1	1	0	1	0	0:	-1-1	277	12	
1:	1	1	4	7	1	1	0	0	1	2:	-1-1	277	12	
1:	1	1	5	5	1	0	0	1	1	1:	-1-1	277	12	
1:	1	1	5	6	1	1	0	1	1	4:	-1-1	277	12	
1:	1	2	2	5	0	0	1	1	0	1:	-1-1	277	4	
1:	1	2	3	4	0	1	2	0	1	0:	-1-1	277	4	
1:	1	2	3	5	1	0	2	0	0	3:	-1-1	277	4	
1:	1	3	3	3	1	0	3	0	0	1:	-1-1	277	4	
1:	2	2	3	3	1	2	-1	2	0	3:	-1-1	277	2	

D= 280

GENUS#	F11	F22	F33	F44	F12	F13	F23	F14	F24	F34	HASSE SYM	LEVEL	AUTOS	MASS
1:	1	1	1	35	1	1	0	0	0	0:	1-1-1	280	96	67/96
1:	1	1	2	9	0	0	0	0	1	0:	1-1-1	280	16	
1:	1	2	3	4	0	1	0	0	0	3:	1-1-1	280	8	
1:	1	2	3	4	0	1	2	0	0	1:	1-1-1	280	8	
1:	2	2	2	3	1	1	-1	1	1	0:	1-1-1	280	4	
1:	2	2	3	3	2	2	0	1	2	0:	1-1-1	280	8	
2:	1	1	2	10	0	1	0	0	0	0:	-1-1 1	280	16	67/96
2:	1	1	2	14	1	1	0	0	0	0:	-1-1 1	280	24	
2:	1	1	3	7	0	1	1	0	0	0:	-1-1 1	280	32	
2:	1	2	2	6	1	0	1	0	2	0:	-1-1 1	280	4	
2:	1	2	2	6	0	0	2	0	0	1:	-1-1 1	280	8	
2:	1	2	3	4	0	0	2	0	2	1:	-1-1 1	280	16	
2:	2	2	3	3	0	2	1	0	2	3:	-1-1 1	280	8	
3:	1	1	2	12	1	0	0	1	0	0:	1 1 1	280	24	67/96
3:	1	1	4	5	0	1	1	0	0	0:	1 1 1	280	32	
3:	1	1	5	5	0	0	0	1	0	5:	1 1 1	280	16	
3:	1	2	2	5	1	0	0	0	0	0:	1 1 1	280	16	
3:	1	2	2	7	1	0	2	0	0	0:	1 1 1	280	16	
3:	1	2	3	3	0	0	0	0	0	1:	1 1 1	280	16	
3:	1	2	3	4	1	1	0	1	0	0:	1 1 1	280	4	
3:	1	2	3	4	0	0	0	1	2	2:	1 1 1	280	8	
4:	1	1	1	18	0	0	0	1	1	0:	-1 1-1	280	32	67/96
4:	1	1	5	5	1	1	0	0	0	0:	-1 1-1	280	24	
4:	1	2	2	5	0	0	1	0	-1	1:	-1 1-1	280	8	
4:	1	2	2	5	0	0	0	0	2	1:	-1 1-1	280	8	
4:	1	2	3	4	0	1	1	-1	-1	1:	-1 1-1	280	4	
4:	1	3	3	3	1	0	2	1	1	2:	-1 1-1	280	8	

D= 281

GENUS#	F11	F22	F33	F44	F12	F13	F23	F14	F24	F34	HASSE SYM	LEVEL	AUTOS	MASS
1:	1	1	1	24	1	0	0	1	0	1:	-1-1	281	24	25/8

GENUS#:	F11	F22	F33	F44	F12	F13	F23	F14	F24	F34:	HASSE SYM	LEVEL	AUTOS	MASS
1:	1	1	2	12	0	1	1	1	0	0:	-1-1	281	8	
1:	1	1	3	7	0	1	0	0	1	2:	-1-1	281	8	
1:	1	1	4	5	0	1	0	0	1	1:	-1-1	281	8	
1:	1	1	4	6	0	1	0	0	1	4:	-1-1	281	8	
1:	1	1	4	7	1	1	0	0	0	3:	-1-1	281	12	
1:	1	2	2	6	1	1	0	0	0	1:	-1-1	281	4	
1:	1	2	2	6	1	0	1	1	2	0:	-1-1	281	4	
1:	1	2	3	4	1	1	0	1	1	0:	-1-1	281	4	
1:	1	2	3	4	1	0	0	0	2	1:	-1-1	281	4	
1:	1	2	3	4	0	0	1	1	2	2:	-1-1	281	4	
1:	1	2	4	4	1	0	2	1	0	3:	-1-1	281	4	
1:	2	2	2	3	1	0	0	1	2	1:	-1-1	281	2	
1:	2	2	2	2	1	2	1	0	0	1	2: -1-1	281	2	

D= 284

GENUS#:	F11	F22	F33	F44	F12	F13	F23	F14	F24	F34:	HASSE SYM	LEVEL	AUTOS	MASS
1:	1	1	1	18	0	0	0	1	0	0:	-1-1	284	32	29/24
1:	1	1	1	24	1	0	0	0	0	0:	-1-1	284	24	
1:	1	1	1	36	1	1	0	1	0	0:	-1-1	284	96	
1:	1	1	2	12	0	1	1	0	0	1:	-1-1	284	16	
1:	1	1	2	9	0	0	0	0	0	1:	-1-1	284	16	
1:	1	1	3	6	0	0	0	0	0	1:	-1-1	284	16	
1:	1	1	3	8	0	1	1	0	0	3:	-1-1	284	16	
1:	1	1	4	5	0	0	0	0	0	3:	-1-1	284	16	
1:	1	1	4	5	0	1	0	0	0	2:	-1-1	284	8	
1:	1	1	4	5	0	0	0	1	1	1:	-1-1	284	16	
1:	1	1	4	5	0	1	0	1	0	0:	-1-1	284	8	
1:	1	2	3	4	1	1	0	0	1	1:	-1-1	284	4	
1:	2	2	2	3	1	0	0	0	1	2:	-1-1	284	4	
2:	1	1	5	6	1	1	0	0	1	3:	1 1	284	12	29/24
2:	1	2	2	6	1	0	1	0	0	2:	1 1	284	4	
2:	1	2	3	4	1	0	0	1	0	2:	1 1	284	4	
2:	1	2	3	4	0	1	0	1	2	0:	1 1	284	8	
2:	1	2	3	5	1	0	2	1	0	2:	1 1	284	4	
2:	2	2	3	3	2	1	0	2	2	2:	1 1	284	4	

D= 285

GENUS#:	F11	F22	F33	F44	F12	F13	F23	F14	F24	F34:	HASSE SYM	LEVEL	AUTOS	MASS
1:	1	1	1	24	1	0	0	0	0	1:	-1 1-1 1	285	48	1/2
1:	1	1	1	36	1	1	0	0	1	0:	-1 1-1 1	285	48	
1:	1	1	3	9	1	1	0	0	0	1:	-1 1-1 1	285	12	
1:	1	1	4	6	1	0	0	0	0	1:	-1 1-1 1	285	24	
1:	1	1	5	6	1	0	0	0	0	5:	-1 1-1 1	285	48	
1:	1	2	2	5	0	0	1	1	0	0:	-1 1-1 1	285	16	
1:	1	3	3	3	0	0	3	1	0	1:	-1 1-1 1	285	8	
1:	2	2	3	3	1	2	-1	2	2	1:	-1 1-1 1	285	8	
2:	1	1	2	11	0	1	0	0	1	2:	-1-1-1-1	285	8	1/2
2:	1	3	3	3	1	1	-1	1	2	0:	-1-1-1-1	285	8	
2:	2	2	2	3	1	1	-1	0	0	1:	-1-1-1-1	285	4	
3:	1	1	2	12	1	0	0	0	0	1:	-1 1 1-1	285	24	1/2
3:	1	1	2	15	1	1	0	1	1	2:	-1 1 1-1	285	24	
3:	1	1	3	7	0	1	0	1	1	0:	-1 1 1-1	285	8	
3:	1	1	3	8	1	0	0	0	0	1:	-1 1 1-1	285	24	
3:	1	1	3	9	1	0	0	1	0	3:	-1 1 1-1	285	24	
3:	1	1	4	7	1	1	0	1	0	2:	-1 1 1-1	285	12	
3:	1	3	3	4	1	0	3	1	3	0:	-1 1 1-1	285	8	
4:	1	1	3	8	0	1	1	1	0	3:	-1-1 1 1	285	16	1/2
4:	1	1	4	5	0	1	0	0	1	0:	-1-1 1 1	285	16	
4:	1	2	3	4	0	1	1	0	1	-2:	-1-1 1 1	285	4	
4:	1	2	3	5	1	1	0	1	2	3:	-1-1 1 1	285	8	

D= 288

GENUS#:	F11	F22	F33	F44	F12	F13	F23	F14	F24	F34:	HASSE SYM	LEVEL	AUTOS	MASS
1:	1	1	1	18	0	0	0	0	0	0:	1 1	72	96	5/32
1:	1	2	2	5	0	0	0	0	0	2:	1 1	72	16	
1:	2	2	3	3	2	2	0	0	0	2:	1 1	72	12	
2:	1	1	1	24	1	0	0	0	0	0:	1 1	96	48	5/24
2:	1	1	3	7	0	0	0	1	0	3:	1 1	96	16	
2:	1	2	4	4	1	0	2	0	1	4:	1 1	96	8	
3:	1	1	1	36	1	1	0	0	0	0:	1 1	144	96	5/32
3:	1	1	5	6	1	1	0	0	0	4:	1 1	144	12	
3:	1	2	3	4	0	1	2	0	0	0:	1 1	144	16	

	F11	F22	F33	F44	F12	F13	F23	F14	F24	F34:	HASSE	SYM	LEVEL	AUTOS	MASS
4:	1	1	2	10	0	0	0	1	1	2:	1 1		144	32	15/32
4:	1	2	3	4	1	0	1	1	0	1:	1 1		144	4	
4:	1	2	4	4	0	0	0	1	2	4:	1 1		144	16	
4:	2	2	2	3	0	0	0	2	1	1:	1 1		144	8	
5:	1	1	2	15	1	1	0	0	1	1:	1 1		288	12	5/8
5:	1	1	3	10	1	1	0	1	1	3:	1 1		288	24	
5:	1	2	3	4	0	0	2	1	1	1:	1 1		288	4	
5:	1	3	3	3	0	0	2	1	2	2:	1 1		288	8	
5:	2	2	3	3	0	2	2	1	-1	-2:	1 1		288	8	
6:	1	1	2	9	0	0	0	0	0	0:	1 1		72	32	5/32
6:	1	2	3	4	0	0	2	0	0	2:	1 1		72	8	
7:	1	1	3	6	0	0	0	0	0	0:	1 1		24	32	5/96
7:	2	2	2	3	2	0	0	0	0	0:	1 1		24	48	
8:	1	1	3	8	1	0	0	0	0	0:	-1-1		96	48	5/24
8:	1	3	3	3	0	0	0	1	3	0:	-1-1		96	16	
8:	2	2	3	3	1	2	-1	2	2	0:	-1-1		96	8	
9:	1	1	3	7	0	1	0	1	0	2:	1 1		288	8	5/8
9:	1	1	3	9	1	1	0	0	0	0:	1 1		288	24	
9:	1	1	4	7	1	1	0	0	1	1:	1 1		288	12	
9:	1	2	3	4	1	0	1	0	1	2:	1 1		288	4	
9:	1	3	3	3	0	0	2	0	1	3:	1 1		288	8	
10:	1	1	4	7	1	0	0	0	0	4:	1 1		96	48	5/24
10:	1	2	2	7	1	1	0	1	2	2:	1 1		96	16	
10:	1	2	3	4	1	0	0	0	2	0:	1 1		96	8	
11:	1	1	4	6	0	0	0	1	1	4:	1 1		144	32	15/32
11:	1	2	2	8	1	0	2	0	2	0:	1 1		144	8	
11:	2	2	2	4	2	1	0	2	0	0:	1 1		144	4	
11:	2	2	2	5	2	2	0	1	0	2:	1 1		144	16	
12:	1	2	2	6	0	0	2	0	0	0:	-1-1		24	48	5/96
12:	1	2	3	3	0	0	0	0	0	0:	-1-1		24	32	
13:	1	1	4	5	0	0	0	1	1	0:	1 1		144	32	5/32
13:	1	2	3	5	0	1	2	1	2	0:	1 1		144	8	
14:	1	1	5	5	1	0	0	0	0	2:	-1-1		96	48	5/24
14:	1	2	2	5	0	0	1	0	1	1:	-1-1		96	8	
14:	1	3	3	4	1	0	3	1	2	3:	-1-1		96	16	
15:	1	1	2	12	0	1	1	0	0	0:	1 1		48	32	5/32
15:	1	2	2	6	1	1	0	0	0	0:	1 1		48	16	
15:	2	2	2	4	2	0	0	1	2	2:	1 1		48	16	
16:	1	1	2	12	1	0	0	0	0	0:	-1-1		48	48	5/96
16:	2	3	3	3	2	0	3	0	3	0:	-1-1		48	32	
17:	1	1	4	5	0	0	0	0	1	2:	1 1		288	8	5/8
17:	1	2	3	4	1	1	0	0	1	-1:	1 1		288	4	
17:	2	2	2	3	1	1	1	-1	0	1:	1 1		288	4	
18:	1	1	4	6	1	0	0	0	0	0:	1 1		48	48	5/96
18:	1	1	5	5	0	1	1	1	1	4:	⊥ ⊥		48	32	
19:	1	2	2	5	0	0	0	1	1	1:	1 1		288	8	5/8
19:	1	2	2	6	1	0	1	1	0	1:	1 1		288	4	
19:	2	2	3	3	1	1	-1	0	2	2:	1 1		288	4	
20:	1	2	3	4	0	0	0	1	0	3:	-1-1		48	16	5/32
20:	2	2	2	4	2	1	-1	0	0	0:	-1-1		48	32	
20:	2	2	3	3	2	1	-1	1	-1	0:	-1-1		48	16	

D= 289

GENUS#:	F11	F22	F33	F44	F12	F13	F23	F14	F24	F34:	HASSE	SYM	LEVEL	AUTOS	MASS
1:	1	1	6	6	1	1	0	1	1	6:	-1-1		17	72	2/9
1:	1	2	3	5	1	0	2	0	1	3:	-1-1		17	8	
1:	2	2	3	3	2	1	0	1	1	3:	-1-1		17	12	

D= 292

GENUS#:	F11	F22	F33	F44	F12	F13	F23	F14	F24	F34:	HASSE	SYM	LEVEL	AUTOS	MASS
1:	1	1	1	19	0	0	0	1	1	1:	1 1		146	96	11/48
1:	1	1	1	37	1	1	0	1	0	0:	1 1		146	96	
1:	1	1	2	13	1	0	0	1	0	2:	1 1		146	24	
1:	1	2	2	7	0	0	2	1	2	0:	1 1		146	24	
1:	1	2	3	5	0	1	2	1	0	3:	1 1		146	8	
2:	1	1	2	11	0	1	0	0	0	2:	-1-1		292	8	11/4
2:	1	1	2	15	1	1	0	1	0	0:	-1-1		292	12	
2:	1	1	3	7	0	1	0	0	0	2:	-1-1		292	8	
2:	1	1	4	7	1	1	0	1	0	0:	-1-1		292	12	
2:	1	1	5	6	1	0	0	1	0	4:	-1-1		292	12	
2:	1	2	2	5	0	0	1	0	0	1:	-1-1		292	4	
2:	1	2	2	6	1	0	1	1	1	1:	-1-1		292	4	

	F11	F22	F33	F44	F12	F13	F23	F14	F24	F34:	HASSE SYM	LEVEL	AUTOS	MASS
2:	1	2	3	4	0	1	1	0	2	0:	-1-1	292	4	
2:	1	2	3	4	0	0	2	0	-1	1:	-1-1	292	4	
2:	1	2	3	4	0	1	1	1	1	2:	-1-1	292	4	
2:	1	2	3	5	1	0	2	0	2	0:	-1-1	292	4	
2:	1	3	3	3	1	1	-1	0	0	2:	-1-1	292	4	
2:	2	2	3	3	1	1	-1	0	1	3:	-1-1	292	2	
3:	1	1	4	6	0	1	1	1	1	3:	-1-1	146	16	33/16
3:	1	2	2	6	1	0	0	1	0	2:	-1-1	146	8	
3:	1	2	2	8	1	0	2	0	0	2:	-1-1	146	8	
3:	1	2	3	4	1	0	1	0	0	2:	-1-1	146	4	
3:	1	2	3	4	0	0	2	1	0	1:	-1-1	146	8	
3:	1	2	3	4	1	0	1	1	1	1:	-1-1	146	4	
3:	1	2	4	4	0	0	2	1	0	4:	-1-1	146	8	
3:	2	2	2	3	1	1	0	-1	1	-1:	-1-1	146	2	
3:	2	2	2	4	2	1	0	0	2	0:	-1-1	146	4	
3:	2	2	3	3	2	1	0	1	2	2:	-1-1	146	4	

D= 293

GENUS#:	F11	F22	F33	F44	F12	F13	F23	F14	F24	F34:	HASSE SYM	LEVEL	AUTOS	MASS
1:	1	1	1	25	1	0	0	1	0	1:	-1-1	293	24	85/48
1:	1	1	1	37	1	0	0	0	1	0:	-1-1	293	48	
1:	1	1	2	11	0	1	0	1	1	0:	-1-1	293	8	
1:	1	1	2	15	1	1	0	0	1	0:	-1-1	293	12	
1:	1	1	3	10	1	1	0	0	1	2:	-1-1	293	12	
1:	1	1	3	7	0	1	0	0	1	1:	-1-1	293	8	
1:	1	1	4	7	1	1	0	0	1	0:	-1-1	293	12	
1:	1	1	4	7	1	0	0	1	1	3:	-1-1	293	12	
1:	1	1	5	5	0	1	1	0	1	4:	-1-1	293	8	
1:	1	2	2	6	1	0	1	0	1	-1:	-1-1	293	4	
1:	1	2	3	4	1	1	0	0	1	0:	-1-1	293	4	
1:	1	2	3	4	0	0	1	1	1	-2:	-1-1	293	4	
1:	1	3	3	3	1	1	-1	0	1	-2:	-1-1	293	4	

D= 296

GENUS#:	F11	F22	F33	F44	F12	F13	F23	F14	F24	F34:	HASSE SYM	LEVEL	AUTOS	MASS
1:	1	1	1	19	0	0	0	1	1	0:	1 1	296	32	41/32
1:	1	1	1	25	1	0	0	1	0	0:	1 1	296	24	
1:	1	1	3	7	0	1	0	1	0	0:	1 1	296	8	
1:	1	1	3	7	0	0	0	1	1	2:	1 1	296	16	
1:	1	1	3	8	0	1	1	1	1	0:	1 1	296	16	
1:	1	1	4	7	1	1	0	0	0	2:	1 1	296	12	
1:	1	2	2	5	0	0	0	1	0	1:	1 1	296	8	
1:	1	2	3	4	1	0	0	0	1	-2:	1 1	296	4	
1:	2	2	2	3	1	0	0	2	0	0:	1 1	296	4	
1:	2	2	3	3	2	1	0	0	2	1:	1 1	296	4	
2:	1	1	1	37	1	1	0	0	0	0:	-1-1	296	96	41/32
2:	1	1	2	10	0	0	0	0	1	2:	-1-1	296	16	
2:	1	1	2	13	0	1	1	1	1	0:	-1-1	296	16	
2:	1	1	4	5	0	1	0	0	0	1:	-1-1	296	8	
2:	1	1	4	6	0	1	0	0	0	4:	-1-1	296	8	
2:	1	1	5	5	0	0	0	1	1	4:	-1-1	296	16	
2:	1	1	5	6	1	1	0	1	1	3:	-1-1	296	12	
2:	1	2	3	4	0	0	1	0	0	3:	-1-1	296	4	
2:	1	2	3	5	1	0	2	1	2	0:	-1-1	296	4	
2:	2	2	2	4	2	1	0	1	2	1:	-1-1	296	4	

D= 297

GENUS#:	F11	F22	F33	F44	F12	F13	F23	F14	F24	F34:	HASSE SYM	LEVEL	AUTOS	MASS
1:	1	1	1	25	1	0	0	0	0	1:	-1 1-1	99	48	3/16
1:	1	2	2	7	1	1	0	0	2	1:	-1 1-1	99	8	
1:	2	2	2	4	1	1	1	-2	-2	1:	-1 1-1	99	24	
2:	1	1	2	11	0	1	0	0	1	1:	-1-1 1	297	8	27/16
2:	1	1	5	5	0	1	0	0	1	4:	-1-1 1	297	16	
2:	1	2	2	6	1	0	1	1	0	0:	-1-1 1	297	4	
2:	1	2	2	8	1	0	2	1	0	1:	-1-1 1	297	4	
2:	1	2	3	4	1	1	0	0	0	1:	-1-1 1	297	4	
2:	2	2	2	3	1	1	0	1	-1	0:	-1-1 1	297	2	
2:	2	2	2	4	2	1	0	1	1	2:	-1-1 1	297	4	
3:	1	1	2	13	0	1	1	1	0	2:	-1-1 1	99	16	3/16
3:	1	2	4	4	1	0	2	1	2	3:	-1-1 1	99	8	
4:	1	1	2	15	1	1	0	0	0	1:	-1 1-1	297	12	27/16
4:	1	1	3	7	0	1	0	0	1	0:	-1 1-1	297	16	

GENUS#	F11	F22	F33	F44	F12	F13	F23	F14	F24	F34	HASSE SYM	LEVEL	AUTOS	MASS
4:	1	1	4	6	0	1	1	0	1	3:	-1 1-1	297	8	
4:	1	1	4	8	1	1	0	1	1	4:	-1 1-1	297	24	
4:	1	2	2	6	1	0	0	0	2	1:	-1 1-1	297	4	
4:	1	2	3	4	0	0	1	1	2	0:	-1 1-1	297	4	
4:	1	2	3	4	1	0	1	0	-1	1:	-1 1-1	297	4	
4:	1	2	3	4	1	0	1	1	0	0:	-1 1-1	297	4	
4:	1	2	4	4	1	0	1	1	1	4:	-1 1-1	297	8	
4:	2	2	3	3	0	2	1	-1	-2	1:	-1 1-1	297	4	
5:	1	1	3	9	1	0	0	0	0	3:	-1 1-1	33	48	1/16
5:	1	3	3	4	0	0	3	1	3	0:	-1 1-1	33	24	
6:	2	2	2	5	2	1	-1	1	2	1:	-1-1 1	33	16	1/16
7:	1	1	5	5	1	0	0	0	0	1:	-1 1-1	99	48	3/8
7:	1	2	2	6	0	0	1	1	2	2:	-1 1-1	99	8	
7:	1	3	3	3	1	0	0	0	0	3:	-1 1-1	99	48	
7:	2	2	2	3	1	1	1	1	1	1:	-1 1-1	99	12	
7:	2	2	3	3	1	2	-1	-1	-1	2:	-1 1-1	99	8	
8:	1	2	3	4	0	1	1	1	0	2:	-1-1 1	99	4	3/8
8:	2	2	3	3	2	1	-1	2	1	0:	-1-1 1	99	8	

D= 300

GENUS#	F11	F22	F33	F44	F12	F13	F23	F14	F24	F34	HASSE SYM	LEVEL	AUTOS	MASS
1:	1	1	1	25	1	0	0	0	0	0:	-1-1 1	300	48	65/96
1:	1	1	2	11	0	1	0	1	0	0:	-1-1 1	300	8	
1:	1	1	3	7	0	0	0	0	0	3:	-1-1 1	300	32	
1:	1	2	3	4	0	1	0	0	2	1:	-1-1 1	300	8	
1:	2	2	2	3	1	1	1	1	0	1:	-1-1 1	300	4	
1:	2	2	2	5	2	1	-1	0	2	0:	-1-1 1	300	8	
2:	1	1	1	38	1	1	0	1	0	0:	1 1 1	300	96	65/96
2:	1	1	2	10	0	0	0	1	1	1:	1 1 1	300	16	
2:	1	1	6	6	1	0	0	1	0	6:	1 1 1	300	24	
2:	1	2	2	7	0	0	2	1	1	2:	1 1 1	300	16	
2:	1	2	3	5	1	0	2	1	1	2:	1 1 1	300	4	
2:	2	2	3	3	0	2	2	2	1	1:	1 1 1	300	4	
3:	1	1	2	15	1	1	0	0	0	0:	1 1 1	60	24	13/96
3:	1	2	3	4	0	0	2	1	0	0:	1 1 1	60	16	
3:	2	2	3	3	2	1	-1	1	-1	1:	1 1 1	60	32	
4:	1	1	3	8	0	1	1	1	1	1:	1-1-1	60	32	13/96
4:	2	2	2	3	1	1	-1	0	0	0:	1-1-1	60	24	
4:	2	2	2	3	1	0	0	0	0	2:	1-1-1	60	16	
5:	1	1	2	13	0	1	1	1	1	1:	1 1 1	300	32	65/96
5:	1	1	3	9	1	0	0	1	0	2:	1 1 1	300	12	
5:	1	1	4	6	0	1	1	0	0	3:	1 1 1	300	16	
5:	1	2	2	6	1	0	0	0	1	2:	1 1 1	300	8	
5:	1	2	3	4	0	1	1	1	1	0:	1 1 1	300	4	
5:	1	3	3	3	1	0	2	0	2	0:	1 1 1	300	8	
6:	1	1	5	5	1	0	0	0	0	0:	-1-1 1	60	96	13/96
6:	1	2	2	5	0	0	1	0	0	0:	-1-1 1	60	16	
6:	1	2	2	8	1	0	2	0	1	2:	-1-1 1	60	16	
7:	1	1	4	5	0	1	0	0	0	0:	-1 1-1	60	16	13/96
7:	1	1	5	5	0	0	0	0	0	5:	-1 1-1	60	96	
7:	1	3	3	4	1	1	0	1	3	3:	-1 1-1	60	16	
8:	1	1	1	19	0	0	0	0	1	0:	-1-1 1	300	32	65/96
8:	1	1	2	13	1	0	0	0	0	2:	-1-1 1	300	48	
8:	1	1	4	5	0	0	0	1	0	1:	-1-1 1	300	8	
8:	1	2	3	4	0	0	0	0	1	3:	-1-1 1	300	8	
8:	1	2	3	5	1	1	0	0	2	2:	-1-1 1	300	4	
8:	2	2	3	3	2	0	0	1	0	3:	-1-1 1	300	8	

D= 301

GENUS#	F11	F22	F33	F44	F12	F13	F23	F14	F24	F34	HASSE SYM	LEVEL	AUTOS	MASS
1:	1	1	1	38	1	1	0	0	1	0:	-1 1-1	301	48	31/24
1:	1	1	2	11	0	1	0	0	1	0:	-1 1-1	301	16	
1:	1	1	3	10	1	1	0	1	1	2:	-1 1-1	301	12	
1:	1	2	3	4	0	1	0	1	1	2:	-1 1-1	301	4	
1:	1	2	3	5	0	1	2	0	1	3:	-1 1-1	301	4	
1:	1	3	3	3	1	1	0	1	2	1:	-1 1-1	301	8	
1:	1	3	3	3	1	1	0	0	2	1:	-1 1-1	301	4	
1:	2	2	3	3	1	2	0	2	1	3:	-1 1-1	301	4	
2:	1	1	2	13	1	0	0	1	0	1:	-1 1-1	301	12	31/24
2:	1	1	3	8	0	1	1	0	1	2:	-1 1-1	301	8	
2:	1	1	5	6	1	1	0	0	1	2:	-1 1-1	301	12	
2:	1	2	2	6	1	0	1	0	1	1:	-1 1-1	301	4	

GENUS#	F11	F22	F33	F44	F12	F13	F23	F14	F24	F34	HASSE SYM	LEVEL	AUTOS	MASS
2:	1	2	3	4	0	1	1	0	1	2:	-1 1-1	301	4	
2:	1	2	3	5	1	0	2	0	2	1:	-1 1-1	301	4	
2:	1	3	3	4	1	0	3	0	3	1:	-1 1-1	301	4	

D= 304

GENUS#	F11	F22	F33	F44	F12	F13	F23	F14	F24	F34	HASSE SYM	LEVEL	AUTOS	MASS
1:	1	1	1	19	0	0	0	0	0	0:	-1-1	76	96	19/96
1:	1	1	4	5	0	0	0	0	0	2:	-1-1	76	16	
1:	1	2	3	4	0	0	0	0	2	2:	-1-1	76	8	
2:	1	1	1	38	1	1	0	0	0	0:	1 1	152	96	19/96
2:	1	1	3	8	0	1	1	0	0	2:	1 1	152	16	
2:	1	2	3	5	0	1	2	1	2	1:	1 1	152	8	
3:	1	1	2	11	0	1	0	0	0	1:	-1-1	304	8	19/12
3:	1	1	3	7	0	1	0	0	0	1:	-1-1	304	8	
3:	1	1	4	5	0	0	0	0	1	0:	-1-1	304	16	
3:	1	1	4	6	0	0	0	0	1	4:	-1-1	304	16	
3:	1	1	4	8	1	1	0	0	1	3:	-1-1	304	12	
3:	1	2	2	5	0	0	0	0	1	1:	-1-1	304	8	
3:	1	2	3	4	1	1	0	0	0	0:	-1-1	304	8	
3:	1	2	3	4	0	0	1	0	2	2:	-1-1	304	4	
3:	1	3	3	4	1	0	3	1	3	1:	-1-1	304	4	
3:	1	3	3	4	1	1	-1	1	3	2:	-1-1	304	8	
3:	2	2	2	3	1	1	0	0	1	1:	-1-1	304	4	
4:	1	1	2	13	1	0	0	1	0	0:	-1-1	152	24	19/96
4:	1	2	2	5	0	0	0	1	0	0:	-1-1	152	32	
4:	1	3	3	3	1	1	0	1	2	0:	-1-1	152	8	
5:	1	1	3	10	1	1	0	0	1	1:	1 1	304	12	19/12
5:	1	2	2	6	1	0	1	0	1	0:	1 1	304	4	
5:	1	2	3	4	0	1	1	0	0	2:	1 1	304	4	
5:	1	2	3	4	0	1	1	1	1	1:	1 1	304	4	
5:	1	3	3	3	1	0	0	0	-2	2:	1 1	304	4	
5:	2	2	3	3	1	2	-1	2	0	0:	1 1	304	2	
6:	1	1	2	10	0	0	0	0	0	2:	1 1	76	32	19/96
6:	1	2	2	7	0	0	2	0	2	0:	1 1	76	24	
6:	2	2	3	3	0	2	2	0	2	0:	1 1	76	8	
7:	1	1	2	10	0	0	0	1	1	0:	1 1	152	32	19/32
7:	1	1	4	6	0	1	1	1	1	0:	1 1	152	16	
7:	1	2	2	6	1	0	0	0	2	0:	1 1	152	8	
7:	2	2	2	3	0	0	0	2	1	0:	1 1	152	8	
7:	2	2	4	2	1	0	0	0	2	2:	1 1	152	4	
8:	1	2	2	6	0	0	0	1	2	2:	-1-1	152	32	19/32
8:	1	2	3	4	0	0	0	1	2	1:	-1-1	152	8	
8:	1	2	4	4	1	0	0	1	0	4:	-1-1	152	8	
8:	2	2	2	3	1	1	0	1	1	1:	-1-1	152	8	
8:	2	2	2	5	2	2	0	1	0	0:	-1-1	152	16	

D= 305

GENUS#	F11	F22	F33	F44	F12	F13	F23	F14	F24	F34	HASSE SYM	LEVEL	AUTOS	MASS
1:	1	1	1	26	1	0	0	1	0	2:	-1-1 1	305	24	41/24
1:	1	1	2	16	1	1	0	1	1	2:	-1-1 1	305	24	
1:	1	1	4	6	0	1	0	1	1	3:	-1-1 1	305	8	
1:	1	2	2	6	1	0	1	0	0	1:	-1-1 1	305	4	
1:	1	2	2	7	1	1	0	1	1	2:	-1-1 1	305	4	
1:	1	2	3	4	1	0	1	0	1	1:	-1-1 1	305	4	
1:	2	2	2	4	1	1	-1	0	1	2:	-1-1 1	305	4	
1:	2	2	2	4	2	1	0	2	1	0:	-1-1 1	305	2	
2:	1	1	2	13	0	1	1	1	0	0:	-1 1-1	305	8	41/24
2:	1	1	3	8	0	1	0	0	1	3:	-1 1-1	305	8	
2:	1	1	4	7	1	1	0	0	0	1:	-1 1-1	305	12	
2:	1	2	2	8	1	0	2	1	1	1:	-1 1-1	305	8	
2:	1	2	3	4	1	0	0	1	1	1:	-1 1-1	305	4	
2:	1	2	4	4	1	0	2	0	2	3:	-1 1-1	305	8	
2:	1	2	4	4	0	0	1	1	2	4:	-1 1-1	305	4	
2:	1	2	4	4	1	0	2	1	1	3:	-1 1-1	305	4	
2:	1	3	3	3	1	0	2	0	2	1:	-1 1-1	305	8	
2:	2	2	2	3	1	0	0	1	-1	1:	-1 1-1	305	4	

D= 308

GENUS#	F11	F22	F33	F44	F12	F13	F23	F14	F24	F34	HASSE SYM	LEVEL	AUTOS	MASS
1:	1	1	1	20	0	0	0	1	1	1:	1-1-1	154	96	5/16
1:	1	1	4	6	0	1	1	1	1	1:	1-1-1	154	32	
1:	1	2	2	6	1	0	0	0	0	2:	1-1-1	154	16	

GENUS#	F11	F22	F33	F44	F12	F13	F23	F14	F24	F34	HASSE SYM	LEVEL	AUTOS	MASS
1:	2	2	2	4	2	1	0	1	2	0:	1-1-1	154	8	
1:	2	2	2	5	2	2	0	0	1	0:	1-1-1	154	12	
2:	1	1	1	26	1	0	0	1	0	0:	-1 1-1	308	24	5/4
2:	1	1	2	16	1	1	0	0	1	1:	-1 1-1	308	12	
2:	1	1	3	7	0	0	0	0	1	2:	-1 1-1	308	8	
2:	1	2	2	7	1	1	0	0	0	0:	-1 1-1	308	4	
2:	1	2	3	4	1	0	1	0	1	0:	-1 1-1	308	4	
2:	1	2	3	4	0	0	2	0	1	1:	-1 1-1	308	4	
2:	1	2	3	5	1	0	1	1	0	3:	-1 1-1	308	4	
3:	1	1	1	39	1	1	0	1	0	0:	-1-1 1	154	96	5/16
3:	1	1	3	10	1	1	0	0	0	2:	-1-1 1	154	12	
3:	1	1	5	5	0	1	1	1	1	3:	-1-1 1	154	32	
3:	1	2	3	4	0	1	0	0	2	0:	-1-1 1	154	16	
3:	1	3	3	3	1	0	1	-1	1	-1:	-1-1 1	154	8	
4:	1	1	2	10	0	0	0	0	1	1:	-1-1 1	308	8	5/4
4:	1	1	2	11	0	1	0	0	0	0:	-1-1 1	308	16	
4:	1	1	3	10	1	1	0	1	0	0:	-1-1 1	308	12	
4:	1	1	3	7	0	1	0	0	0	0:	-1-1 1	308	16	
4:	1	1	4	7	1	1	0	0	0	0:	-1-1 1	308	24	
4:	1	2	3	4	1	0	0	0	0	2:	-1-1 1	308	8	
4:	1	2	4	4	0	0	2	1	1	4:	-1-1 1	308	8	
4:	1	3	3	3	1	0	2	0	1	2:	-1-1 1	308	4	
4:	1	3	3	3	0	0	2	0	-1	2:	-1-1 1	308	8	
4:	2	2	2	3	1	1	0	0	-1	1:	-1-1 1	308	4	
5:	1	1	2	13	0	1	1	0	0	1:	1 1 1	154	16	5/16
5:	1	2	2	8	1	0	2	1	0	0:	1 1 1	154	8	
5:	2	2	2	4	2	0	0	2	2	1:	1 1 1	154	12	
5:	2	2	3	4	2	2	0	2	2	3:	1 1 1	154	24	5/16
6:	1	1	3	7	0	0	0	1	1	1:	-1 1-1	154	16	
6:	1	1	4	7	1	0	0	1	0	2:	-1 1-1	154	12	
6:	1	1	5	7	1	1	0	1	1	5:	-1 1-1	154	24	
6:	1	2	3	5	0	0	2	1	0	3:	-1 1-1	154	8	

D= 309

GENUS#	F11	F22	F33	F44	F12	F13	F23	F14	F24	F34	HASSE SYM	LEVEL	AUTOS	MASS
1:	1	1	1	26	1	0	0	0	0	1:	-1-1 1	309	48	59/48
1:	1	1	2	13	0	0	0	0	0	1:	-1-1 1	309	24	
1:	1	1	3	8	0	1	1	0	1	0:	-1-1 1	309	8	
1:	1	1	4	7	1	0	0	0	0	3:	-1-1 1	309	24	
1:	1	2	3	4	0	1	0	0	1	2:	-1-1 1	309	4	
1:	1	2	3	5	1	0	0	0	2	3:	-1-1 1	309	8	
1:	1	3	3	4	1	0	3	0	1	3:	-1-1 1	309	8	
1:	2	2	2	3	1	1	1	0	1	0:	-1-1 1	309	4	
1:	2	2	3	3	1	0	0	2	0	3:	-1-1 1	309	4	
2:	1	1	1	39	1	1	0	0	1	0:	-1 1-1	309	48	59/48
2:	1	1	3	10	1	1	0	0	1	0:	-1 1-1	309	12	
2:	1	1	3	9	1	0	0	1	1	1:	-1 1-1	309	12	
2:	1	1	4	6	0	1	0	0	1	3:	-1 1-1	309	8	
2:	1	1	5	6	1	1	0	1	1	2:	-1 1-1	309	12	
2:	1	1	5	6	1	1	0	0	1	3:	-1 1-1	309	12	
2:	1	2	3	4	0	0	1	1	1	2:	-1 1-1	309	4	
2:	1	2	3	5	1	0	2	1	0	1:	-1 1-1	309	4	
2:	1	3	3	3	1	1	0	0	2	-1:	-1 1-1	309	4	

D= 312

GENUS#	F11	F22	F33	F44	F12	F13	F23	F14	F24	F34	HASSE SYM	LEVEL	AUTOS	MASS
1:	1	1	1	26	1	0	0	0	0	0:	1 1 1	312	48	23/32
1:	1	1	3	7	0	0	0	1	1	0:	1 1 1	312	32	
1:	1	1	3	9	1	0	0	0	0	2:	1 1 1	312	24	
1:	1	1	3	9	1	0	0	1	0	0:	1 1 1	312	24	
1:	1	1	5	6	1	1	0	0	1	1:	1 1 1	312	12	
1:	1	2	3	5	1	0	2	0	0	2:	1 1 1	312	4	
1:	1	3	3	3	0	0	0	0	1	3:	1 1 1	312	8	
1:	2	2	3	3	0	0	0	2	1	3:	1 1 1	312	8	
2:	1	1	1	39	1	1	0	0	0	0:	-1 1-1	312	96	23/32
2:	1	1	2	10	0	0	0	0	1	0:	-1 1-1	312	16	
2:	1	1	2	13	1	0	0	0	0	0:	-1 1-1	312	48	
2:	1	1	2	16	1	1	0	1	0	0:	-1 1-1	312	12	
2:	1	1	5	6	1	0	0	0	0	4:	-1 1-1	312	24	
2:	1	2	2	6	0	0	1	0	2	2:	-1 1-1	312	8	
2:	1	2	3	4	1	0	0	1	0	0:	-1 1-1	312	8	
2:	1	2	3	4	0	0	0	1	2	0:	-1 1-1	312	16	

2:	1	2	3	4	0	0	0	0	0	3:	-1 1-1	312	16	
2:	1	2	3	4	0	0	2	0	0	1:	-1 1-1	312	8	
3:	1	1	2	13	0	1	1	0	0	0:	1-1-1	312	32	23/32
3:	1	1	5	5	0	1	0	1	0	4:	1-1-1	312	16	
3:	1	2	3	5	1	1	0	1	1	3:	1-1-1	312	4	
3:	2	2	2	5	2	1	-1	2	0	1:	1-1-1	312	8	
3:	2	2	3	3	2	2	0	0	-1	1:	1-1-1	312	4	
4:	1	1	1	20	0	0	0	1	1	0:	-1-1 1	312	32	23/32
4:	1	2	2	5	0	0	0	0	1	0:	-1-1 1	312	8	
4:	1	2	2	6	1	0	1	0	0	0:	-1-1 1	312	8	
4:	1	2	2	7	0	0	2	0	2	1:	-1-1 1	312	16	
4:	1	2	2	8	1	0	2	0	1	0:	-1-1 1	312	8	
4:	1	2	3	4	0	1	0	1	0	2:	-1-1 1	312	8	
4:	2	2	3	3	2	1	-1	0	0	2:	-1-1 1	312	8	

D= 313

GENUS#:	F11	F22	F33	F44	F12	F13	F23	F14	F24	F34:	HASSE SYM	LEVEL	AUTOS	MASS
1:	1	1	2	12	0	1	0	0	1	2:	-1-1	313	8	25/6
1:	1	1	2	16	1	1	0	0	1	0:	-1-1	313	12	
1:	1	1	4	6	0	1	1	0	1	2:	-1-1	313	8	
1:	1	1	5	6	1	0	0	1	0	3:	-1-1	313	12	
1:	1	2	2	6	1	0	0	1	0	1:	-1-1	313	4	
1:	1	2	2	6	0	0	1	1	0	2:	-1-1	313	4	
1:	1	2	2	8	1	0	2	0	0	1:	-1-1	313	4	
1:	1	2	3	4	0	0	1	1	0	2:	-1-1	313	4	
1:	1	2	3	4	0	1	1	1	0	0:	-1-1	313	4	
1:	1	2	3	4	0	1	1	0	1	-1:	-1-1	313	4	
1:	1	2	3	4	1	0	0	0	0	1:	-1-1	313	4	
1:	1	2	3	5	1	0	2	0	-1	1:	-1-1	313	4	
1:	1	2	3	5	1	0	1	0	2	3:	-1-1	313	4	
1:	2	2	2	3	1	1	0	1	0	1:	-1-1	313	2	
1:	2	2	3	3	2	1	0	2	1	2:	-1-1	313	2	
1:	2	2	3	3	1	1	-1	2	0	3:	-1-1	313	2	

D= 316

GENUS#:	F11	F22	F33	F44	F12	F13	F23	F14	F24	F34:	HASSE SYM	LEVEL	AUTOS	MASS
1:	1	1	1	20	0	0	0	1	0	0:	-1-1	316	32	7/4
1:	1	1	1	40	1	1	0	1	0	0:	-1-1	316	96	
1:	1	1	2	10	0	0	0	0	0	1:	-1-1	316	16	
1:	1	1	3	8	0	1	0	0	0	3:	-1-1	316	8	
1:	1	1	3	8	0	1	1	0	0	1:	-1-1	316	16	
1:	1	1	4	5	0	0	0	0	0	1:	-1-1	316	16	
1:	1	1	4	6	0	0	1	1	0	3:	-1-1	316	16	
1:	1	1	4	8	1	1	0	1	1	3:	-1-1	316	12	
1:	1	1	5	5	0	0	0	0	1	4:	-1-1	316	8	
1:	1	2	3	4	0	0	1	0	-1	2:	-1-1	316	4	
1:	1	2	3	5	0	1	2	0	2	0:	-1-1	316	8	
1:	1	2	3	5	1	1	0	1	2	2:	-1-1	316	4	
1:	2	2	2	4	2	1	0	1	0	-1:	-1-1	316	4	
1:	2	2	3	3	0	2	0	2	1	3:	-1-1	316	4	
2:	1	1	2	14	1	0	0	1	0	2:	1 1	316	24	7/4
2:	1	1	5	6	1	1	0	1	0	0:	1 1	316	12	
2:	1	2	2	7	0	0	2	1	0	1:	1 1	316	8	
2:	1	2	3	4	0	0	0	1	1	2:	1 1	316	4	
2:	1	2	3	5	1	0	2	0	1	2:	1 1	316	4	
2:	1	3	3	3	1	1	0	0	0	2:	1 1	316	4	
2:	2	2	3	3	2	1	0	0	2	0:	1 1	316	4	
2:	2	2	3	3	1	2	0	2	2	0:	1 1	316	2	

D= 317

GENUS#:	F11	F22	F33	F44	F12	F13	F23	F14	F24	F34:	HASSE SYM	LEVEL	AUTOS	MASS
1:	1	1	1	27	1	0	0	1	0	1:	-1-1	317	24	101/48
1:	1	1	1	40	1	1	0	0	1	0:	-1-1	317	48	
1:	1	1	2	16	1	1	0	0	0	1:	-1-1	317	12	
1:	1	1	3	10	1	1	0	0	0	1:	-1-1	317	12	
1:	1	1	4	7	1	0	0	1	0	1:	-1-1	317	12	
1:	1	1	5	5	0	1	1	1	0	3:	-1-1	317	8	
1:	1	1	5	6	1	1	0	0	1	0:	-1-1	317	12	
1:	1	1	5	7	1	1	0	0	1	4:	-1-1	317	12	
1:	1	2	2	7	1	0	1	1	0	2:	-1-1	317	4	
1:	1	2	3	4	0	1	0	1	1	1:	-1-1	317	4	
1:	1	2	3	4	1	0	0	0	1	1:	-1-1	317	4	

	F11	F22	F33	F44	F12	F13	F23	F14	F24	F34:	HASSE	SYM	LEVEL	AUTOS	MASS
1:	1	2	3	5	1	0	2	1	1	1:	-1-1		317	4	
1:	1	2	3	5	1	1	0	0	0	3:	-1-1		317	4	
1:	1	3	3	3	1	0	1	0	2	-1:	-1-1		317	4	

D= 320

GENUS#:	F11	F22	F33	F44	F12	F13	F23	F14	F24	F34:	HASSE	SYM	LEVEL	AUTOS	MASS
1:	1	1	1	20	0	0	0	0	0	0:	1 1		80	96	1/24
1:	1	1	4	5	0	0	0	0	0	0:	1 1		80	32	
2:	1	1	1	27	1	0	0	1	0	0:	-1-1		320	24	2/3
2:	1	1	3	7	0	0	0	0	1	1:	-1-1		320	8	
2:	1	2	4	4	1	0	0	0	1	4:	-1-1		320	8	
2:	1	3	3	4	1	1	-1	0	3	1:	-1-1		320	8	
2:	2	2	2	3	1	1	0	1	0	0:	-1-1		320	4	
3:	1	1	1	40	1	1	0	0	0	0:	-1-1		160	96	1/6
3:	1	1	2	11	0	0	0	1	1	2:	-1-1		160	32	
3:	1	1	3	8	0	1	1	0	0	0:	-1-1		160	32	
3:	1	1	5	5	0	1	1	1	1	0:	-1-1		160	32	
3:	2	2	3	3	0	2	0	2	2	1:	-1-1		160	16	
4:	1	1	2	10	0	0	0	0	0	0:	-1-1		40	32	1/16
4:	2	2	2	3	0	0	0	2	0	0:	-1-1		40	32	
5:	1	1	2	12	0	1	0	0	0	2:	-1-1		320	8	2/3
5:	1	1	4	6	0	1	0	1	0	3:	-1-1		320	8	
5:	1	1	4	8	1	0	0	1	0	4:	-1-1		320	24	
5:	1	2	2	7	1	1	0	1	0	0:	-1-1		320	8	
5:	1	2	3	4	0	0	1	0	2	0:	-1-1		320	4	
6:	1	1	2	14	0	1	1	1	1	0:	1 1		160	16	1/2
6:	1	1	4	6	0	1	1	0	0	2:	1 1		160	16	
6:	1	2	2	6	1	0	0	1	0	0:	1 1		160	8	
6:	2	2	2	3	1	0	0	-1	1	0:	1 1		160	8	
6:	2	2	2	4	2	0	0	1	1	2:	1 1		160	8	
7:	1	1	2	16	1	1	0	0	0	0:	-1-1		320	24	2/3
7:	1	2	2	7	1	0	1	0	2	2:	-1-1		320	4	
7:	1	2	3	5	0	0	2	1	-1	2:	-1-1		320	8	
7:	2	2	3	3	1	2	0	1	2	2:	-1-1		320	4	
8:	1	1	3	10	1	1	0	0	0	0:	-1-1		160	24	1/6
8:	1	2	3	4	0	1	0	0	0	2:	-1-1		160	8	
9:	1	1	3	11	1	1	0	1	1	3:	-1-1		40	24	1/6
9:	1	3	3	3	0	0	2	-1	-1	1:	-1-1		40	8	
10:	1	1	4	7	1	0	0	1	0	0:	-1-1		80	24	1/6
10:	1	3	3	3	1	1	-1	0	-1	1:	-1-1		80	8	
11:	1	1	6	6	1	1	0	1	1	5:	-1-1		320	24	2/3
11:	1	2	3	5	1	0	1	1	1	3:	-1-1		320	8	
11:	1	2	3	5	1	0	2	1	0	0:	-1-1		320	4	
11:	2	2	2	4	1	1	-1	1	2	1:	-1-1		320	4	
12:	1	1	4	6	0	0	0	0	0	4:	1 1		40	32	1/16
12:	1	2	2	5	0	0	0	0	0	0:	1 1		40	32	
13:	1	2	2	8	1	0	2	0	0	0:	1 1		160	16	1/2
13:	1	2	4	4	0	1	2	1	2	0:	1 1		160	16	
13:	1	2	4	4	1	0	2	1	0	2:	1 1		160	4	
13:	2	2	3	3	2	1	0	1	0	-2:	1 1		160	8	
14:	1	1	3	7	0	0	0	0	0	2:	-1-1		80	16	1/8
14:	1	2	4	4	0	0	0	0	2	4:	-1-1		80	16	
15:	1	2	2	6	0	0	0	0	2	2:	1 1		80	32	1/24
15:	2	2	2	5	2	2	0	0	0	0:	1 1		80	96	
16:	1	2	2	7	1	1	0	0	1	1:	-1-1		80	8	1/2
16:	1	2	3	4	1	0	1	0	0	0:	-1-1		80	8	
16:	2	2	2	4	1	1	0	0	2	2:	-1-1		80	4	
17:	1	2	3	4	0	0	2	0	0	0:	-1-1		80	16	1/8
17:	2	2	3	3	0	2	2	0	0	2:	-1-1		80	16	
18:	1	2	4	4	1	0	2	0	2	0:	1 1		40	8	1/6
18:	2	2	3	3	1	1	-1	1	-1	-2:	1 1		40	24	
19:	1	3	3	3	0	0	2	0	2	2:	-1-1		20	24	1/24

D= 321

GENUS#:	F11	F22	F33	F44	F12	F13	F23	F14	F24	F34:	HASSE	SYM	LEVEL	AUTOS	MASS
1:	1	1	1	27	1	0	0	0	0	1:	-1 1-1		321	48	33/16
1:	1	1	3	10	1	0	0	1	0	3:	-1 1-1		321	24	
1:	1	1	3	8	0	1	0	1	1	2:	-1 1-1		321	8	
1:	1	1	3	9	1	0	0	0	0	1:	-1 1-1		321	24	
1:	1	1	4	6	0	1	1	1	0	0:	-1 1-1		321	8	
1:	1	1	4	8	1	1	0	0	1	2:	-1 1-1		321	12	
1:	1	2	2	6	1	0	0	0	1	-1:	-1 1-1		321	4	

	F11	F22	F33	F44	F12	F13	F23	F14	F24	F34:	HASSE SYM	LEVEL	AUTOS	MASS
1:	1	2	2	7	1	1	0	1	1	0:	-1 1-1	321	8	
1:	1	2	3	4	0	0	1	-1	-1	1:	-1 1-1	321	4	
1:	1	3	3	4	1	0	0	0	3	3:	-1 1-1	321	8	
1:	1	3	3	4	0	0	3	1	1	3:	-1 1-1	321	8	
1:	1	3	3	4	1	0	2	1	0	3:	-1 1-1	321	4	
1:	2	2	2	3	1	0	0	1	1	1:	-1 1-1	321	4	
1:	2	2	3	3	1	1	1	2	2	-1:	-1 1-1	321	4	33/16
2:	1	1	2	12	0	1	0	1	1	0:	-1-1 1	321	8	
2:	1	1	2	14	0	1	1	1	0	2:	-1-1 1	321	16	
2:	1	1	4	6	0	1	0	1	1	2:	-1-1 1	321	8	
2:	1	2	3	4	0	1	1	0	1	1:	-1-1 1	321	4	
2:	1	2	4	4	1	0	0	1	2	3:	-1-1 1	321	4	
2:	2	2	2	3	1	1	0	0	0	1:	-1-1 1	321	2	
2:	2	2	2	4	2	1	0	-1	-1	1:	-1-1 1	321	2	
2:	2	2	2	5	2	1	-1	1	2	0:	-1-1 1	321	4	

D= 324

GENUS#:	F11	F22	F33	F44	F12	F13	F23	F14	F24	F34:	HASSE SYM	LEVEL	AUTOS	MASS
1:	1	1	1	21	0	0	0	1	1	1:	1 1	162	96	9/32
1:	1	1	3	9	0	1	1	0	0	3:	1 1	162	16	
1:	1	2	3	5	0	1	0	0	2	3:	1 1	162	8	
1:	2	2	3	3	2	0	0	2	2	1:	1 1	162	12	
2:	1	1	1	27	1	0	0	0	0	0:	-1-1	108	48	3/8
2:	1	1	3	7	0	0	0	0	1	0:	-1-1	108	16	
2:	1	1	4	7	1	0	0	0	0	2:	-1-1	108	24	
2:	1	2	2	4	1	0	0	0	1	0:	-1-1	108	8	
2:	1	3	3	4	0	0	3	0	3	0:	-1-1	108	24	
2:	2	2	2	4	1	1	1	2	2	2:	-1-1	108	12	
3:	1	1	1	41	1	1	0	1	0	0:	1 1	162	96	9/32
3:	1	1	5	5	0	0	1	1	1	3:	1 1	162	16	
3:	1	1	5	6	1	1	0	0	0	2:	1 1	162	12	
3:	1	2	3	5	0	1	2	1	0	2:	1 1	162	8	
4:	1	1	3	8	0	0	0	1	1	3:	-1-1	54	32	9/32
4:	1	2	2	7	1	1	0	0	-1	1:	-1-1	54	8	
4:	2	2	2	4	1	0	0	2	2	2:	-1-1	54	8	
5:	1	1	4	6	0	1	0	0	0	3:	-1-1	108	8	3/8
5:	1	2	4	4	1	0	1	1	2	3:	-1-1	108	4	
6:	1	1	6	6	1	0	0	0	0	6:	1 1	18	144	1/144
7:	2	2	2	3	1	1	1	0	0	0:	-1-1	36	24	1/24
8:	2	2	2	5	2	1	-1	-1	1	-1:	-1-1	18	32	1/32
9:	1	1	2	14	0	0	0	0	0	2:	1 1	54	48	1/16
9:	2	3	3	3	0	0	3	2	3	0:	1 1	54	24	
10:	1	1	2	14	0	1	1	1	1	1:	-1-1	54	32	9/32
10:	1	2	4	4	1	0	2	1	2	0:	-1-1	54	8	
10:	2	2	2	4	2	0	0	2	1	1:	-1-1	54	8	
11:	1	1	3	9	1	0	0	0	0	0:	-1-1	36	48	1/24
11:	1	3	3	3	0	0	0	0	0	3:	-1-1	36	48	
12:	1	1	5	5	0	1	1	1	1	1:	1 1	18	64	1/64
13:	1	2	2	7	0	0	2	1	0	0:	1 1	54	48	1/16
13:	1	3	3	3	1	1	0	1	0	0:	1 1	54	24	
14:	1	2	3	5	0	1	2	0	2	1:	1 1	18	16	1/16
15:	1	3	3	4	0	0	0	1	3	3:	-1-1	18	32	1/32
16:	2	2	3	3	2	0	0	0	0	3:	1 1	18	144	1/144

D= 325

GENUS#:	F11	F22	F33	F44	F12	F13	F23	F14	F24	F34:	HASSE SYM	LEVEL	AUTOS	MASS
1:	1	1	1	41	1	1	0	0	1	0:	-1 1-1	325	48	65/48
1:	1	1	2	12	0	1	0	0	1	1:	-1 1-1	325	8	
1:	1	1	3	11	1	1	0	0	1	2:	-1 1-1	325	12	
1:	1	2	2	7	1	0	1	1	1	2:	-1 1-1	325	8	
1:	1	2	3	5	0	1	2	1	1	2:	-1 1-1	325	4	
1:	1	3	3	3	1	1	-1	0	1	0:	-1 1-1	325	4	
1:	2	2	3	3	1	2	0	2	2	1:	-1 1-1	325	2	
2:	1	1	2	17	1	1	0	1	1	2:	-1 1-1	65	24	13/48
2:	1	1	5	7	1	0	0	1	0	5:	-1 1-1	65	24	
2:	1	2	2	6	0	0	1	-1	-1	1:	-1 1-1	65	8	
2:	2	3	3	3	1	1	-3	-1	3	-1:	-1 1-1	65	16	
3:	1	1	2	14	1	0	0	1	1	1:	-1 1-1	325	12	65/48
3:	1	1	3	8	0	1	0	0	1	2:	-1 1-1	325	8	
3:	1	1	4	8	1	1	0	0	0	3:	-1 1-1	325	12	
3:	1	1	5	5	0	1	0	0	1	3:	-1 1-1	325	16	
3:	1	2	3	4	0	1	1	0	1	0:	-1 1-1	325	4	

										HASSE	SYM	LEVEL	AUTOS	MASS
3:	1	2	3	5	0	1	1	1	2	3: -1 1-1		325	4	
3:	1	2	3	5	1	1	0	0	2	1: -1 1-1		325	4	
3:	1	3	3	4	1	0	3	1	2	2: -1 1-1		325	4	
4:	1	1	3	9	0	1	1	1	0	3: -1-1 1		65	16	13/48
4:	1	3	3	3	1	1	0	1	1	0: -1-1 1		65	24	
4:	1	3	3	4	1	1	0	0	3	2: -1-1 1		65	8	
4:	2	2	2	4	1	1	-1	1	-1	-1: -1-1 1		65	24	

D= 328

GENUS#:	F11	F22	F33	F44	F12	F13	F23	F14	F24	F34:	HASSE	SYM	LEVEL	AUTOS	MASS
1:	1	1	1	21	0	0	0	1	1	0:	1 1		328	32	27/16
1:	1	1	1	41	1	1	0	0	0	0:	1 1		328	96	
1:	1	1	2	11	0	0	0	1	2:	1 1		328	16		
1:	1	1	2	17	1	1	0	0	1	1:	1 1		328	12	
1:	1	2	2	6	0	0	0	1	2	1:	1 1		328	8	
1:	1	2	2	9	1	0	2	0	2	0:	1 1		328	8	
1:	1	2	3	4	0	1	1	0	0	1:	1 1		328	4	
1:	1	2	3	5	0	0	2	0	0	3:	1 1		328	8	
1:	1	2	3	5	0	1	2	0	0	2:	1 1		328	8	
1:	1	2	3	5	0	1	1	0	2	3:	1 1		328	4	
1:	2	2	3	3	1	1	0	2	1	3:	1 1		328	2	
2:	1	1	2	12	0	1	0	1	0	0:	-1-1		328	8	27/16
2:	1	1	2	14	1	0	0	1	0	0:	-1-1		328	24	
2:	1	1	4	7	0	1	1	0	0	4:	-1-1		328	16	
2:	1	1	5	6	1	0	0	1	0	2:	-1-1		328	12	
2:	1	2	2	6	0	0	1	0	2	0:	-1-1		328	4	
2:	1	2	2	6	1	0	0	0	1	0:	-1-1		328	8	
2:	1	2	2	7	0	0	2	0	1	0:	-1-1		328	8	
2:	1	2	3	4	0	1	0	1	0	0:	-1-1		328	8	
2:	1	2	3	4	0	0	0	1	0	2:	-1-1		328	8	
2:	1	2	3	4	0	0	0	0	2	1:	-1-1		328	8	
2:	1	2	3	5	1	0	2	0	1	0:	-1-1		328	4	
2:	2	2	3	3	2	1	0	1	2	1:	-1-1		328	4	

D= 329

GENUS#:	F11	F22	F33	F44	F12	F13	F23	F14	F24	F34:	HASSE	SYM	LEVEL	AUTOS	MASS
1:	1	1	1	28	1	0	0	1	0	1:	-1 1-1		329	24	47/24
1:	1	1	2	12	0	1	0	0	1	0:	-1 1-1		329	16	
1:	1	1	2	14	0	1	1	0	1	0:	-1 1-1		329	8	
1:	1	1	4	7	0	1	1	1	0	4:	-1 1-1		329	16	
1:	1	1	6	6	1	1	0	0	1	4:	-1 1-1		329	24	
1:	1	2	2	6	1	0	0	0	0	1:	-1 1-1		329	8	
1:	1	2	2	7	1	1	0	0	1	0:	-1 1-1		329	4	
1:	1	2	3	4	1	0	0	0	0	1:	-1 1-1		329	8	
1:	1	2	3	5	1	0	2	0	1	1: -1 1-1		329	4	
1:	1	2	4	4	1	0	2	0	0	3:	-1 1-1		329	4	
1:	1	2	4	4	1	0	2	0	2	1:	-1 1-1		329	8	
1:	2	2	2	4	1	1	0	-1	2	1:	-1 1-1		329	4	
1:	2	2	3	3	2	1	0	1	1	-2:	-1 1-1		329	4	
2:	1	1	3	8	0	1	0	1	1	0:	-1-1 1		329	8	47/24
2:	1	1	4	6	0	1	0	0	1	2:	-1-1 1		329	8	
2:	1	1	4	6	0	1	0	1	1	0:	-1-1 1		329	8	
2:	1	1	4	8	1	1	0	1	1	2:	-1-1 1		329	12	
2:	1	2	3	4	0	0	1	1	1	1:	-1-1 1		329	4	
2:	1	3	3	4	1	0	2	0	3	3:	-1-1 1		329	4	
2:	2	2	2	3	1	0	0	0	1	1:	-1-1 1		329	2	
2:	2	2	2	4	2	1	0	0	-1	1:	-1-1 1		329	2	

D= 332

GENUS#:	F11	F22	F33	F44	F12	F13	F23	F14	F24	F34:	HASSE	SYM	LEVEL	AUTOS	MASS
1:	1	1	1	21	0	0	0	0	0	1:	-1-1		332	32	43/32
1:	1	1	1	28	1	0	0	1	0	0:	-1-1		332	24	
1:	1	1	2	12	0	1	0	0	0	1:	-1-1		332	8	
1:	1	1	3	7	0	0	0	0	0	1:	-1-1		332	16	
1:	1	1	3	8	0	1	0	1	0	2:	-1-1		332	8	
1:	1	1	4	6	0	0	0	1	0	3:	-1-1		332	8	
1:	1	1	4	8	1	1	0	0	1	1:	-1-1		332	12	
1:	1	2	2	9	1	0	2	0	0	2:	-1-1		332	8	
1:	1	2	3	4	0	0	1	0	1	2:	-1-1		332	4	
1:	1	2	4	4	1	1	0	0	1	3:	-1-1		332	4	
1:	1	2	4	4	0	1	2	0	2	3:	-1-1		332	8	
2:	1	1	1	42	1	1	0	1	0	0:	1 1		332	96	43/32

GENUS#	F11	F22	F33	F44	F12	F13	F23	F14	F24	F34	HASSE SYM	LEVEL	AUTOS	MASS
2:	1	1	2	11	0	0	0	1	1	1:	1 1	332	16	
2:	1	1	2	14	0	1	1	0	0	1:	1 1	332	16	
2:	1	1	2	17	1	1	0	1	0	0:	1 1	332	12	
2:	1	1	4	6	0	1	1	0	0	1:	1 1	332	16	
2:	1	1	5	6	0	1	1	0	0	5:	1 1	332	16	
2:	1	2	2	7	1	0	0	0	2	2:	1 1	332	8	
2:	1	2	2	7	1	0	1	0	2	0:	1 1	332	4	
2:	1	2	3	5	0	0	2	1	2	0:	1 1	332	8	
2:	2	2	2	4	2	1	0	1	0	1:	1 1	332	4	
2:	2	2	3	3	0	1	0	2	2	2:	1 1	332	4	

D= 333

GENUS#	F11	F22	F33	F44	F12	F13	F23	F14	F24	F34	HASSE SYM	LEVEL	AUTOS	MASS
1:	1	1	1	28	1	0	0	0	0	1:	-1-1 1	111	48	5/24
1:	1	1	3	10	1	0	0	0	0	3:	-1-1 1	111	48	
1:	1	1	4	7	1	0	0	0	0	1:	-1-1 1	111	24	
1:	1	2	3	5	1	0	0	1	0	3:	-1-1 1	111	8	
2:	1	1	1	42	1	1	0	0	1	0:	-1 1-1	333	48	5/4
2:	1	1	2	17	1	1	0	0	1	0:	-1 1-1	333	12	
2:	1	1	3	11	1	1	0	1	1	2:	-1 1-1	333	12	
2:	1	1	5	6	1	1	0	0	0	1:	-1 1-1	333	12	
2:	1	1	5	6	0	1	1	1	0	5:	-1 1-1	333	16	
2:	1	1	5	7	1	1	0	1	1	4:	-1 1-1	333	12	
2:	1	1	6	6	1	0	0	1	0	5:	-1 1-1	333	12	
2:	1	2	3	4	0	0	1	1	1	0:	-1 1-1	333	4	
2:	1	2	3	5	1	0	2	0	0	1:	-1 1-1	333	4	
2:	1	3	3	3	0	0	1	1	2	1:	-1 1-1	333	4	
3:	1	1	2	14	1	0	0	0	0	1:	-1 1-1	111	24	5/6
3:	1	1	5	6	1	0	0	0	0	3:	-1 1-1	111	24	
3:	1	2	2	6	0	0	1	1	1	1:	-1 1-1	111	8	
3:	1	2	3	5	0	1	2	0	1	2:	-1 1-1	111	4	
3:	1	3	3	4	1	0	3	1	2	0:	-1 1-1	111	8	
3:	2	3	3	3	1	0	0	2	3	3:	-1 1-1	111	4	
4:	1	1	5	5	0	1	0	1	1	2:	-1 1-1	333	8	5/4
4:	1	2	2	7	1	0	1	1	2	0:	-1 1-1	333	4	
4:	1	2	3	4	0	1	0	0	-1	1:	-1 1-1	333	4	
4:	1	2	3	5	1	1	0	1	2	1:	-1 1-1	333	4	
4:	1	2	3	5	0	1	2	1	1	1:	-1 1-1	333	8	
4:	2	2	3	3	1	2	0	1	0	3:	-1 1-1	333	4	

D= 336

GENUS#	F11	F22	F33	F44	F12	F13	F23	F14	F24	F34	HASSE SYM	LEVEL	AUTOS	MASS
1:	1	1	1	21	0	0	0	0	0	0:	1-1-1	84	96	5/96
1:	2	2	3	3	2	0	0	2	0	0:	1-1-1	84	24	
2:	1	1	1	28	1	0	0	0	0	0:	-1-1 1	336	48	5/6
2:	1	1	2	12	0	1	0	0	0	0:	-1-1 1	336	16	
2:	1	1	3	8	0	0	0	1	0	3:	-1-1 1	336	16	
2:	1	1	5	5	0	1	0	1	0	3:	-1-1 1	336	16	
2:	1	2	3	5	1	1	0	0	2	0:	-1-1 1	336	4	
2:	1	2	4	4	0	0	2	0	-1	3:	-1-1 1	336	8	
2:	2	2	2	4	1	1	1	2	0	2:	-1-1 1	336	4	
3:	1	1	1	42	1	1	0	0	0	0:	-1 1-1	168	96	5/24
3:	1	1	2	11	0	0	0	1	1	0:	-1 1-1	168	32	
3:	1	1	5	6	1	1	0	0	0	0:	-1 1-1	168	24	
3:	1	3	3	3	1	1	0	0	1	1:	-1 1-1	168	8	
4:	1	1	3	11	1	1	0	0	1	1:	-1 1-1	84	12	5/24
4:	1	3	3	3	1	0	1	0	1	2:	-1 1-1	84	8	
5:	1	1	3	7	0	0	0	0	0	0:	-1-1 1	84	32	5/32
5:	1	2	3	4	0	0	0	0	2	0:	-1-1 1	84	16	
5:	2	2	3	3	0	2	0	0	2	0:	-1-1 1	84	16	
6:	1	1	2	14	1	0	0	0	0	0:	-1-1 1	160	48	5/24
6:	1	1	3	9	0	1	1	1	1	0:	-1-1 1	168	16	
6:	1	2	3	5	0	0	0	1	2	0:	-1-1 1	168	16	
6:	2	2	3	3	0	0	0	0	2	3:	-1-1 1	168	16	
7:	1	1	4	7	1	0	0	0	0	0:	-1-1 1	336	48	5/6
7:	1	2	2	7	1	0	1	0	0	2:	-1-1 1	336	4	
7:	1	2	2	7	1	1	0	0	0	0:	-1-1 1	336	16	
7:	1	2	3	4	1	0	0	0	0	0:	-1-1 1	336	16	
7:	1	3	3	3	1	0	0	1	2	0:	-1-1 1	336	16	
7:	2	2	2	3	1	1	0	0	0	0:	-1-1 1	336	8	
7:	2	2	3	3	1	0	0	2	2	2:	-1-1 1	336	4	
8:	1	2	2	8	1	1	0	1	2	2:	-1-1 1	84	16	5/24

#	F11	F22	F33	F44	F12	F13	F23	F14	F24	F34	HASSE	SYM	LEVEL	AUTOS	MASS
8	1	2	4	4	1	0	0	0	0	4	-1-1	1	84	48	
8	2	2	2	4	1	1	0	1	2	2	-1-1	1	84	8	
9	1	1	2	11	0	0	0	0	0	2	-1 1	-1	84	32	5/32
9	1	3	3	3	0	0	2	0	0	2	-1 1	-1	84	8	
10	1	1	2	14	0	1	1	0	0	0	1 1	1	168	32	5/8
10	1	1	4	6	0	1	1	0	0	0	1 1	1	168	32	
10	1	2	2	6	1	0	0	0	0	0	1 1	1	168	16	
10	1	2	4	4	1	0	2	1	1	2	1 1	1	168	4	
10	1	2	4	4	0	1	2	1	2	1	1 1	1	168	16	
10	2	2	2	4	2	0	0	2	1	0	1 1	1	168	16	
10	2	2	3	3	0	2	1	2	1	0	1 1	1	168	8	
11	1	1	3	10	1	0	0	1	1	2	-1 1	-1	336	12	5/6
11	1	2	3	4	0	1	1	0	0	0	-1 1	-1	336	8	
11	1	2	4	4	0	0	1	1	1	4	-1 1	-1	336	8	
11	1	3	3	3	0	0	0	1	-2	1	-1 1	-1	336	4	
11	2	2	3	3	0	1	1	2	2	0	-1 1	-1	336	4	
12	1	1	3	8	0	1	0	0	0	2	-1 1	-1	336	8	5/6
12	1	1	4	6	0	1	0	1	0	2	-1 1	-1	336	8	
12	1	1	4	8	1	1	0	1	0	0	-1 1	-1	336	12	
12	1	2	3	4	0	0	1	0	0	2	-1 1	-1	336	4	
12	1	3	3	4	1	0	1	1	3	3	-1 1	-1	336	4	
13	1	1	4	6	0	0	0	1	1	2	1-1	-1	168	16	5/8
13	1	2	2	6	0	0	0	1	2	0	1-1	-1	168	16	
13	2	2	2	3	1	0	0	1	1	0	1-1	-1	168	8	
13	2	2	2	4	2	1	0	1	0	0	1-1	-1	168	4	
13	2	2	2	5	2	1	-1	1	1	1	1-1	-1	168	8	
14	1	1	4	8	1	0	0	0	0	4	-1-1	1	84	48	5/24
14	1	3	3	4	1	1	-1	0	2	2	-1-1	1	84	16	
14	1	3	3	4	1	0	3	0	2	0	-1-1	1	84	8	
15	1	1	5	5	0	0	0	0	0	4	1 1	1	84	32	5/96
15	1	2	2	7	0	0	2	0	0	0	1 1	1	84	48	
16	1	3	3	4	0	0	2	1	3	3	-1 1	-1	84	8	5/24
16	2	2	2	3	0	0	0	1	1	1	-1 1	-1	84	12	

D= 337

GENUS#	F11	F22	F33	F44	F12	F13	F23	F14	F24	F34	HASSE	SYM	LEVEL	AUTOS	MASS
1	1	1	2	17	1	1	0	0	0	1	-1-1		337	12	19/4
1	1	1	3	8	0	1	0	0	1	1	-1-1		337	8	
1	1	1	4	8	1	1	0	0	1	0	-1-1		337	12	
1	1	1	5	6	0	1	0	0	1	5	-1-1		337	8	
1	1	1	5	6	1	0	0	1	0	1	-1-1		337	12	
1	1	2	2	6	0	0	1	1	0	1	-1-1		337	4	
1	1	2	2	9	1	0	2	0	2	1	-1-1		337	4	
1	1	2	3	4	0	0	1	1	0	1	-1-1		337	4	
1	1	2	3	5	1	0	1	0	0	3	-1-1		337	4	
1	1	2	3	5	1	0	1	1	2	2	-1-1		337	4	
1	1	2	3	5	1	1	0	0	1	2	-1-1		337	4	
1	1	2	3	6	1	0	2	1	1	3	-1-1		337	4	
1	1	2	4	4	0	1	0	0	1	4	-1-1		337	4	
1	1	3	3	3	1	0	0	0	2	1	-1-1		337	4	
1	2	2	2	4	2	1	0	1	1	0	-1-1		337	2	
1	2	2	3	3	0	1	1	1	2	3	-1-1		337	2	
1	2	2	3	3	1	1	-1	2	2	1	-1-1		337	2	
1	2	2	3	3	2	1	-1	0	1	-1	-1-1		337	2	

D= 340

GENUS#	F11	F22	F33	F44	F12	F13	F23	F14	F24	F34	HASSE	SYM	LEVEL	AUTOS	MASS
1	1	1	1	22	0	0	0	1	1	1	1 1	1	170	96	15/32
1	1	2	4	4	0	1	0	1	2	3	1 1	1	170	8	
1	1	2	4	4	1	1	0	1	1	3	1 1	1	170	8	
1	2	2	2	6	2	2	0	2	0	1	1 1	1	170	12	
1	2	2	3	3	2	1	-1	1	1	0	1 1	1	170	8	
2	1	1	1	43	1	1	0	1	0	0	-1-1	1	170	96	15/32
2	1	1	3	11	1	1	0	1	0	0	-1-1	1	170	12	
2	1	2	3	5	0	1	2	1	0	0	-1-1	1	170	8	
2	1	3	3	3	1	0	1	1	1	1	-1-1	1	170	8	
2	2	3	3	3	2	0	2	2	1	3	-1-1	1	170	8	
3	1	1	2	15	1	0	0	1	0	2	-1 1	-1	170	24	15/32
3	1	1	3	9	0	1	1	1	1	1	-1 1	-1	170	32	
3	1	1	5	7	1	1	0	0	1	3	-1 1	-1	170	12	
3	1	2	3	5	0	0	2	1	2	1	-1 1	-1	170	16	
3	1	3	3	4	1	0	3	0	0	2	-1 1	-1	170	4	

	F11	F22	F33	F44	F12	F13	F23	F14	F24	F34:	HASSE SYM	LEVEL	AUTOS	MASS
4:	1	1	2	17	1	1	0	0	0	0:	-1 1-1	340	24	15/8
4:	1	1	3	8	0	1	0	1	0	0:	-1 1-1	340	8	
4:	1	1	4	8	1	1	0	0	0	2:	-1 1-1	340	12	
4:	1	2	3	4	0	0	0	0	1	2:	-1 1-1	340	4	
4:	1	2	3	4	0	0	0	1	-1	1:	-1 1-1	340	4	
4:	1	2	3	5	0	0	2	0	-1	2:	-1 1-1	340	8	
4:	1	2	3	5	1	0	1	0	2	2:	-1 1-1	340	4	
4:	1	3	3	4	1	0	2	-1	1	-2:	-1 1-1	340	4	
4:	1	3	3	4	1	0	3	1	2	1:	-1 1-1	340	4	
4:	2	2	3	3	0	2	0	1	1	3:	-1 1-1	340	4	
5:	1	1	2	11	0	0	0	0	1	1:	-1-1 1	340	8	15/8
5:	1	1	3	11	1	1	0	0	0	2:	-1-1 1	340	12	
5:	1	1	5	6	1	0	0	1	0	0:	-1-1 1	340	24	
5:	1	2	2	6	0	0	1	0	1	-1:	-1-1 1	340	8	
5:	1	2	2	7	1	0	1	1	0	1:	-1-1 1	340	4	
5:	1	2	3	5	1	0	2	0	0	0:	-1-1 1	340	8	
5:	1	3	3	3	0	0	1	0	2	2:	-1-1 1	340	8	
5:	1	3	3	3	1	0	1	0	0	2:	-1-1 1	340	4	
5:	2	2	2	4	1	1	-1	0	0	0:	-1-1 1	340	4	
5:	2	2	3	3	1	2	-1	0	1	-2:	-1-1 1	340	2	
6:	1	1	5	6	0	0	0	1	1	5:	1-1-1	170	32	15/32
6:	1	2	2	8	0	0	2	1	2	0:	1-1-1	170	24	
6:	1	2	2	9	1	0	2	0	1	2:	1-1-1	170	16	
6:	2	2	2	4	2	1	0	0	0	1:	1-1-1	170	4	
6:	2	2	3	4	2	2	0	0	2	1:	1-1-1	170	12	

D= 341

GENUS#:	F11	F22	F33	F44	F12	F13	F23	F14	F24	F34:	HASSE SYM	LEVEL	AUTOS	MASS
1:	1	1	1	29	1	0	0	1	0	1:	-1-1 1	341	24	59/48
1:	1	1	3	8	0	1	0	0	1	0:	-1-1 1	341	16	
1:	1	1	4	6	0	1	0	0	1	1:	-1-1 1	341	8	
1:	1	1	4	7	0	1	0	0	1	4:	-1-1 1	341	8	
1:	1	1	4	8	1	0	0	1	0	3:	-1-1 1	341	12	
1:	1	1	4	9	1	1	0	1	1	4:	-1-1 1	341	24	
1:	1	2	3	4	0	1	0	0	1	0:	-1-1 1	341	8	
1:	1	2	3	5	0	0	1	1	2	3:	-1-1 1	341	4	
1:	1	2	4	4	1	0	1	0	1	-3:	-1-1 1	341	8	
1:	2	2	3	3	1	2	0	0	2	1:	-1-1 1	341	4	
2:	1	1	1	43	1	1	0	0	1	0:	-1 1-1	341	48	59/48
2:	1	1	2	13	0	1	0	0	1	2:	-1 1-1	341	8	
2:	1	1	3	11	1	1	0	0	1	0:	-1 1-1	341	12	
2:	1	1	3	9	0	1	1	1	0	2:	-1 1-1	341	8	
2:	1	1	5	5	0	1	1	0	1	0:	-1 1-1	341	8	
2:	1	2	3	5	1	1	0	1	1	2:	-1 1-1	341	4	
2:	1	3	3	4	1	1	-1	1	3	1:	-1 1-1	341	4	
2:	1	3	3	4	1	0	1	0	3	3:	-1 1-1	341	4	

D= 344

GENUS#:	F11	F22	F33	F44	F12	F13	F23	F14	F24	F34:	HASSE SYM	LEVEL	AUTOS	MASS
1:	1	1	1	29	1	0	0	1	0	0:	1 1	344	24	155/96
1:	1	1	1	43	1	1	0	0	0	0:	1 1	344	96	
1:	1	1	2	11	0	0	0	0	1	0:	1 1	344	16	
1:	1	1	3	8	0	0	0	1	1	2:	1 1	344	16	
1:	1	1	5	5	0	0	0	1	0	3:	1 1	344	8	
1:	1	1	5	5	0	0	0	1	1	2:	1 1	344	16	
1:	1	2	3	4	0	1	0	0	0	1:	1 1	344	8	
1:	1	2	3	5	1	1	0	1	0	2:	1 1	344	4	
1:	1	2	4	4	1	1	0	0	-1	3:	1 1	344	4	
1:	1	2	4	4	0	0	0	1	2	3:	1 1	344	8	
1:	2	2	2	3	1	0	0	1	0	0:	1 1	344	4	
1:	2	2	3	4	2	2	0	2	1	3:	1 1	344	4	
2:	1	1	1	22	0	0	0	1	1	0:	-1-1	344	32	155/96
2:	1	1	2	15	0	1	1	1	1	0:	-1-1	344	16	
2:	1	1	3	9	0	1	1	0	0	2:	-1-1	344	16	
2:	1	1	4	6	0	1	0	1	0	0:	-1-1	344	8	
2:	1	1	4	6	0	1	0	0	0	2:	-1-1	344	8	
2:	1	1	5	7	1	1	0	0	0	4:	-1-1	344	12	
2:	1	2	2	6	0	0	0	1	2	2:	-1-1	344	8	
2:	1	2	2	7	1	0	1	1	1	1:	-1-1	344	4	
2:	1	2	3	4	0	0	1	0	1	-1:	-1-1	344	4	
2:	1	2	3	5	1	0	0	1	2	2:	-1-1	344	4	
2:	2	2	3	3	0	2	0	1	2	2:	-1-1	344	4	

D= 345

GENUS#:	F11	F22	F33	F44	F12	F13	F23	F14	F24	F34:	HASSE SYM	LEVEL	AUTOS	MASS
1:	1	1	1	29	1	0	0	0	0	1:	-1-1-1-1	345	48	55/48
1:	1	2	2	7	1	0	1	0	1	-1:	-1-1-1-1	345	4	
1:	1	2	2	8	1	1	0	0	2	1:	-1-1-1-1	345	8	
1:	2	2	2	4	1	1	1	2	2	1:	-1-1-1-1	345	4	
1:	2	2	2	4	1	1	-1	1	0	2:	-1-1-1-1	345	4	
1:	2	2	2	5	2	1	-1	0	-1	1:	-1-1-1-1	345	4	
2:	1	1	2	15	0	1	1	1	0	2:	-1 1 1-1	345	16	55/48
2:	1	1	3	10	1	0	0	1	0	1:	-1 1 1-1	345	12	
2:	1	2	4	4	1	1	0	1	0	3:	-1 1 1-1	345	8	
2:	1	2	4	4	0	1	2	-1	-1	1:	-1 1 1-1	345	4	
2:	1	2	4	4	1	0	2	1	0	1:	-1 1 1-1	345	4	
2:	1	3	3	4	0	0	1	1	3	3:	-1 1 1-1	345	8	
2:	1	3	3	4	1	0	2	1	1	3:	-1 1 1-1	345	4	
3:	1	1	2	18	1	1	0	1	1	2:	-1 1-1 1	345	24	55/48
3:	1	1	4	6	0	1	0	0	1	0:	-1 1-1 1	345	16	
3:	1	1	6	6	1	1	0	1	1	4:	-1 1-1 1	345	24	
3:	1	2	3	4	0	0	1	1	0	0:	-1 1-1 1	345	8	
3:	1	2	3	5	1	0	1	1	0	2:	-1 1-1 1	345	4	
3:	1	2	3	6	1	0	2	0	0	3:	-1 1-1 1	345	4	
3:	2	2	3	3	0	2	1	1	2	2:	-1 1-1 1	345	4	
3:	2	2	3	4	2	1	-1	1	2	2:	-1 1-1 1	345	8	
4:	1	1	5	5	0	1	0	0	1	2:	-1-1 1 1	345	16	55/48
4:	1	1	5	7	1	0	0	0	0	5:	-1-1 1 1	345	48	
4:	1	2	2	6	0	0	1	1	0	0:	-1-1 1 1	345	16	
4:	1	2	2	9	1	0	2	1	1	1:	-1-1 1 1	345	8	
4:	1	2	3	5	1	1	0	0	-1	2:	-1-1 1 1	345	4	
4:	1	2	3	5	1	0	0	0	1	3:	-1-1 1 1	345	8	
4:	2	2	2	3	1	0	0	0	0	1:	-1-1 1 1	345	8	
4:	2	2	3	3	1	0	0	1	-1	-3:	-1-1 1 1	345	8	
4:	2	2	3	3	2	1	0	1	1	2:	-1-1 1 1	345	4	

D= 348

GENUS#:	F11	F22	F33	F44	F12	F13	F23	F14	F24	F34:	HASSE SYM	LEVEL	AUTOS	MASS
1:	1	1	1	29	1	0	0	0	0	0:	-1 1-1	348	48	13/16
1:	1	1	1	44	1	1	0	1	0	0:	-1 1-1	348	96	
1:	1	1	2	11	0	0	0	0	0	1:	-1 1-1	348	16	
1:	1	1	3	8	0	0	0	0	0	3:	-1 1-1	348	32	
1:	1	1	3	8	0	1	0	0	0	1:	-1 1-1	348	8	
1:	1	1	4	7	0	1	1	1	1	3:	-1 1-1	348	16	
1:	1	1	4	9	1	1	0	0	1	3:	-1 1-1	348	12	
1:	1	1	5	6	1	0	0	0	0	2:	-1 1-1	348	24	
1:	1	2	2	6	0	0	1	0	1	1:	-1 1-1	348	8	
1:	1	2	2	7	1	0	0	1	0	2:	-1 1-1	348	8	
1:	1	3	3	4	0	0	3	0	1	3:	-1 1-1	348	8	
2:	1	1	1	22	0	0	0	0	1	0:	-1-1 1	348	32	13/16
2:	1	1	2	13	0	1	0	0	0	2:	-1-1 1	348	8	
2:	1	1	2	15	0	1	1	1	1	1:	-1-1 1	348	32	
2:	1	1	4	6	0	0	0	0	0	3:	-1-1 1	348	16	
2:	1	1	4	6	0	0	0	1	1	1:	-1-1 1	348	16	
2:	1	2	4	4	0	1	0	0	2	3:	-1-1 1	348	8	
2:	2	2	2	5	2	1	0	1	2	2:	-1-1 1	348	4	
2:	2	2	3	3	2	0	0	2	1	1:	-1-1 1	348	8	
3:	1	1	2	15	1	0	0	0	0	2:	1 1 1	348	48	13/16
3:	1	1	2	18	1	1	0	0	1	1:	1 1 1	348	12	
3:	1	1	3	10	1	0	0	1	0	0:	1 1 1	348	24	
3:	1	1	3	10	1	0	0	0	0	2:	1 1 1	348	24	
3:	1	2	3	4	0	0	0	1	1	0:	1 1 1	348	8	
3:	1	2	3	5	0	0	2	1	0	2:	1 1 1	348	8	
3:	1	3	3	3	0	0	0	1	2	0:	1 1 1	348	8	
3:	2	2	3	3	1	1	1	2	2	2:	1 1 1	348	4	
4:	1	2	2	8	0	0	2	-1	1	-1:	1-1-1	348	16	13/16
4:	1	2	2	9	1	0	2	1	0	0:	1-1-1	348	8	
4:	1	2	3	5	0	1	1	0	0	3:	1-1-1	348	4	
4:	2	2	2	5	2	1	-1	1	1	0:	1-1-1	348	8	
4:	2	2	3	3	1	2	0	2	0	0:	1-1-1	348	4	

D= 349

GENUS#:	F11	F22	F33	F44	F12	F13	F23	F14	F24	F34:	HASSE SYM	LEVEL	AUTOS	MASS
1:	1	1	1	44	1	1	0	0	1	0:	-1-1	349	48	151/48
1:	1	1	2	13	0	1	0	1	1	0:	-1-1	349	8	

GENUS#	F11	F22	F33	F44	F12	F13	F23	F14	F24	F34	HASSE SYM	LEVEL	AUTOS
1:	1	1	2	15	1	0	0	1	0	1:	-1-1	349	12
1:	1	1	3	11	1	1	0	0	0	1:	-1-1	349	12
1:	1	1	3	9	0	1	0	0	1	3:	-1-1	349	8
1:	1	1	3	9	0	1	1	1	0	0:	-1-1	349	8
1:	1	1	4	8	1	1	0	0	0	1:	-1-1	349	12
1:	1	2	2	7	1	0	1	1	0	0:	-1-1	349	4
1:	1	2	3	5	0	1	2	0	1	0:	-1-1	349	4
1:	1	2	3	5	0	1	1	1	0	3:	-1-1	349	4
1:	1	2	3	5	0	1	1	1	2	0:	-1-1	349	4
1:	1	3	3	3	1	1	0	0	1	0:	-1-1	349	4
1:	1	3	3	4	1	1	-1	0	3	0:	-1-1	349	4
1:	1	3	3	4	1	1	0	1	2	3:	-1-1	349	4
1:	1	3	3	4	1	0	3	0	-1	1:	-1-1	349	4
1:	2	2	3	3	1	2	-1	0	1	1:	-1-1	349	2

D= 352

GENUS#	F11	F22	F33	F44	F12	F13	F23	F14	F24	F34	HASSE SYM	LEVEL	AUTOS	MASS
1:	1	1	1	22	0	0	0	0	0	0:	-1-1	88	96	23/96
1:	1	2	2	6	0	0	0	0	2	0:	-1-1	88	16	
1:	1	2	2	8	0	0	2	0	2	0:	-1-1	88	24	
1:	1	2	2	5	0	0	2	0	2	0:	-1-1	88	8	
2:	1	1	1	44	1	1	0	0	0	0:	1 1	176	96	23/96
2:	1	1	2	15	1	0	0	1	0	0:	1 1	176	24	
2:	1	2	3	4	0	1	0	0	0	0:	1 1	176	16	
2:	1	2	3	5	0	1	2	0	0	1:	1 1	176	8	
3:	1	1	2	11	0	0	0	0	0	0:	1 1	88	32	23/96
3:	1	2	3	4	0	0	0	0	0	2:	1 1	88	8	
3:	2	2	3	4	2	2	0	2	0	0:	1 1	88	12	
4:	1	1	2	12	0	0	0	1	1	2:	1 1	176	32	23/32
4:	1	2	2	9	1	0	2	0	1	0:	1 1	176	8	
4:	1	2	3	4	0	0	0	1	0	1:	1 1	176	8	
4:	1	2	4	4	0	0	0	1	0	4:	1 1	176	16	
4:	2	2	2	3	0	0	0	1	1	0:	1 1	176	8	
4:	2	2	3	4	2	1	-1	1	1	3:	1 1	176	4	
5:	1	1	2	18	1	1	0	1	0	0:	-1-1	352	12	23/24
5:	1	2	2	6	0	0	0	1	1	1:	-1-1	352	8	
5:	1	2	3	5	0	0	2	1	1	2:	-1-1	352	4	
5:	2	2	3	3	1	2	0	2	1	0:	-1-1	352	4	
5:	2	3	3	3	2	2	0	1	3	2:	-1-1	352	4	
6:	1	1	3	11	1	1	0	0	0	0:	1 1	352	24	23/24
6:	1	1	6	6	1	1	0	0	1	3:	1 1	352	24	
6:	1	2	3	6	1	0	2	1	0	2:	1 1	352	4	
6:	1	3	3	3	1	0	0	0	0	2:	1 1	352	16	
6:	1	3	3	4	1	0	3	1	1	1:	1 1	352	8	
6:	1	3	3	4	0	0	2	-1	-2	2:	1 1	352	16	
6:	2	2	3	3	1	1	-1	2	2	0:	1 1	352	4	
6:	2	2	3	3	0	1	1	2	2	1:	1 1	352	8	
7:	1	1	3	12	1	1	0	1	1	3:	1 1	352	24	23/24
7:	1	1	3	8	0	1	0	0	0	0:	1 1	352	16	
7:	1	1	4	8	1	1	0	0	0	0:	1 1	352	24	
7:	1	1	5	5	0	1	0	1	0	2:	1 1	352	16	
7:	1	2	3	5	1	1	0	0	0	2:	1 1	352	4	
7:	1	3	3	3	0	0	2	0	1	-1:	1 1	352	8	
7:	1	3	3	4	1	0	3	0	1	2:	1 1	352	8	
7:	2	2	3	3	1	1	0	0	1	3:	1 1	352	4	
8:	1	1	4	7	0	0	0	1	1	4:	-1-1	176	32	23/96
8:	1	1	5	7	1	1	0	1	1	3:	-1-1	176	12	
8:	2	3	3	3	0	0	2	2	3	3:	-1-1	176	8	
9:	1	1	4	6	0	0	0	1	1	0:	-1-1	176	32	23/32
9:	1	2	3	5	1	0	1	0	-1	2:	-1-1	176	4	
9:	2	2	2	4	2	1	0	0	0	0:	-1-1	176	8	
9:	2	2	2	6	2	2	0	2	1	1:	-1-1	176	16	
9:	2	2	3	3	2	1	0	1	0	2:	-1-1	176	8	
9:	2	2	3	3	0	2	1	2	1	2:	-1-1	176	8	
10:	1	1	4	6	0	0	0	0	1	2:	-1-1	352	8	23/24
10:	1	1	5	7	1	0	0	1	0	4:	-1-1	352	12	
10:	1	2	2	6	0	0	1	0	0	1:	-1-1	352	4	
10:	1	2	3	4	0	0	1	0	1	1:	-1-1	352	4	
10:	1	2	3	5	1	0	1	0	2	0:	-1-1	352	4	

D= 353
GENUS#: F11 F22 F33 F44 F12 F13 F23 F14 F24 F34: HASSE SYM LEVEL AUTOS MASS

GENUS#	F11	F22	F33	F44	F12	F13	F23	F14	F24	F34	HASSE	SYM	LEVEL	AUTOS	MASS
1:	1	1	1	30	1	0	0	1	0	1:		-1-1	353	24	4/1
1:	1	1	2	13	0	1	0	0	1	1:		-1-1	353	8	
1:	1	1	2	15	0	1	1	1	0	0:		-1-1	353	8	
1:	1	1	2	18	1	1	0	0	1	0:		-1-1	353	12	
1:	1	1	4	7	0	1	1	0	1	3:		-1-1	353	8	
1:	1	2	2	7	1	0	0	1	2	1:		-1-1	353	4	
1:	1	2	2	7	1	0	1	0	1	1:		-1-1	353	4	
1:	1	2	2	8	1	1	0	1	2	1:		-1-1	353	4	
1:	1	2	2	9	1	0	2	0	1	1:		-1-1	353	4	
1:	1	2	3	5	1	0	1	1	1	2:		-1-1	353	4	
1:	1	2	3	5	1	0	1	1	2	0:		-1-1	353	4	
1:	1	2	4	4	1	0	0	1	0	3:		-1-1	353	4	
1:	1	2	4	4	1	0	1	1	2	2:		-1-1	353	4	
1:	1	2	4	4	0	0	1	1	2	3:		-1-1	353	4	
1:	1	2	4	4	1	1	0	1	2	1:		-1-1	353	4	
1:	2	2	2	4	1	1	0	2	1	2:		-1-1	353	2	
1:	2	2	3	3	1	1	-1	2	1	2:		-1-1	353	2	

D= 356

GENUS#	F11	F22	F33	F44	F12	F13	F23	F14	F24	F34	HASSE	SYM	LEVEL	AUTOS	MASS
1:	1	1	1	23	0	0	0	1	1	1:		1 1	178	96	13/48
1:	1	1	1	45	1	1	0	1	0	0:		1 1	178	96	
1:	1	1	3	9	0	1	1	0	0	1:		1 1	178	16	
1:	1	1	5	5	0	0	0	1	1	1:		1 1	178	16	
1:	1	2	3	5	0	1	0	1	2	2:		1 1	178	8	
2:	1	1	1	30	1	0	0	1	0	0:		-1-1	356	24	13/4
2:	1	1	2	13	0	1	0	1	0	0:		-1-1	356	8	
2:	1	1	3	8	0	0	0	0	1	2:		-1-1	356	8	
2:	1	1	4	6	0	1	0	0	0	1:		-1-1	356	8	
2:	1	1	4	7	0	1	0	0	0	4:		-1-1	356	8	
2:	1	1	4	8	1	0	0	1	1	2:		-1-1	356	12	
2:	1	1	5	6	0	1	0	0	0	5:		-1-1	356	8	
2:	1	2	2	7	1	0	1	0	1	0:		-1-1	356	4	
2:	1	2	3	4	0	0	1	0	1	0:		-1-1	356	4	
2:	1	2	3	5	1	1	0	1	0	0:		-1-1	356	4	
2:	1	2	4	4	1	0	2	0	1	2:		-1-1	356	4	
2:	1	2	4	4	0	1	1	0	2	3:		-1-1	356	4	
2:	1	2	4	5	1	0	2	1	1	4:		-1-1	356	4	
2:	1	3	3	4	1	1	-1	1	0	3:		-1-1	356	4	
2:	1	3	3	4	1	0	2	0	3	0:		-1-1	356	4	
2:	2	2	3	3	1	2	0	0	1	2:		-1-1	356	2	
3:	1	1	2	15	0	1	1	0	0	1:		-1-1	178	16	39/16
3:	1	1	3	8	0	0	0	1	1	1:		-1-1	178	16	
3:	1	1	4	7	0	1	1	0	0	3:		-1-1	178	16	
3:	1	2	2	7	1	0	0	0	1	2:		-1-1	178	8	
3:	1	2	2	8	1	1	0	0	2	0:		-1-1	178	4	
3:	1	2	4	4	1	0	0	1	2	2:		-1-1	178	4	
3:	1	2	4	4	1	1	0	0	2	0:		-1-1	178	4	
3:	1	2	4	4	0	0	2	1	2	0:		-1-1	178	8	
3:	2	2	2	4	2	0	0	1	1	1:		-1-1	178	4	
3:	2	2	2	4	1	0	0	0	2	2:		-1-1	178	4	
3:	2	2	2	4	1	1	0	2	2	0:		-1-1	178	2	
3:	2	2	3	3	0	1	0	-1	-2	2:		-1-1	178	4	

D= 357

GENUS#	F11	F22	F33	F44	F12	F13	F23	F14	F24	F34	HASSE	SYM	LEVEL	AUTOS	MASS
1:	1	1	1	30	1	0	0	0	0	1:		-1 1 1-1	357	48	11/16
1:	1	1	2	15	1	0	0	0	0	1:		-1 1 1-1	357	24	
1:	1	1	2	18	1	1	0	0	0	1:		-1 1 1-1	357	12	
1:	1	1	4	8	1	0	0	0	0	3:		-1 1 1-1	357	24	
1:	1	1	5	7	1	1	0	0	1	2:		-1 1 1-1	357	12	
1:	1	1	6	7	1	1	0	1	1	6:		-1 1 1-1	357	24	
1:	1	2	3	6	1	0	2	0	1	3:		-1 1 1-1	357	8	
1:	1	3	3	4	0	0	3	1	2	2:		-1 1 1-1	357	8	
1:	1	3	3	4	1	0	3	1	0	0:		-1 1 1-1	357	8	
2:	1	1	1	45	1	1	0	0	1	0:		-1 1-1 1	357	48	11/16
2:	1	1	3	10	1	0	0	0	0	1:		-1 1-1 1	357	24	
2:	1	1	3	11	1	0	0	1	0	3:		-1 1-1 1	357	24	
2:	1	1	3	12	1	1	0	0	1	2:		-1 1-1 1	357	12	
2:	1	1	5	6	1	0	0	0	0	1:		-1 1-1 1	357	24	
2:	1	1	6	6	1	0	0	0	0	5:		-1 1-1 1	357	48	
2:	1	2	2	7	0	0	1	1	2	2:		-1 1-1 1	357	8	

```
            2:   1   2   3   5   1   0   0   0   0   3: -1 1-1 1    357    16
            2:   1   3   3   3   1   0   1   0  -1   1: -1 1-1 1    357     8
            2:   2   3   3   3   2   1  -1   1   2   3: -1 1-1 1    357     8
            3:   1   1   2  13   0   1   0   0   1   0: -1-1-1-1    357    16   11/16
            3:   1   2   3   5   0   1   0   1   1   3: -1-1-1-1    357     4
            3:   1   3   3   4   1   1  -1   1   2   2: -1-1-1-1    357     8
            3:   2   2   3   3   1   2   0   1   2   1: -1-1-1-1    357     4
            4:   1   1   5   5   0   1   0   0   1   1: -1-1 1 1    357    16   11/16
            4:   1   2   2   7   1   0   1   0   0   1: -1-1 1 1    357     4
            4:   1   2   3   5   1   1   0   1   1   0: -1-1 1 1    357     4
            4:   1   2   3   5   0   1   1   1   2   1: -1-1 1 1    357     8

D= 360
GENUS#: F11 F22 F33 F44 F12 F13 F23 F14 F24 F34: HASSE SYM   LEVEL AUTOS   MASS
     1:   1   1   1  30   1   0   0   0   0   0:  1-1-1     120    48    7/48
     1:   1   1   3   8   0   0   0   1   1   0:  1-1-1     120    32
     1:   2   2   2   3   1   0   0   0   0   0:  1-1-1     120    16
     1:   2   2   2   5   2   1  -1   0   0   0:  1-1-1     120    32
     2:   1   1   1  45   1   1   0   0   0   0: -1 1-1     360    96    7/8
     2:   1   1   2  12   0   0   0   0   1   2: -1 1-1     360    16
     2:   1   1   5   5   0   1   1   0   0   0: -1 1-1     360    32
     2:   1   1   5   6   0   0   0   0   1   5: -1 1-1     360    16
     2:   1   1   6   6   1   0   0   1   1   4: -1 1-1     360    12
     2:   1   2   3   4   0   0   1   0   0   1: -1 1-1     360     4
     2:   1   2   3   6   1   0   2   0   2   0: -1 1-1     360     4
     2:   2   2   3   3   0   2   1   2   1   1: -1 1-1     360     8
     3:   1   1   2  13   0   1   0   0   0   1: -1 1-1     360     8    7/8
     3:   1   1   5   6   0   1   1   1   1   4: -1 1-1     360    16
     3:   1   2   2   9   1   0   2   0   0   0: -1 1-1     360    16
     3:   1   2   3   5   0   1   2   0   0   0: -1 1-1     360    16
     3:   1   2   4   4   0   0   2   0   2   3: -1 1-1     360    16
     3:   2   2   2   4   1   1  -1  -1   0   0: -1 1-1     360     4
     3:   2   2   2   5   2   1   0   0   2   1: -1 1-1     360     4
     4:   1   1   2  18   1   1   0   0   0   0:  1 1 1     360    24    7/8
     4:   1   1   3   9   0   1   0   0   0   3:  1 1 1     360     8
     4:   1   1   4   7   0   1   1   1   1   0:  1 1 1     360    16
     4:   1   1   4   9   1   1   0   1   1   3:  1 1 1     360    12
     4:   1   2   2   7   1   0   0   0   2   0:  1 1 1     360     8
     4:   1   2   3   5   0   0   2   0   2   1:  1 1 1     360    16
     4:   1   2   4   4   0   1   1   1   1  -2:  1 1 1     360     8
     4:   1   3   3   4   1   0   2   1   3   0:  1 1 1     360     4
     5:   1   1   2  15   0   1   1   0   0   0: -1-1 1     120    32    7/48
     5:   1   1   3  10   1   0   0   0   0   0: -1-1 1     120    48
     5:   1   1   4   6   0   1   0   0   0   0: -1-1 1     120    16
     5:   1   3   3   4   0   0   0   0   3   3: -1-1 1     120    32
     6:   1   1   1  23   0   0   0   0   1   1:  1 1 1     360    32    7/8
     6:   1   1   3   9   0   1   1   0   0   0:  1 1 1     360    32
     6:   1   1   5   5   0   1   0   1   0   0:  1 1 1     360    16
     6:   1   2   2   6   0   0   0   1   1   0:  1 1 1     360     8
     6:   1   2   3   5   1   1   0   0   1   1:  1 1 1     360     4
     6:   2   2   3   3   0   2   0   1   0   3:  1 1 1     360     8
     6:   2   2   3   3   2   1   0   0   0   2:  1 1 1     360     4
     7:   1   1   2  15   1   0   0   0   0   0:  1 1 1     120    48    7/12
     7:   1   2   3   4   0   0   0   1   0   0:  1 1 1     120    16
     7:   1   2   3   5   0   1   1   0   2   2:  1 1 1     120     4
     7:   1   2   3   5   0   0   0   0   2   3:  1 1 1     120    16
     7:   2   2   3   3   2   0   0   1   2   0:  1 1 1     120    16
     7:   2   3   3   3   1   1   0   0   3   3:  1 1 1     120     8
     8:   1   1   5   6   1   0   0   0   0   0: -1 1-1     120    48    7/12
     8:   1   2   2   6   0   0   1   0   0   0: -1 1-1     120    16
     8:   1   2   2   8   0   0   2   0   2   1: -1 1-1     120    16
     8:   1   2   3   6   0   1   2   1   2   0: -1 1-1     120     8
     8:   1   3   3   3   1   1   0   0   0   0: -1 1-1     120    16
     8:   2   2   3   3   1   2  -1   1   1   1: -1 1-1     120     4

D= 361
GENUS#: F11 F22 F33 F44 F12 F13 F23 F14 F24 F34: HASSE SYM   LEVEL AUTOS   MASS
     1:   1   1   5   5   0   1   0   0   1   0: -1-1        19    32    9/32
     1:   1   2   3   6   1   1   0   1   2   3: -1-1        19     8
     1:   2   2   3   3   0   2   1   1   2   1: -1-1        19     8
```

D= 364

GENUS#:	F11	F22	F33	F44	F12	F13	F23	F14	F24	F34:	HASSE SYM	LEVEL	AUTOS	MASS
1:	1	1	1	46	1	1	0	1	0	0:	1-1-1	364	96	103/96
1:	1	1	2	12	0	0	0	1	1	1:	1-1-1	364	16	
1:	1	2	3	5	0	1	1	1	1	-1:	1-1-1	364	4	
1:	1	2	3	6	0	1	2	1	0	3:	1-1-1	364	8	
1:	2	2	3	3	1	2	0	0	2	0:	1-1-1	364	4	
1:	2	2	3	3	0	2	2	0	1	1:	1-1-1	364	4	
1:	2	2	3	4	2	2	0	0	1	2:	1-1-1	364	8	
2:	1	1	2	13	0	1	0	0	0	0:	-1 1-1	364	16	103/96
2:	1	1	2	16	1	0	0	1	0	2:	-1 1-1	364	24	
2:	1	1	5	5	0	1	0	0	0	2:	-1 1-1	364	8	
2:	1	1	5	5	0	0	0	0	0	3:	-1 1-1	364	32	
2:	1	2	3	4	0	0	0	0	1	-1:	-1 1-1	364	4	
2:	1	2	3	5	1	1	0	0	-1	1:	-1 1-1	364	4	
2:	1	2	4	4	0	0	2	1	2	1:	-1 1-1	364	16	
2:	1	2	4	5	1	1	0	1	2	4:	-1 1-1	364	8	
2:	2	2	3	3	2	1	0	1	0	-1:	-1 1-1	364	8	
3:	1	1	3	10	0	1	1	0	0	3:	1 1 1	364	16	103/96
3:	1	1	4	7	0	1	1	1	1	1:	1 1 1	364	32	
3:	1	1	6	6	1	1	0	1	1	3:	1 1 1	364	24	
3:	1	2	2	7	1	0	1	0	0	0:	1 1 1	364	8	
3:	1	2	2	7	1	0	0	0	0	2:	1 1 1	364	16	
3:	1	2	2	8	0	0	2	1	1	0:	1 1 1	364	8	
3:	1	2	3	6	1	0	2	1	2	0:	1 1 1	364	4	
3:	2	2	3	3	0	2	0	2	1	2:	1 1 1	364	4	
3:	2	3	3	3	2	2	0	2	3	2:	1 1 1	364	8	
4:	1	1	1	23	0	0	0	0	1	0:	-1-1 1	364	32	103/96
4:	1	1	4	6	0	0	0	0	1	1:	-1-1 1	364	8	
4:	1	1	5	8	1	1	0	1	1	5:	-1-1 1	364	24	
4:	1	2	3	5	0	1	0	0	2	2:	-1-1 1	364	8	
4:	1	2	3	5	0	0	1	0	2	3:	-1-1 1	364	4	
4:	1	3	3	4	1	1	0	0	1	3:	-1-1 1	364	4	
4:	2	2	3	3	1	1	0	1	1	3:	-1-1 1	364	4	

D= 365

GENUS#:	F11	F22	F33	F44	F12	F13	F23	F14	F24	F34:	HASSE SYM	LEVEL	AUTOS	MASS
1:	1	1	1	31	1	0	0	1	0	1:	-1-1 1	365	24	65/48
1:	1	1	1	46	1	1	0	0	1	0:	-1-1 1	365	48	
1:	1	1	3	12	1	1	0	1	1	2:	-1-1 1	365	12	
1:	1	1	4	8	1	0	0	1	1	1:	-1-1 1	365	12	
1:	1	1	5	6	0	1	1	0	1	4:	-1-1 1	365	8	
1:	1	2	3	5	0	0	1	1	0	3:	-1-1 1	365	4	
1:	1	2	4	4	1	0	1	0	1	3:	-1-1 1	365	8	
1:	1	3	3	3	1	0	1	0	1	1:	-1-1 1	365	8	
1:	2	2	2	4	1	1	-1	1	-1	1:	-1-1 1	365	4	
1:	2	2	3	3	1	2	0	2	0	1:	-1-1 1	365	4	65/48
2:	1	1	2	19	1	1	0	1	1	2:	-1 1-1	365	24	
2:	1	1	3	10	0	1	1	1	0	3:	-1 1-1	365	16	
2:	1	1	3	9	0	1	0	1	1	2:	-1 1-1	365	8	
2:	1	1	4	7	0	1	0	1	1	3:	-1 1-1	365	8	
2:	1	1	4	9	1	1	0	0	1	2:	-1 1-1	365	12	
2:	1	1	5	7	1	1	0	0	0	3:	-1 1-1	365	12	
2:	1	1	5	7	1	1	0	1	1	2:	-1 1-1	365	12	
2:	1	2	3	5	1	0	0	1	2	1:	-1 1-1	365	4	
2:	1	2	4	4	0	1	1	0	1	-3:	-1 1-1	365	4	
2:	1	3	3	4	1	0	2	0	0	3:	-1 1-1	365	4	

D= 368

GENUS#:	F11	F22	F33	F44	F12	F13	F23	F14	F24	F34:	HASSE SYM	LEVEL	AUTOS	MASS
1:	1	1	1	23	0	0	0	0	0	0:	-1-1	92	96	5/24
1:	1	1	2	12	0	0	0	0	0	2:	-1-1	92	32	
1:	1	1	3	8	0	0	0	0	0	2:	-1-1	92	16	
1:	1	1	4	6	0	0	0	0	0	2:	-1-1	92	16	
1:	2	2	2	5	2	0	0	2	0	2:	-1-1	92	24	
2:	1	1	1	31	1	0	0	1	0	0:	-1-1	368	24	5/3
2:	1	1	3	12	1	1	0	1	0	1:	-1-1	368	12	
2:	1	1	3	8	0	0	0	1	0	1:	-1-1	368	8	
2:	1	1	4	6	0	0	0	1	0	0:	-1-1	368	16	
2:	1	1	4	7	0	0	0	1	0	4:	-1-1	368	16	
2:	1	1	4	8	1	0	0	1	0	0:	-1-1	368	24	
2:	1	2	2	6	0	0	0	0	-1	1:	-1-1	368	8	

	F11	F22	F33	F44	F12	F13	F23	F14	F24	F34:	HASSE	SYM	LEVEL	AUTOS	MASS
2:	1	2	2	8	1	1	0	0	1	1:	-1-1		368	8	
2:	1	2	3	4	0	0	1	0	0	0:	-1-1		368	8	
2:	1	2	4	4	0	0	0	0	1	4:	-1-1		368	8	
2:	1	2	4	4	1	0	1	0	2	0:	-1-1		368	4	
2:	1	2	4	5	1	0	2	0	0	4:	-1-1		368	4	
2:	1	3	3	3	0	0	1	0	1	2:	-1-1		368	8	
2:	1	3	3	4	0	0	2	0	3	3:	-1-1		368	8	
3:	1	1	1	46	1	1	0	0	0	0:	-1-1		184	96	5/24
3:	1	1	4	9	1	0	0	1	0	4:	-1-1		184	24	
3:	1	1	5	6	0	1	1	0	0	4:	-1-1		184	16	
3:	1	2	2	7	0	0	0	1	2	2:	-1-1		184	32	
3:	2	2	3	3	0	0	0	2	2	1:	-1-1		184	16	
4:	1	1	2	12	0	0	0	1	1	0:	-1-1		184	32	5/8
4:	1	1	2	16	0	1	1	1	0	0:	-1-1		184	16	
4:	1	2	2	6	0	0	0	1	0	0:	-1-1		184	32	
4:	1	2	2	8	1	1	0	1	0	0:	-1-1		184	8	
4:	2	2	2	3	0	0	0	1	0	0:	-1-1		184	16	
4:	2	2	2	4	0	0	0	2	2	1:	-1-1		184	16	
4:	2	2	2	4	2	0	0	0	1	0:	-1-1		184	8	
4:	2	2	2	6	2	2	0	0	1	-1:	-1-1		184	16	
4:	2	2	3	4	0	2	2	0	0	3:	-1-1		184	16	
5:	1	1	2	19	1	1	0	0	1	1:	1 1		368	12	5/3
5:	1	1	6	7	1	1	0	0	1	5:	1 1		368	12	
5:	1	2	3	5	0	0	2	1	1	0:	1 1		368	4	
5:	1	2	3	5	1	0	1	1	0	1:	1 1		368	4	
5:	1	2	3	5	1	0	0	1	0	2:	1 1		368	4	
5:	1	2	3	6	1	0	2	1	1	2:	1 1		368	4	
5:	2	2	3	3	1	2	0	1	-1	2:	1 1		368	2	
6:	1	1	5	7	1	1	0	0	1	1:	1 1		184	12	5/24
6:	1	2	3	5	0	1	0	0	0	3:	1 1		184	8	
7:	1	2	2	10	1	0	2	0	2	0:	1 1		184	8	5/8
7:	1	2	3	5	1	0	1	0	1	2:	1 1		184	4	
7:	2	2	3	4	2	1	0	1	2	3:	1 1		184	4	
8:	1	2	3	5	0	0	2	0	0	2:	1 1		92	8	5/24
8:	2	2	3	4	2	2	0	0	2	0:	1 1		92	12	

D= 369

GENUS#:	F11	F22	F33	F44	F12	F13	F23	F14	F24	F34:	HASSE	SYM	LEVEL	AUTOS	MASS
1:	1	1	1	31	1	0	0	0	0	1:	-1 1-1		123	48	5/6
1:	1	1	2	16	0	1	1	1	0	2:	-1 1-1		123	16	
1:	1	1	4	7	0	1	0	0	1	3:	-1 1-1		123	8	
1:	1	2	2	8	1	1	0	1	1	0:	-1 1-1		123	8	
1:	1	2	4	4	1	0	1	1	2	0:	-1 1-1		123	4	
1:	2	2	2	4	1	0	0	-2	1	-1:	-1 1-1		123	4	
2:	1	1	2	14	0	1	0	0	1	2:	-1 1-1		369	8	5/2
2:	1	1	5	6	0	1	0	1	0	4:	-1 1-1		369	8	
2:	1	2	2	8	1	0	1	1	0	2:	-1 1-1		369	4	
2:	1	2	3	5	1	1	0	0	1	0:	-1 1-1		369	4	
2:	1	2	4	4	1	0	0	0	-1	3:	-1 1-1		369	4	
2:	2	2	2	4	1	1	0	0	1	2:	-1 1-1		369	2	
2:	2	2	2	4	1	1	1	1	0	2:	-1 1-1		369	4	
2:	2	2	2	5	2	1	0	2	1	2:	-1 1-1		369	2	
2:	2	2	3	3	2	1	0	0	1	1:	-1 1-1		369	4	
3:	1	1	1	11	1	0	0	0	0	3:	-1-1 1		123	48	5/6
3:	1	2	3	5	0	1	1	0	1	-2:	-1-1 1		123	4	
3:	1	3	3	4	1	0	0	1	2	3:	-1-1 1		123	8	
3:	2	2	2	6	2	1	-1	1	2	1:	-1-1 1		123	16	
3:	2	2	3	3	1	0	0	1	1	3:	-1-1 1		123	8	
3:	2	2	3	3	1	1	1	2	2	1:	-1-1 1		123	4	
4:	1	1	3	9	0	1	0	0	1	2:	-1 1-1		369	8	5/2
4:	1	1	4	7	0	1	1	0	1	2:	-1 1-1		369	8	
4:	1	1	4	9	1	1	0	0	0	3:	-1 1-1		369	12	
4:	1	1	6	6	1	1	0	0	1	2:	-1 1-1		369	24	
4:	1	2	2	7	1	0	0	1	1	1:	-1 1-1		369	4	
4:	1	2	3	6	1	0	2	0	2	1:	-1 1-1		369	4	
4:	1	2	4	4	1	1	0	1	1	2:	-1 1-1		369	8	
4:	1	2	4	4	0	1	2	1	1	0:	-1 1-1		369	4	
4:	1	2	4	4	1	1	0	0	1	2:	-1 1-1		369	4	
4:	1	3	3	4	0	0	3	1	2	1:	-1 1-1		369	8	
4:	1	3	3	4	1	0	2	0	3	1:	-1 1-1		369	4	
4:	1	3	4	4	1	0	3	1	1	4:	-1 1-1		369	8	
4:	2	2	3	3	0	2	1	1	-1	2:	-1 1-1		369	2	

D= 372

GENUS#:	F11	F22	F33	F44	F12	F13	F23	F14	F24	F34:	HASSE SYM	LEVEL	AUTOS	MASS
1:	1	1	1	24	0	0	0	1	1	1:	1-1-1	186	96	15/32
1:	1	2	2	10	1	0	2	0	0	2:	1-1-1	186	8	
1:	2	2	2	6	2	1	-1	2	0	0:	1-1-1	186	8	
1:	2	2	2	6	2	0	1	1	0	0:	1-1-1	186	12	
1:	2	2	3	4	2	0	0	2	0	3:	1-1-1	186	24	
1:	2	2	3	4	2	2	0	2	0	1:	1-1-1	186	12	
2:	1	1	1	31	1	0	0	0	0	0:	-1-1 1	372	48	15/8
2:	1	1	3	8	0	0	0	0	1	0:	-1-1 1	372	16	
2:	1	1	5	7	1	0	0	0	0	4:	-1-1 1	372	24	
2:	1	2	2	7	0	0	1	0	2	2:	-1-1 1	372	8	
2:	1	2	2	8	1	0	1	0	2	2:	-1-1 1	372	4	
2:	1	2	2	8	1	1	0	0	1	-1:	-1-1 1	372	8	
2:	1	2	3	4	0	0	0	0	1	0:	-1-1 1	372	8	
2:	1	2	3	5	1	0	0	0	2	0:	-1-1 1	372	8	
2:	1	2	3	5	0	0	0	1	1	3:	-1-1 1	372	8	
2:	1	3	3	4	1	0	3	0	1	0:	-1-1 1	372	8	
2:	2	2	2	4	1	1	1	2	0	0:	-1-1 1	372	4	
2:	2	2	3	3	1	2	0	0	-1	2:	-1-1 1	372	2	
3:	1	1	1	47	1	1	0	1	0	0:	-1 1-1	186	96	15/32
3:	1	1	3	12	1	1	0	0	0	2:	-1 1-1	186	12	
3:	1	1	5	7	1	0	0	1	0	0:	-1 1-1	186	12	
3:	1	1	6	7	1	0	0	1	0	6:	-1 1-1	186	24	
3:	1	2	3	5	0	0	2	1	0	1:	-1 1-1	186	8	
3:	1	3	3	3	0	0	0	1	1	1:	-1 1-1	186	8	
4:	1	1	2	12	0	0	0	0	1	1:	-1 1-1	372	8	15/8
4:	1	1	2	19	1	1	0	1	0	0:	-1 1-1	372	12	
4:	1	1	3	11	1	0	0	1	0	2:	-1 1-1	372	12	
4:	1	1	3	12	1	1	0	1	0	0:	-1 1-1	372	12	
4:	1	2	3	5	0	0	2	0	1	2:	-1 1-1	372	4	
4:	1	2	3	5	1	0	1	1	1	1:	-1 1-1	372	4	
4:	1	2	3	5	1	0	1	0	0	2:	-1 1-1	372	4	
4:	1	3	3	3	0	0	1	0	0	2:	-1 1-1	372	4	
4:	1	3	3	4	1	0	1	1	0	3:	-1 1-1	372	4	
4:	1	3	3	4	0	0	2	1	3	0:	-1 1-1	372	4	
5:	1	1	2	16	1	0	0	0	0	1:	-1-1 1	186	48	15/32
5:	1	1	3	9	0	0	0	1	1	3:	-1-1 1	186	32	
5:	1	1	4	8	1	0	0	0	0	2:	-1-1 1	186	24	
5:	1	2	3	5	0	1	0	1	2	0:	-1-1 1	186	8	
5:	1	3	3	5	1	0	3	0	3	0:	-1-1 1	186	8	
5:	2	3	3	3	2	0	3	0	1	3:	-1-1 1	186	8	
6:	1	1	2	16	0	1	1	1	1	1:	1 1 1	186	32	15/32
6:	1	2	2	8	0	0	2	1	0	0:	1 1 1	186	48	
6:	1	2	4	4	1	1	0	1	0	2:	1 1 1	186	8	
6:	1	2	4	4	0	1	0	1	2	2:	1 1 1	186	8	
6:	2	2	2	4	2	0	0	0	0	1:	1 1 1	186	24	
6:	2	2	3	4	2	1	-1	0	0	3:	1 1 1	186	8	

D= 373

GENUS#:	F11	F22	F33	F44	F12	F13	F23	F14	F24	F34:	HASSE SYM	LEVEL	AUTOS	MASS
1:	1	1	1	47	1	1	0	1	0	1:	-1-1	373	48	161/48
1:	1	1	2	16	1	0	0	1	0	1:	-1-1	373	12	
1:	1	1	2	19	1	1	0	0	1	0:	-1-1	373	12	
1:	1	1	3	12	1	1	0	0	1	0:	-1-1	373	12	
1:	1	1	3	9	0	1	0	1	1	0:	-1-1	373	8	
1:	1	1	4	9	1	1	0	1	1	2:	-1-1	373	12	
1:	1	1	5	6	0	1	0	0	1	4:	-1-1	373	8	
1:	1	1	5	7	1	0	0	1	1	3:	-1-1	373	12	
1:	1	1	5	7	1	1	0	0	1	0:	-1-1	373	12	
1:	1	1	5	8	1	1	0	0	1	4:	-1-1	373	12	
1:	1	2	2	7	0	0	1	1	2	0:	-1-1	373	4	
1:	1	2	3	5	0	0	1	1	2	2:	-1-1	373	4	
1:	1	2	3	5	1	1	0	0	0	1:	-1-1	373	4	
1:	1	2	3	6	0	1	2	1	1	-1:	-1-1	373	4	
1:	1	3	3	3	1	0	0	0	1	1:	-1-1	373	4	
1:	1	3	3	4	1	0	3	0	0	1:	-1-1	373	4	
1:	2	2	3	3	1	1	0	2	2	1:	-1-1	373	2	
1:	2	3	3	3	1	0	1	2	3	3:	-1-1	373	2	

D= 376

GENUS#:	F11	F22	F33	F44	F12	F13	F23	F14	F24	F34:	HASSE SYM	LEVEL	AUTOS	MASS

#	F11	F22	F33	F44	F12	F13	F23	F14	F24	F34	HASSE SYM	LEVEL	AUTOS	MASS
1:	1	1	1	24	0	0	0	1	1	0:	-1-1	376	32	53/24
1:	1	1	1	47	1	1	0	0	0	0:	-1-1	376	96	
1:	1	1	2	12	0	0	0	1	0	0:	-1-1	376	16	
1:	1	1	2	14	0	1	0	0	0	2:	-1-1	376	8	
1:	1	1	2	16	1	0	0	1	0	0:	-1-1	376	24	
1:	1	1	4	7	0	1	1	0	0	2:	-1-1	376	16	
1:	1	2	2	6	0	0	0	0	1	0:	-1-1	376	8	
1:	1	2	2	7	1	0	0	1	0	0:	-1-1	376	8	
1:	1	2	2	8	0	0	2	0	0	1:	-1-1	376	8	
1:	1	2	3	4	0	0	0	0	0	1:	-1-1	376	8	
1:	1	2	3	5	0	1	1	1	1	2:	-1-1	376	4	
1:	1	2	3	5	0	0	0	1	2	2:	-1-1	376	8	
1:	1	2	3	5	0	1	1	0	2	0:	-1-1	376	4	
1:	1	2	3	6	0	1	2	1	2	1:	-1-1	376	8	
1:	1	2	4	4	0	0	0	0	2	3:	-1-1	376	8	
1:	2	2	3	3	1	1	1	0	2	2:	-1-1	376	2	
2:	1	1	3	10	0	1	1	1	1	0:	1 1	376	16	53/24
2:	1	1	3	9	0	1	0	1	0	2:	1 1	376	8	
2:	1	1	4	9	1	1	0	0	1	1:	1 1	376	12	
2:	1	1	5	5	0	0	0	0	1	1:	1 1	376	8	
2:	1	1	5	6	0	0	0	1	1	4:	1 1	376	16	
2:	1	2	3	6	1	1	0	0	2	2:	1 1	376	4	
2:	1	3	3	4	1	1	0	1	3	1:	1 1	376	4	
2:	1	3	3	4	1	0	0	1	3	2:	1 1	376	4	
2:	2	2	2	5	2	1	0	2	0	0:	1 1	376	4	
2:	2	3	3	0	2	0	1	2	0	0:	1 1	376	4	
2:	2	2	3	3	1	1	0	2	2	0:	1 1	376	2	

D= 377

GENUS#:	F11	F22	F33	F44	F12	F13	F23	F14	F24	F34:	HASSE SYM	LEVEL	AUTOS	MASS
1:	1	1	1	32	1	0	0	1	0	1:	-1 1-1	377	24	53/24
1:	1	1	4	7	0	1	1	0	1	0:	-1 1-1	377	8	
1:	1	1	6	6	1	1	0	1	1	2:	-1 1-1	377	24	
1:	1	2	2	10	1	0	2	0	2	1:	-1 1-1	377	4	
1:	1	2	2	7	1	0	0	0	-1	0:	-1 1-1	377	4	
1:	1	2	2	8	1	1	0	0	1	0:	-1 1-1	377	4	
1:	1	2	3	6	1	0	2	1	0	1:	-1 1-1	377	4	
1:	1	2	4	4	0	1	1	1	2	1:	-1 1-1	377	8	
1:	1	2	4	4	1	0	1	1	1	2:	-1 1-1	377	4	
1:	1	3	3	4	1	0	2	0	1	3:	-1 1-1	377	8	
1:	2	2	3	3	1	1	-1	0	1	2:	-1 1-1	377	2	
2:	1	1	2	14	0	1	0	1	1	0:	-1-1 1	377	8	53/24
2:	1	1	2	16	0	1	1	0	1	0:	-1-1 1	377	8	
2:	1	1	2	19	1	1	0	0	0	1:	-1-1 1	377	12	
2:	1	2	2	8	1	0	1	1	1	2:	-1-1 1	377	8	
2:	1	2	3	5	1	0	1	1	0	0:	-1-1 1	377	4	
2:	1	2	3	5	1	0	1	0	-1	1:	-1-1 1	377	4	
2:	1	2	4	4	1	0	2	0	0	1:	-1-1 1	377	4	
2:	1	2	4	5	1	0	2	1	0	3:	-1-1 1	377	4	
2:	2	2	2	4	1	1	0	1	2	1:	-1-1 1	377	2	
2:	2	2	3	3	2	1	0	1	1	1:	-1-1 1	377	4	

D= 380

GENUS#:	F11	F22	F33	F44	F12	F13	F23	F14	F24	F34:	HASSE SYM	LEVEL	AUTOS	MASS
1:	1	1	1	32	1	0	0	1	0	0:	-1-1 1	380	24	43/48
1:	1	1	1	48	1	1	0	1	0	0:	-1-1 1	380	96	
1:	1	1	2	12	0	0	0	0	0	1:	-1-1 1	380	16	
1:	1	1	3	10	0	1	1	1	1	1:	-1-1 1	380	32	
1:	1	1	3	8	0	0	0	0	0	1:	-1-1 1	380	16	
1:	1	1	4	7	0	1	0	1	0	3:	-1-1 1	380	8	
1:	1	1	5	6	0	1	1	1	1	3:	-1-1 1	380	16	
1:	1	2	4	4	1	0	1	0	-1	2:	-1-1 1	380	8	
1:	2	2	2	5	2	1	0	0	2	0:	-1-1 1	380	4	
1:	2	2	3	3	0	2	1	2	0	1:	-1-1 1	380	8	
2:	1	1	1	24	0	0	0	1	0	0:	-1 1-1	380	32	43/48
2:	1	1	2	16	0	1	1	0	0	1:	-1 1-1	380	16	
2:	1	1	3	9	0	1	0	0	0	2:	-1 1-1	380	8	
2:	1	1	4	6	0	0	0	0	0	1:	-1 1-1	380	16	
2:	1	1	4	7	0	0	0	1	1	3:	-1 1-1	380	16	
2:	1	1	4	9	1	1	0	1	0	0:	-1 1-1	380	12	
2:	1	1	5	5	0	0	0	0	1	0:	-1 1-1	380	16	
2:	1	1	5	6	0	0	0	0	0	5:	-1 1-1	380	32	

GENUS#:	F11	F22	F33	F44	F12	F13	F23	F14	F24	F34:	HASSE SYM	LEVEL	AUTOS	MASS
2:	1	2	3	5	1	1	0	0	0	0:	-1 1-1	380	8	
2:	1	3	3	4	0	0	1	0	3	3:	-1 1-1	380	8	
2:	2	2	2	4	1	0	0	1	-1	2:	-1 1-1	380	8	
3:	1	1	2	19	1	1	0	0	0	0:	1-1-1	380	24	43/48
3:	1	1	6	6	1	1	0	0	1	1:	1-1-1	380	24	
3:	1	2	3	5	0	0	2	1	0	0:	1-1-1	380	16	
3:	1	2	3	6	1	0	2	0	0	2:	1-1-1	380	4	
3:	2	2	2	4	1	1	-1	1	1	0:	1-1-1	380	4	
3:	2	2	3	4	2	2	0	1	2	1:	1-1-1	380	4	
4:	1	1	5	7	1	1	0	0	0	2:	1 1 1	380	12	43/48
4:	1	2	2	10	1	0	2	0	1	2:	1 1 1	380	16	
4:	1	2	3	5	1	0	0	0	1	2:	1 1 1	380	4	
4:	1	2	4	4	0	1	1	1	1	3:	1 1 1	380	8	
4:	1	3	3	4	1	0	2	1	0	2:	1 1 1	380	4	
4:	2	2	3	3	2	1	0	1	0	1:	1 1 1	380	8	

D= 381

GENUS#:	F11	F22	F33	F44	F12	F13	F23	F14	F24	F34:	HASSE SYM	LEVEL	AUTOS	MASS
1:	1	1	1	32	1	0	0	0	0	1:	-1-1 1	381	48	77/48
1:	1	1	2	14	0	1	0	0	1	1:	-1-1 1	381	8	
1:	1	1	2	16	0	0	0	0	0	1:	-1-1 1	381	24	
1:	1	1	3	10	0	1	1	1	0	2:	-1-1 1	381	8	
1:	1	1	4	8	1	0	0	0	0	1:	-1-1 1	381	24	
1:	1	3	3	4	1	1	-1	0	1	2:	-1-1 1	381	4	
1:	1	3	3	4	1	0	0	0	2	3:	-1-1 1	381	8	
1:	1	3	3	5	1	0	3	1	3	0:	-1-1 1	381	8	
1:	2	2	2	4	1	1	1	1	1	-1:	-1-1 1	381	4	
1:	2	2	3	3	1	0	0	2	2	1:	-1-1 1	381	4	
1:	2	2	3	3	1	0	0	1	0	2:	-1-1 1	381	4	
2:	1	1	1	48	1	1	0	0	1	0:	-1 1-1	381	48	77/48
2:	1	1	3	11	1	0	0	1	1	1:	-1 1-1	381	12	
2:	1	1	3	12	1	1	0	0	0	1:	-1 1-1	381	12	
2:	1	1	3	9	0	1	0	0	1	1:	-1 1-1	381	8	
2:	1	1	4	7	0	1	0	1	1	2:	-1 1-1	381	8	
2:	1	1	4	9	1	1	0	0	1	0:	-1 1-1	381	12	
2:	1	1	6	6	1	0	0	1	0	3:	-1 1-1	381	12	
2:	1	2	3	5	0	1	1	1	0	2:	-1 1-1	381	4	
2:	1	2	3	6	1	0	2	0	-1	1:	-1 1-1	381	4	
2:	1	3	3	4	1	1	0	0	0	3:	-1 1-1	381	4	
2:	1	3	3	4	1	0	1	1	3	2:	-1 1-1	381	4	

D= 384

GENUS#:	F11	F22	F33	F44	F12	F13	F23	F14	F24	F34:	HASSE SYM	LEVEL	AUTOS	MASS
1:	1	1	1	24	0	0	0	0	0	0:	-1-1	96	96	1/32
1:	2	2	3	3	2	0	0	0	0	2:	-1-1	96	40	
2:	1	1	1	32	1	0	0	0	0	0:	1 1	384	48	1/1
2:	1	1	3	9	1	0	1	0	0	0:	1 1	384	8	
2:	1	1	3	9	0	0	0	0	1	3:	1 1	384	16	
2:	1	1	4	9	1	1	0	0	0	2:	1 1	384	12	
2:	1	1	4	9	1	0	0	0	0	4:	1 1	384	48	
2:	1	2	2	9	1	1	0	1	2	2:	1 1	384	16	
2:	1	3	3	4	0	0	3	0	2	2:	1 1	384	8	
2:	1	3	3	4	1	0	1	0	3	2:	1 1	384	4	
2:	1	3	4	4	1	0	3	0	0	4:	1 1	384	8	
2:	2	2	3	3	0	1	1	0	0	3:	1 1	384	8	
3:	1	1	1	48	1	1	0	0	0	0:	1 1	192	96	1/8
3:	1	1	2	13	0	0	0	1	1	2:	1 1	192	32	
3:	1	1	6	6	1	0	0	0	0	4:	1 1	192	48	
3:	1	3	3	5	1	0	3	1	2	3:	1 1	192	16	
4:	1	1	2	12	0	0	0	0	0	0:	1 1	48	32	3/32
4:	1	2	3	4	0	0	0	0	0	0:	1 1	48	16	
5:	1	1	2	14	0	1	0	1	0	0:	-1-1	384	8	1/1
5:	1	1	4	7	0	1	0	0	0	3:	-1-1	384	8	
5:	1	2	4	4	0	0	1	0	2	3:	-1-1	384	4	
5:	1	2	4	5	1	0	0	0	2	4:	-1-1	384	8	
5:	1	3	3	4	1	1	-1	1	2	-1:	-1-1	384	8	
5:	2	2	2	4	1	1	0	-1	1	1:	-1-1	384	4	
6:	1	1	3	10	0	1	1	0	0	2:	-1-1	192	16	1/8
6:	2	2	3	3	0	2	0	0	2	1:	-1-1	192	16	
7:	1	1	3	12	1	1	0	0	0	0:	1 1	96	24	1/8
7:	1	1	4	8	1	0	0	0	0	0:	1 1	96	48	
7:	1	3	3	4	1	0	3	0	0	0:	1 1	96	16	

```
 8:  1 1 2 16  1 0  0  0  0  0: 1 1      192  48  1/8
 8:  1 1 3 13  1 1  0  1  1  3: 1 1      192  24
 8:  1 2 3  5  0 0  0  1  0  3: 1 1      192  16
 9:  1 1 3  8  0 0  0  0  0  0: 1 1       96  32  3/32
 9:  1 3 3  4  0 0  2  0  2 -2: 1 1       96  16
10:  1 1 4  6  0 0  0  0  0  0: -1-1      48  32  3/32
10:  2 2 3  3  0 2  0  2  0  0: -1-1      48  16
11:  1 1 5  5  0 0  0  0  0  2: 1 1       96  32  1/32
12:  1 2 2  6  0 0  0  0  0  0: -1-1      24  32  1/32
13:  1 1 3 11  1 0  0  1  0  0: 1 1      384  24  1/1
13:  1 1 3 11  1 0  0  0  0  2: 1 1      384  24
13:  1 1 6  6  1 1  0  1  0  0: 1 1      384  24
13:  1 2 3  6  1 0  2  0  1  2: 1 1      384   4
13:  1 3 3  3  0 0  0  1  1  0: 1 1      384   8
13:  1 3 3  4  0 0  0  1  3  2: 1 1      384   8
13:  1 3 3  4  1 0  1  1  1  3: 1 1      384   8
13:  2 2 3  4  1 2 -1  1  1  3: 1 1      384   4
14:  1 1 2 16  0 1  1  0  0  0: -1-1     192  32  3/8
14:  1 1 6  6  0 1  1  0  0  6: -1-1     192  32
14:  2 2 2  5  2 0  0  1  2  2: -1-1     192  16
14:  2 2 2  5  2 1  0  1  2  1: -1-1     192   4
15:  1 1 4  7  0 0  0  0  0  4: -1-1      96  32  3/32
15:  1 2 4  4  0 0  2  0  2  0: -1-1      96  16
16:  1 1 4  8  0 1  0  0  0  4: 1 1      192  16  3/8
16:  1 2 2  7  1 0  0  0  1  0: 1 1      192   8
16:  2 2 4  4  1 0  0  0  2  0: 1 1      192   8
16:  2 2 3  4  0 2  2  2  0  3: 1 1      192  16
17:  1 2 2  7  0 0  0  0  2  2: -1-1      96  32  3/32
17:  2 3 3  3  2 0  2  2  0  2: -1-1      96  16
18:  1 2 2  8  1 0  1  0  2  0: -1-1     384   4  1/1
18:  1 2 3  5  0 1  1  1  1  0: -1-1     384   4
18:  2 2 3  3  1 0  0  2  1  2: -1-1     384   2
19:  1 2 2  8  0 0  2  0  0  0: -1-1      96  48  1/32
19:  2 2 2  7  2 2  0  2  0  0: -1-1      96  96
20:  1 2 2  8  1 1  0  0  0  0: 1 1       96  16  3/8
20:  1 3 3  4  0 0  2  1  3  1: 1 1       96   8
20:  1 3 4  4  0 0  0  1  3  4: 1 1       96  16
20:  2 2 3  3  0 1  1 -1 -1  2: 1 1       96   8
21:  1 2 4  4  1 1  0  1  0  0: 1 1      192   8  3/8
21:  1 2 4  4  0 0  0  1  2  2: 1 1      192   8
21:  2 2 3  3  1 1  1 -1 -1  2: 1 1      192   8
22:  1 2 4  4  0 0  0  0  0  4: 1 1       24  48  1/32
22:  2 2 2  3  0 0  0  0  0  0: 1 1       24  96
23:  1 2 4  4  1 0  0  0  2  0: -1-1      96   8  3/8
23:  2 2 2  4  1 1  0  2  0  0: -1-1      96   4
24:  1 2 4  5  1 0  2  0  1  4: -1-1     192   8  3/8
24:  2 2 2  6  2 1 -1  0  0  2: -1-1     192   8
24:  2 2 3  3  2 1  0  1  0  0: -1-1     192   8
25:  1 3 3  3  0 0  2  0  0  0: 1 1       96  16  3/32
25:  2 2 3  4  0 2  2  2  2  2: 1 1       96  32
26:  1 3 3  4  1 1  0  0  2  2: -1-1     192   8  1/8
27:  1 3 3  4  1 1 -1  0 -2  2: -1-1      96   8  1/8
28:  2 2 3  3  0 0  0  2  2  0: 1 1       96  32  1/32
```

D= 385

GENUS#	F11	F22	F33	F44	F12	F13	F23	F14	F24	F34:	HASSE SYM	LEVEL	AUTOS	MASS
1:	1	1	2	14	0	1	0	0	1	0:	-1-1 1 1	385	16	71/48
1:	1	1	6	6	1	1	0	0	1	0:	-1-1 1 1	385	24	
1:	1	2	2	10	1	0	2	1	1	1:	-1-1 1 1	385	8	
1:	1	2	2	8	1	0	1	1	2	0:	-1-1 1 1	385	4	
1:	1	2	3	6	1	0	2	1	1	0:	-1-1 1 1	385	4	
1:	1	2	4	4	1	0	0	0	0	3:	-1-1 1 1	385	16	
1:	1	3	3	3	1	0	0	0	0	1:	-1-1 1 1	385	16	
1:	2	2	2	4	1	1	0	2	1	0:	-1-1 1 1	385	4	
1:	2	2	3	3	1	1	-1	-1	1	1:	-1-1 1 1	385	8	
1:	2	2	3	4	1	2	-1	1	2	2:	-1-1 1 1	385	4	
2:	1	1	2	20	1	1	0	1	1	1:	-1 1 1-1	385	24	71/48
2:	1	1	4	8	0	1	1	1	0	4:	-1 1 1-1	385	16	
2:	1	2	2	7	1	0	0	0	0	1:	-1 1 1-1	385	8	
2:	1	2	3	5	1	0	1	0	1	1:	-1 1 1-1	385	4	
2:	1	2	4	4	0	1	2	0	-1	1:	-1 1 1-1	385	4	
2:	1	2	4	4	1	1	0	1	1	0:	-1 1 1-1	385	8	

GENUS#	F11	F22	F33	F44	F12	F13	F23	F14	F24	F34	HASSE SYM	LEVEL	AUTOS	MASS
2	1	2	4	5	1	1	0	0	2	3	-1 1 1-1	385	4	
2	2	2	3	3	2	1	0	1	1	0	-1 1 1-1	385	4	
2	2	3	3	3	1	1	0	-1	3	2	-1 1 1-1	385	8	
3	1	1	3	9	0	1	0	0	1	0	-1 1-1 1	385	16	71/48
3	1	1	4	10	1	1	0	1	1	4	-1 1-1 1	385	24	
3	1	2	3	5	0	1	1	0	1	2	-1 1-1 1	385	4	
3	1	2	4	4	0	1	0	1	1	3	-1 1-1 1	385	4	
3	1	3	3	4	1	0	2	-1	1	-1	-1 1-1 1	385	4	
3	1	3	4	4	1	1	-1	1	3	3	-1 1-1 1	385	8	
3	2	2	3	3	0	1	0	0	1	3	-1 1-1 1	385	4	
3	2	2	3	4	2	1	-1	0	1	2	-1 1-1 1	385	4	
4	1	1	5	8	1	0	0	1	0	5	-1-1-1-1	385	24	71/48
4	1	1	6	6	0	1	1	1	0	6	-1-1-1-1	385	16	
4	1	2	2	7	0	0	1	1	1	-1	-1-1-1-1	385	8	
4	1	2	3	5	0	0	1	-1	1	-2	-1-1-1-1	385	4	
4	2	2	2	4	1	1	-1	0	1	0	-1-1-1-1	385	4	
4	2	2	2	5	2	1	0	1	1	2	-1-1-1-1	385	4	
4	2	2	3	3	1	1	1	2	1	2	-1-1-1-1	385	2	

D= 388

GENUS#	F11	F22	F33	F44	F12	F13	F23	F14	F24	F34	HASSE SYM	LEVEL	AUTOS	MASS
1	1	1	1	25	0	0	0	1	1	1	1 1	194	96	17/48
1	1	1	1	49	1	1	0	1	0	0	1 1	194	96	
1	1	1	2	17	1	0	0	1	0	2	1 1	194	24	
1	1	2	2	9	0	0	2	1	2	0	1 1	194	24	
1	1	2	3	5	0	1	0	0	2	1	1 1	194	8	
1	1	2	3	6	0	1	2	0	2	0	1 1	194	8	
2	1	1	2	14	0	1	0	0	0	1	-1-1	388	8	17/4
2	1	1	2	20	1	1	0	0	1	1	-1-1	388	12	
2	1	1	5	6	0	1	0	1	0	4	-1-1	388	8	
2	1	1	5	7	1	0	0	1	1	2	-1-1	388	12	
2	1	1	6	7	1	1	0	1	1	5	-1-1	388	12	
2	1	2	2	7	0	0	1	0	0	2	-1-1	388	4	
2	1	2	2	8	1	0	1	0	0	2	-1-1	388	4	
2	1	2	3	5	0	0	2	0	1	0	-1-1	388	4	
2	1	2	3	5	0	0	1	0	0	3	-1-1	388	4	
2	1	2	3	5	0	1	1	0	0	2	-1-1	388	4	
2	1	2	3	5	0	1	1	1	1	1	-1-1	388	4	
2	1	2	3	6	1	1	0	1	1	3	-1-1	388	4	
2	1	2	3	6	1	0	2	1	0	0	-1-1	388	4	
2	1	2	4	4	0	1	1	0	2	2	-1-1	388	4	
2	1	3	3	4	1	0	2	1	2	2	-1-1	388	4	
2	1	3	3	4	1	1	-1	1	2	1	-1-1	388	4	
2	2	2	3	3	1	1	-1	2	0		-1-1	388	2	
2	2	3	3	3	2	1	0	0	2	3	-1-1	388	2	
3	1	1	4	7	0	1	1	0	0	1	-1-1	194	16	51/16
3	1	2	2	10	1	0	2	1	0	0	-1-1	194	8	
3	1	2	2	8	1	0	0	0	2	2	-1-1	194	4	
3	1	2	3	5	1	0	1	0	1	0	-1-1	194	8	
3	1	2	3	6	0	0	2	1	0	3	-1-1	194	8	
3	1	2	3	6	1	0	1	1	0	3	-1-1	194	4	
3	1	2	4	4	0	1	0	1	2	0	-1-1	194	8	
3	1	2	4	4	1	1	0	0	0	2	-1-1	194	4	
3	1	2	4	4	0	1	0	0	2	2	-1-1	194	8	
3	2	2	2	4	1	1	0	0	0	2	-1-1	194	2	
3	2	2	3	3	0	1	0	1	2	2	-1-1	194	4	
3	2	2	3	3	1	1	1	-1	1	-2	-1-1	194	2	
3	2	2	3	4	2	1	0	0	2	2	-1-1	194	4	
3	2	2	3	4	2	1	-1	2	0	0	-1-1	194	4	

D= 389

GENUS#	F11	F22	F33	F44	F12	F13	F23	F14	F24	F34	HASSE SYM	LEVEL	AUTOS	MASS
1	1	1	1	33	1	0	0	1	0	1	-1-1	389	24	151/48
1	1	1	1	49	1	1	0	0	1	0	-1-1	389	48	
1	1	1	3	10	0	1	1	1	0	0	-1-1	389	8	
1	1	1	3	13	1	1	0	0	1	2	-1-1	389	12	
1	1	1	4	7	0	1	0	1	1	0	-1-1	389	8	
1	1	1	4	7	0	1	0	0	1	2	-1-1	389	8	
1	1	1	4	9	1	0	0	1	0	3	-1-1	389	12	
1	1	1	5	6	0	1	1	0	1	3	-1-1	389	8	
1	1	1	5	7	1	1	0	0	0	1	-1-1	389	12	
1	1	1	5	8	1	1	0	1	1	4	-1-1	389	12	

GENUS#	F11	F22	F33	F44	F12	F13	F23	F14	F24	F34:	HASSE SYM	LEVEL	AUTOS	MASS
1:	1	2	3	5	0	1	0	-1	-1	1:	-1-1	389	4	
1:	1	2	3	5	1	0	0	1	1	1:	-1-1	389	4	
1:	1	2	3	5	0	0	1	1	2	0:	-1-1	389	4	
1:	1	2	4	4	0	1	1	1	0	3:	-1-1	389	4	
1:	1	2	4	5	1	0	1	1	0	4:	-1-1	389	4	
1:	1	3	3	4	1	0	2	0	2	-1:	-1-1	389	4	
1:	1	3	3	4	1	0	1	0	2	3:	-1-1	389	4	
1:	2	2	3	3	1	1	0	2	1	2:	-1-1	389	2	

D= 392

GENUS#	F11	F22	F33	F44	F12	F13	F23	F14	F24	F34:	HASSE SYM	LEVEL	AUTOS	MASS
1:	1	1	1	33	1	0	0	1	0	0:	1 1	392	24	7/4
1:	1	1	2	17	0	1	1	1	1	0:	1 1	392	16	
1:	1	1	2	20	1	1	0	1	0	0:	1 1	392	12	
1:	1	1	3	9	0	0	0	1	1	2:	1 1	392	16	
1:	1	1	5	6	0	1	0	0	0	4:	1 1	392	8	
1:	1	2	2	10	1	0	2	0	1	0:	1 1	392	8	
1:	1	2	2	8	1	0	1	1	0	1:	1 1	392	4	
1:	1	2	3	5	0	0	2	0	0	1:	1 1	392	8	
1:	1	2	3	6	1	1	0	1	2	2:	1 1	392	4	
1:	1	2	4	4	1	0	1	1	0	1:	1 1	392	4	
1:	1	2	4	5	0	1	2	0	0	4:	1 1	392	8	
1:	2	2	3	3	0	0	0	2	1	2:	1 1	392	4	
2:	1	1	1	25	0	0	0	1	1	0:	1 1	392	32	7/4
2:	1	1	1	49	1	1	0	0	0	0:	1 1	392	96	
2:	1	1	2	13	0	0	0	0	1	2:	1 1	392	16	
2:	1	1	3	9	0	1	0	0	0	1:	1 1	392	8	
2:	1	1	4	10	1	1	0	0	1	3:	1 1	392	12	
2:	1	1	5	6	0	1	1	1	0	0:	1 1	392	16	
2:	1	2	2	7	0	0	0	1	2	1:	1 1	392	8	
2:	1	2	4	4	0	0	0	1	-1	3:	1 1	392	4	
2:	1	3	3	4	1	0	2	0	1	-2:	1 1	392	4	
2:	2	2	2	5	2	1	0	0	1	2:	1 1	392	4	
2:	2	2	3	3	1	2	0	1	0	2:	1 1	392	2	
3:	1	1	2	14	0	1	0	0	0	0:	-1-1	56	16	3/16
3:	1	1	4	7	0	1	1	0	0	0:	-1-1	56	32	
3:	1	2	2	7	1	0	0	0	0	0:	-1-1	56	16	
3:	1	2	4	4	0	0	2	0	2	1:	-1-1	56	32	
4:	1	1	5	7	1	1	0	0	0	0:	1 1	56	24	1/3
4:	1	2	3	5	1	0	0	0	0	2:	1 1	56	8	
4:	1	3	3	3	0	0	1	0	1	-1:	1 1	56	24	
4:	2	2	3	4	2	2	0	2	1	1:	1 1	56	8	

D= 393

GENUS#	F11	F22	F33	F44	F12	F13	F23	F14	F24	F34:	HASSE SYM	LEVEL	AUTOS	MASS
1:	1	1	1	33	1	0	0	0	0	1:	-1 1-1	393	48	43/16
1:	1	1	2	20	1	1	0	0	1	0:	-1 1-1	393	12	
1:	1	1	3	10	0	1	0	0	1	3:	-1 1-1	393	8	
1:	1	1	3	11	1	0	0	0	0	1:	-1 1-1	393	24	
1:	1	1	3	12	1	0	0	1	0	3:	-1 1-1	393	24	
1:	1	1	4	9	1	1	0	0	0	1:	-1 1-1	393	12	
1:	1	1	5	7	1	0	0	0	0	3:	-1 1-1	393	24	
1:	1	2	2	7	0	0	1	1	1	1:	-1 1-1	393	8	
1:	1	2	2	9	1	1	0	0	2	1:	-1 1-1	393	8	
1:	1	2	3	5	1	0	1	0	0	1:	-1 1-1	393	4	
1:	1	2	3	6	1	0	0	0	2	3:	-1 1-1	393	8	
1:	1	2	3	6	1	0	1	0	2	3:	-1 1-1	393	4	
1:	1	3	3	4	1	0	0	1	0	3:	-1 1-1	393	8	
1:	1	3	3	4	0	0	3	1	1	1:	-1 1-1	393	8	
1:	1	3	3	4	0	0	1	1	2	3:	-1 1-1	393	4	
1:	1	3	4	4	0	0	3	1	0	4:	-1 1-1	393	8	
1:	2	2	3	3	0	1	1	1	2	2:	-1 1-1	393	2	
1:	2	2	3	4	1	2	-1	0	0	3:	-1 1-1	393	4	
2:	1	1	2	17	0	1	1	1	0	2:	-1-1 1	393	16	43/16
2:	1	1	5	6	0	1	0	1	1	3:	-1-1 1	393	8	
2:	1	2	2	10	1	0	2	0	1	1:	-1-1 1	393	4	
2:	1	2	3	6	1	1	0	0	0	3:	-1-1 1	393	4	
2:	1	2	4	4	0	0	1	1	0	3:	-1-1 1	393	4	
2:	1	2	4	5	1	0	2	1	2	3:	-1-1 1	393	4	
2:	2	2	2	5	2	1	0	2	1	0:	-1-1 1	393	2	
2:	2	2	2	6	2	1	-1	2	1	0:	-1-1 1	393	2	
2:	2	2	3	3	1	2	0	1	-1	1:	-1-1 1	393	2	

```
  2:   2   2   3   4   2   1  -1   1   2   1: -1-1 1      393       4

D= 396
GENUS#: F11 F22 F33 F44 F12 F13 F23 F14 F24 F34: HASSE SYM  LEVEL  AUTOS   MASS
  1:    1   1   1  33   1   0   0   0   0   0: -1 1-1      132      48    35/96
  1:    1   1   3   9   0   0   0   0   0   3: -1 1-1      132      32
  1:    1   1   4   7   0   1   0   1   0   2: -1 1-1      132       8
  1:    1   2   4   4   1   0   1   0   1   2: -1 1-1      132       8
  1:    2   2   3   4   2   0   0  -1   1   3: -1 1-1      132      16
  2:    1   1   1  50   1   1   0   1   0   0:  1 1 1      396      96    35/32
  2:    1   1   2  13   0   0   0   1   1   1:  1 1 1      396      16
  2:    1   1   6   6   1   1   0   0   0   2:  1 1 1      396      12
  2:    1   2   3   6   1   0   2   0   1   0:  1 1 1      396       4
  2:    1   2   3   6   0   1   2   0   2   1:  1 1 1      396      16
  2:    1   2   4   4   0   1   1   1   1  -1:  1 1 1      396       8
  2:    1   3   3   4   1   0   2   1   1   2:  1 1 1      396       4
  2:    2   2   3   4   0   2   2   1   0   3:  1 1 1      396       4
  3:    1   1   3  10   0   1   1   0   0   1:  1 1 1      396      16    35/32
  3:    1   1   5   6   0   1   1   1   1   1:  1 1 1      396      32
  3:    1   2   2   8   1   0   1   1   1   1:  1 1 1      396       4
  3:    1   2   3   6   0   1   2   1   0   2:  1 1 1      396       8
  3:    2   2   2   5   2   1   0   1   2   0:  1 1 1      396       8
  3:    2   2   3   3   0   2   0   0   1   2:  1 1 1      396       4
  3:    2   2   3   4   2   2   0   1   1   2:  1 1 1      396       4
  4:    1   1   3  11   1   0   0   0   0   0:  1-1-1      132      48    35/96
  4:    1   2   2   9   0   0   2   1   1   2:  1-1-1      132      16
  4:    1   3   3   3   0   0   0   1   0   0:  1-1-1      132      32
  4:    2   2   3   3   1   0   0   2   2   0:  1-1-1      132       8
  4:    2   3   3   3   2   2   0   0   3   0:  1-1-1      132       8
  5:    1   1   3   9   0   1   0   0   0   0: -1 1-1      396      16    35/32
  5:    1   1   4   9   1   1   0   0   0   0: -1 1-1      396      24
  5:    1   1   5   5   0   0   0   0   0   1: -1 1-1      396      32
  5:    1   1   5   6   0   0   0   0   1   4: -1 1-1      396       8
  5:    1   1   5   8   1   1   0   0   1   3: -1 1-1      396      12
  5:    1   2   3   5   0   0   1   0   2   2: -1 1-1      396       4
  5:    1   2   4   4   1   0   1   1   1   1: -1 1-1      396       8
  5:    1   2   4   5   0   1   2   1   0   4: -1 1-1      396       8
  5:    1   3   3   4   0   0   3   0   2   1: -1 1-1      396       8
  6:    1   1   2  17   0   1   1   1   1   1:  1 1 1      132      32    35/96
  6:    1   1   5   6   0   1   1   0   0   3:  1 1 1      132      16
  6:    1   1   6   7   1   0   0   0   0   6:  1 1 1      132      48
  6:    1   2   3   5   1   0   0   1   0   0:  1 1 1      132       8
  6:    2   2   2   4   1   0   0   1   1   2:  1 1 1      132       8
  7.    1   1   2  17   1   0   0   0   0   2: -1-1 1      132      48    35/96
  7:    1   2   3   5   0   0   0   0   1   3: -1-1 1      132       8
  7:    1   3   3   4   1   1   0   1   2   2: -1-1 1      132       8
  7:    2   2   2   6   2   1  -1  -1   1  -1: -1-1 1      132      32
  7:    2   3   3   3   1   1   0   1   3   3: -1-1 1      132      16
  8:    1   1   1  25   0   0   0   0   0   1: -1 1-1      396      32    35/32
  8:    1   1   4   7   0   0   1   0   0   3: -1 1-1      396       8
  8:    1   2   3   5   0   1   0   0   2   0: -1 1-1      396      16
  8:    1   2   4   4   0   0   1   0   1  -3: -1 1-1      396       8
  8:    1   2   4   5   1   0   1   0   2   4: -1 1-1      396       4
  8:    2   2   2   4   1   1   1  -1   1   0: -1 1-1      396       4
  8:    2   2   3   3   2   0   0   0   1   1: -1 1-1      396       4

D= 397
GENUS#: F11 F22 F33 F44 F12 F13 F23 F14 F24 F34: HASSE SYM  LEVEL  AUTOS   MASS
  1:    1   1   1  50   1   1   0   0   0   0: -1-1       397      48    57/16
  1:    1   1   2  15   0   1   0   0   1   2: -1-1       397       8
  1:    1   1   2  17   1   0   0   1   0   1: -1-1       397      12
  1:    1   1   2  20   1   1   0   0   0   1: -1-1       397      12
  1:    1   1   3  13   1   1   0   1   1   2: -1-1       397      12
  1:    1   1   5   7   1   0   0   1   0   1: -1-1       397      12
  1:    1   1   6   7   1   1   0   0   1   4: -1-1       397      12
  1:    1   2   2   7   0   0   1   1   1   0: -1-1       397       4
  1:    1   2   2   8   1   0   1   0  -1   1: -1-1       397       4
  1:    1   2   3   5   0   1   1   1   0   0: -1-1       397       4
  1:    1   2   3   5   0   1   0   0   1   2: -1-1       397       4
  1:    1   2   3   5   0   1   1   0  -1   1: -1-1       397       4
  1:    1   2   3   6   0   1   2   1   1   0: -1-1       397       4
  1:    1   2   3   6   1   0   2   0   1   1: -1-1       397       4
```

Continuation of D= 397:

#	F11	F22	F33	F44	F12	F13	F23	F14	F24	F34	HASSE	SYM	LEVEL	AUTOS
1:	1	3	3	4	1	1	-1	1	0	2:	-1	-1	397	4
1:	1	3	3	4	1	0	1	1	-1	2:	-1	-1	397	4
1:	1	3	3	5	1	0	3	0	3	1:	-1	-1	397	4
1:	2	3	3	3	1	1	-1	0	3	2:	-1	-1	397	2

D= 400

GENUS#:	F11	F22	F33	F44	F12	F13	F23	F14	F24	F34:	HASSE	SYM	LEVEL	AUTOS	MASS
1:	1	1	1	25	0	0	0	0	0	0:	1	1 •	100	96	
1:	1	2	3	5	0	0	0	0	2	2:	1	1	100	8	5/32
1:	2	2	2	7	2	2	0	0	2	0:	1	1	100	48	
2:	1	1	1	50	1	1	0	0	0	0:	1	1	200	96	5/8
2:	1	1	2	13	0	0	0	1	1	0:	1	1	200	32	
2:	1	1	3	13	1	1	0	0	1	1:	1	1	200	12	
2:	1	2	3	5	0	1	0	1	0	2:	1	1	200	8	
2:	1	2	2	6	0	1	2	0	0	2:	1	1	200	8	
2:	2	3	3	3	2	0	3	2	2	1:	1	1	200	4	
3:	1	1	2	17	1	0	0	1	0	0:	1	1	200	24	5/8
3:	1	1	4	7	0	0	0	1	1	2:	1	1	200	16	
3:	1	1	5	8	1	1	0	0	0	4:	1	1	200	12	
3:	1	2	2	7	0	0	0	1	2	0:	1	1	200	16	
3:	1	2	3	5	0	0	0	1	2	1:	1	1	200	8	
3:	1	3	3	5	1	0	3	1	3	1:	1	1	200	4	
4:	1	1	2	20	1	1	0	0	0	0:	-1	-1	80	24	1/3
4:	1	2	3	6	0	0	2	1	1	3:	-1	-1	80	8	
4:	2	2	2	4	1	1	-1	0	0	0:	-1	-1	80	24	
4:	2	3	3	3	1	1	2	2	3	3:	-1	-1	80	8	
5:	1	1	3	10	0	1	1	0	0	0:	1	1	40	32	3/16
5:	1	3	3	4	1	1	0	0	2	-2:	1	1	40	8	
5:	2	2	3	3	0	2	0	2	0	1:	1	1	40	32	
6:	1	1	5	5	0	0	0	0	0	0:	1	1	20	64	3/64
6:	2	2	3	3	0	2	0	0	2	0:	1	1	20	32	
7:	1	1	5	7	1	0	0	1	0	0:	-1	-1	80	24	1/3
7:	1	2	2	7	0	0	0	1	0	1:	-1	-1	80	8	
7:	1	2	3	5	1	0	1	0	0	0:	-1	-1	80	8	
7:	1	3	3	3	0	0	1	0	1	1:	-1	-1	80	24	
8:	1	1	7	7	1	1	0	1	1	7:	-1	-1	20	72	1/36
8:	3	3	3	3	2	2	2	-3	-3	1:	-1	-1	20	72	
9:	1	2	2	10	1	0	2	0	0	0:	-1	-1	40	16	1/4
9:	1	2	4	4	0	1	0	1	0	3:	-1	-1	40	16	
9:	2	2	3	4	2	1	-1	2	0	1:	-1	-1	40	8	
10:	1	1	2	13	0	0	0	0	0	2:	1	1	100	32	5/32
10:	1	2	2	9	0	0	2	0	2	0:	1	1	100	24	
10:	2	2	3	4	2	2	0	0	0	2:	1	1	100	12	
11:	1	2	3	5	0	0	2	0	0	0:	-1	-1	20	16	1/16
12:	1	2	3	6	1	0	1	1	1	3:	-1	-1	20	8	1/4
12:	2	2	3	3	1	1	-1	1	-1	2:	-1	-1	20	8	

D= 401

GENUS#:	F11	F22	F33	F44	F12	F13	F23	F14	F24	F34:	HASSE	SYM	LEVEL	AUTOS	MASS
1:	1	1	1	34	1	0	0	1	0	1:	-1	-1	401	24	43/8
1:	1	1	2	17	0	1	1	1	0	0:	-1	-1	401	8	
1:	1	1	4	7	0	1	0	0	1	1:	-1	-1	401	8	
1:	1	1	4	8	0	1	0	0	1	4:	-1	-1	401	8	
1:	1	1	5	6	0	1	0	0	1	3:	-1	-1	401	8	
1:	1	1	6	7	1	1	0	0	0	5:	-1	-1	401	12	
1:	1	2	2	8	1	0	1	1	0	0:	-1	-1	401	4	
1:	1	2	2	9	1	1	0	1	1	2:	-1	-1	401	4	
1:	1	2	3	5	1	0	0	0	1	1:	-1	-1	401	4	
1:	1	2	3	5	0	0	1	1	1	2:	-1	-1	401	4	
1:	1	2	3	6	1	1	0	0	2	1:	-1	-1	401	4	
1:	1	2	3	6	1	0	2	0	0	1:	-1	-1	401	4	
1:	1	2	4	4	0	0	1	1	2	0:	-1	-1	401	4	
1:	1	2	4	4	1	0	1	1	0	0:	-1	-1	401	4	
1:	1	2	4	5	1	0	2	0	2	3:	-1	-1	401	4	
1:	1	2	4	5	1	0	2	1	1	3:	-1	-1	401	4	
1:	1	2	4	5	1	1	0	1	1	4:	-1	-1	401	4	
1:	2	2	2	4	1	0	0	0	2	1:	-1	-1	401	2	
1:	2	2	3	3	0	1	0	2	1	2:	-1	-1	401	2	
1:	2	2	3	4	2	1	0	0	1	3:	-1	-1	401	2	
1:	2	2	3	4	1	2	0	1	2	3:	-1	-1	401	2	

```
D= 404
GENUS#: F11 F22 F33 F44 F12 F13 F23 F14 F24 F34: HASSE SYM  LEVEL AUTOS  MASS
   1:     1   1   1  26   0   0   0   1   1   1:  1  1      202    96   95/96
   1:     1   1   2  17   0   1   1   0   0   1:  1  1      202    16
   1:     1   1   4   8   0   1   1   1   1   3:  1  1      202    16
   1:     1   1   5   6   0   0   0   1   1   3:  1  1      202    16
   1:     1   2   2   8   1   0   0   1   0   2:  1  1      202     8
   1:     2   2   2   4   1   0   0   0   1   2:  1  1      202     4
   1:     2   2   2   5   2   1   0  -1   0   1:  1  1      202     4
   1:     2   2   2   5   2   0   0   2   2   1:  1  1      202    12
   1:     2   2   2   7   2   2   0   1   2   1:  1  1      202    12
   2:     1   1   1  34   1   0   0   1   0   0: -1 -1      404    24   95/24
   2:     1   1   2  13   0   0   0   1   0   1: -1 -1      404     8
   2:     1   1   2  15   0   1   0   0   0   2: -1 -1      404     8
   2:     1   1   3  10   0   1   0   0   0   3: -1 -1      404     8
   2:     1   1   3  13   1   1   0   0   0   2: -1 -1      404    12
   2:     1   1   3   9   0   0   0   0   1   2: -1 -1      404     8
   2:     1   1   4  10   1   1   0   1   1   3: -1 -1      404    12
   2:     1   1   4   7   0   1   0   1   0   0: -1 -1      404     8
   2:     1   1   4   7   0   1   0   0   0   2: -1 -1      404     8
   2:     1   2   2   9   1   1   0   0   2   0: -1 -1      404     4
   2:     1   2   4   4   0   0   2   1   1   0: -1 -1      404     4
   2:     1   2   4   4   1   0   1   0   0   2: -1 -1      404     4
   2:     1   3   3   4   1   0   1   0   3   0: -1 -1      404     4
   2:     1   3   3   4   0   0   2  -1   2  -1: -1 -1      404     4
   2:     1   3   3   4   1   0   2   0   2   2: -1 -1      404     4
   2:     1   3   3   4   0   0   2   0   0   3: -1 -1      404     4
   2:     1   3   4   4   1   0   2   1   0   4: -1 -1      404     4
   2:     2   2   2   4   1   1   0  -1  -1   1: -1 -1      404     2
   2:     2   2   3   3   0   2   0  -1   1  -2: -1 -1      202     2
   3:     1   1   1  51   1   1   0   1   0   0: -1 -1      202    96   95/96
   3:     1   1   3  11   0   1   1   0   0   3: -1 -1      202    16
   3:     1   1   3  13   1   1   0   1   0   0: -1 -1      202    12
   3:     1   1   3   9   0   0   0   1   1   1: -1 -1      202    16
   3:     1   1   4   9   1   0   0   1   0   2: -1 -1      202    12
   3:     1   1   5   7   0   1   1   0   0   5: -1 -1      202    16
   3:     1   2   4   5   0   0   2   1   0   4: -1 -1      202     8
   3:     1   3   3   4   1   0   1   0   2  -2: -1 -1      202     4
   3:     1   3   3   4   1   1  -1   0   0   2: -1 -1      202     4
```

```
D= 405
GENUS#: F11 F22 F33 F44 F12 F13 F23 F14 F24 F34: HASSE SYM  LEVEL AUTOS  MASS
   1:     1   1   1  34   1   0   0   0   0   1: -1 1-1     135    48    3/16
   1:     1   1   4   9   1   0   0   0   0   3: -1 1-1     135    24
   1:     1   3   3   5   0   0   3   1   3   0: -1 1-1     135    24
   1:     2   2   2   4   1   1   1   1   1   1: -1 1-1     405    48   27/16
   2:     1   1   1  51   1   1   0   0   1   0: -1 1-1     405    24
   2:     1   1   2  21   1   1   0   1   1   2: -1 1-1     405    12
   2:     1   1   3  13   1   1   0   0   1   0: -1 1-1     405     8
   2:     1   1   5   6   0   1   1   0   1   2: -1 1-1     405    12
   2:     1   1   6   6   1   1   0   0   0   1: -1 1-1     405    12
   2:     1   1   6   7   1   0   0   1   0   5: -1 1-1     405     4
   2:     1   2   3   5   0   0   1   1   0   2: -1 1-1     405     4
   2:     1   2   3   6   0   1   2   0  -1   1: -1 1-1     405     4
   2:     1   2   3   7   1   0   2   1   1   3: -1 1-1     405     4
   2:     1   3   3   4   1   0   1   1   3   0: -1 1-1     405     4
   2:     2   3   3   3   2   1  -1   2   1   3: -1 1-1     405     8   27/16
   3:     1   1   2  15   0   1   0   1   0   3: -1 1-1     405    16
   3:     1   1   3  11   1   1   0   1   0   1: -1 1-1     405     4
   3:     1   2   2   8   1   0   1   0   1   1: -1 1-1     405     4
   3:     1   2   3   5   0   1   0   1   1   2: -1 1-1     405     4
   3:     1   3   3   4   1   1  -1   0  -1   0: -1 1-1     405     4
   3:     2   2   2   5   1   1  -1   1  -2   1: -1 1-1     405     2
   3:     2   2   3   3   1   1   0   0   2   1: -1-1 1     135    24    3/8
   4:     1   1   2  17   1   0   0   0   0   1: -1-1 1     135    48
   4:     1   1   5   8   1   0   0   0   0   5: -1-1 1     135    16
   4:     1   2   2   7   0   0   1   1   0   0: -1-1 1     135     8
   4:     1   3   3   5   1   0   3   0   1   3: -1-1 1     135    12
   4:     2   3   3   3   1   0   3   1   3   0: -1-1 1     135    24
   4:     3   3   3   3   1   0   0   3   3   3: -1-1 1      45    48    1/48
   5:     1   1   3  12   1   0   0   0   1   0: -1 1-1     135    16    3/16
   6:     1   1   4   7   0   1   0   0   1
```

```
        F11 F22 F33 F44 F12 F13 F23 F14 F24 F34: HASSE SYM   LEVEL AUTOS   MASS
  6:  1   2   4   5   1   0   1   1   1   4: -1 1-1    135     8
  7:  1   1   5   7   0   1   1   1   0   5: -1 1-1     45    16    3/16
  7:  1   2   4   4   1   0   1   0   1  -1: -1 1-1     45     8
  8:  1   1   6   6   1   0   0   0   0   3: -1 1-1     45    48    1/24
  8:  1   3   3   4   0   0   3   1   0   0: -1 1-1     45    48
  9:  1   2   3   5   0   1   1   0   1   1: -1-1 1    135     4    3/8
  9:  1   3   3   4   1   1   0   1   2  -1: -1-1 1    135     8
 10:  1   2   3   6   0   1   2   1   1   1: -1 1-1     45     8    3/16
 10:  2   3   3   3   1   1  -1   1  -3  -1: -1 1-1     45    16
 11:  2   2   2   5   1   1   1  -2  -2   1: -1 1-1     45    24    1/24
 12:  2   2   3   3   1   0   0   0   0   3: -1-1 1     45    48    1/48

D= 408
GENUS#: F11 F22 F33 F44 F12 F13 F23 F14 F24 F34: HASSE SYM   LEVEL AUTOS   MASS
  1:  1   1   1  34   1   0   0   0   0   0:  1-1-1    408    48   103/96
  1:  1   1   2  17   1   0   0   0   0   0:  1-1-1    408    48
  1:  1   1   3   9   0   0   0   1   1   0:  1-1-1    408    32
  1:  1   2   2  11   1   0   2   0   2   0:  1-1-1    408     8
  1:  1   2   3   5   0   1   0   0   0   2:  1-1-1    408     8
  1:  1   2   3   5   0   0   0   0   0   3:  1-1-1    408    16
  1:  1   2   3   5   0   0   0   1   2   0:  1-1-1    408    16
  1:  2   2   2   4   1   1   1   0   1   1:  1-1-1    408     4
  1:  2   2   3   3   0   0   0   0   1   3:  1-1-1    408     8
  1:  2   2   3   4   2   2   0  -1  -1   1:  1-1-1    408     8
  2:  1   1   1  51   1   1   0   0   0   0:  1 1 1    408    96   103/96
  2:  1   1   2  13   0   0   0   0   1   0:  1 1 1    408    16
  2:  1   1   3  12   1   0   0   0   1   0:  1 1 1    408    12
  2:  1   1   6   6   1   0   0   1   0   2:  1 1 1    408    24
  2:  1   2   3   6   1   0   2   0   0   0:  1 1 1    408     8
  2:  1   2   4   4   0   0   0   1   2   1:  1 1 1    408     8
  2:  1   2   4   4   0   1   0   0   0   3:  1 1 1    408     8
  2:  1   2   4   5   1   1   0   0   0   4:  1 1 1    408     4
  2:  1   3   3   3   0   0   0   0  -1   1:  1 1 1    408     8
  2:  2   2   3   4   2   1  -1  -1   1   0:  1 1 1    408     8
  3:  1   1   2  17   0   1   1   0   0   0: -1 1-1    408    32   103/96
  3:  1   1   2  21   1   1   0   0   1   1: -1 1-1    408    12
  3:  1   1   5   8   1   1   0   1   1   3: -1 1-1    408    12
  3:  1   1   6   6   0   1   0   0   0   6: -1 1-1    408    16
  3:  1   2   2   9   0   0   2   0   1  -1: -1 1-1    408    16
  3:  1   2   3   5   0   0   1   0   1  -2: -1 1-1    408     4
  3:  1   2   3   6   0   0   2   0   0   3: -1 1-1    408     8
  3:  1   2   4   4   0   1   1   1   1   2: -1 1-1    408     8
  3:  1   3   3   4   1   0   2   1   2   0: -1 1-1    408     4
  4:  1   1   1  26   0   0   0   1   1   0: -1-1 1    408    32   103/96
  4:  1   1   5   7   1   0   0   0   0   2: -1-1 1    408    24
  4:  1   2   2   7   0   0   0   0   1   2: -1-1 1    408     8
  4:  1   2   2   7   0   0   1   0   1   1: -1-1 1    408     8
  4:  1   2   2   8   1   0   1   0   1   0: -1-1 1    408     4
  4:  1   2   3   5   1   0   0   0   1   0: -1-1 1    408     8
  4:  2   2   2   6   2   1  -1   1   1   1: -1-1 1    408     8
  4:  2   2   3   3   2   0   0   1   1   0: -1-1 1    408     8
  4:  2   3   3   3   2   0   2   0   3   0: -1-1 1    408     8

D= 409
GENUS#: F11 F22 F33 F44 F12 F13 F23 F14 F24 F34: HASSE SYM   LEVEL AUTOS   MASS
  1:  1   1   2  15   0   1   0   0   0   1: -1-1      409     8   79/12
  1:  1   1   3  10   0   1   0   1   1   2: -1-1      409     8
  1:  1   1   4  10   1   1   0   0   1   2: -1-1      409    12
  1:  1   1   4   8   1   0   1   1   0   3: -1-1      409     8
  1:  1   1   5   6   0   1   0   1   1   2: -1-1      409     8
  1:  1   2   2   8   1   0   1   0   0   1: -1-1      409     4
  1:  1   2   2   8   1   0   0   0   2   1: -1-1      409     4
  1:  1   2   3   5   0   1   1   0   1   0: -1-1      409     4
  1:  1   2   3   6   1   1   0   1   2   1: -1-1      409     4
  1:  1   2   3   6   0   1   1   1   2   3: -1-1      409     4
  1:  1   2   4   4   0   1   1   0  -1   2: -1-1      409     4
  1:  1   2   4   4   1   1   0   0   0   1: -1-1      409     4
  1:  1   2   4   5   0   1   2   0   1   4: -1-1      409     4
  1:  1   2   4   5   1   1   0   1   2   3: -1-1      409     4
  1:  1   3   3   4   1   0   2   1   2   1: -1-1      409     4
  1:  1   3   3   4   1   0   0   0   3   1: -1-1      409     4
  1:  1   3   4   4   1   0   3  -1   1  -2: -1-1      409     4
```

GENUS#	F11	F22	F33	F44	F12	F13	F23	F14	F24	F34:	HASSE	SYM	LEVEL	AUTOS	MASS
1:	2	2	2	4	1	1	0	1	0	-1:	-1-1		409	2	
1:	2	2	2	5	2	1	0	1	1	-1:	-1-1		409	2	
1:	2	2	3	3	1	1	1	-1	0	2:	-1-1		409	2	
1:	2	2	3	3	0	1	1	2	1	0:	-1-1		409	2	
1:	2	2	3	3	1	1	1	2	1	0:	-1-1		409	2	
1:	2	2	3	4	1	2	-1	2	0	3:	-1-1		409	2	

D= 412

GENUS#:	F11	F22	F33	F44	F12	F13	F23	F14	F24	F34:	HASSE	SYM	LEVEL	AUTOS	MASS
1:	1	1	1	26	0	0	0	0	0	1:	-1-1		412	32	19/8
1:	1	1	1	52	1	1	0	1	0	0:	-1-1		412	96	
1:	1	1	2	13	0	0	0	0	0	1:	-1-1		412	16	
1:	1	1	2	15	0	1	0	1	0	0:	-1-1		412	8	
1:	1	1	4	7	0	0	0	1	1	1:	-1-1		412	16	
1:	1	1	4	7	0	0	0	0	0	3:	-1-1		412	16	
1:	1	1	4	8	0	1	1	0	0	3:	-1-1		412	16	
1:	1	1	5	6	0	1	0	1	0	3:	-1-1		412	8	
1:	1	1	5	8	1	0	0	1	1	4:	-1-1		412	12	
1:	1	2	2	7	0	0	1	0	1	0:	-1-1		412	4	
1:	1	2	2	8	1	0	0	0	1	2:	-1-1		412	8	
1:	1	2	3	5	0	0	1	0	2	0:	-1-1		412	4	
1:	1	2	3	6	0	1	2	1	0	0:	-1-1		412	8	
1:	1	2	3	6	1	1	0	0	2	0:	-1-1		412	4	
1:	1	2	4	4	0	1	0	0	2	1:	-1-1		412	8	
1:	1	2	4	5	0	1	0	0	2	4:	-1-1		412	8	
1:	2	2	3	3	0	2	1	0	1	-1:	-1-1		412	4	
1:	2	2	3	4	2	1	-1	1	1	2:	-1-1		412	4	
2:	1	1	2	18	1	0	0	1	0	2:	1 1		412	24	19/8
2:	1	1	2	21	1	1	0	1	0	0:	1 1		412	12	
2:	1	2	2	11	0	0	2	0	0	2:	1 1		412	8	
2:	1	2	2	9	0	0	2	1	1	0:	1 1		412	8	
2:	1	2	3	5	0	1	1	0	0	1:	1 1		412	4	
2:	1	2	3	5	0	0	0	-1	-1	2:	1 1		412	4	
2:	1	2	3	6	0	1	1	0	2	3:	1 1		412	4	
2:	1	2	3	6	0	1	0	0	2	3:	1 1		412	8	
2:	1	2	3	6	0	0	2	1	2	0:	1 1		412	8	
2:	2	2	3	3	1	1	1	2	0	0:	1 1		412	2	
2:	2	3	3	3	1	1	-3	2	2	0:	1 1		412	2	

D= 413

GENUS#:	F11	F22	F33	F44	F12	F13	F23	F14	F24	F34:	HASSE	SYM	LEVEL	AUTOS	MASS
1:	1	1	1	35	1	0	0	1	0	1:	-1 1-1		413	24	73/48
1:	1	1	2	15	0	1	0	0	1	0:	-1 1-1		413	16	
1:	1	1	2	21	1	0	0	0	1	0:	-1 1-1		413	12	
1:	1	1	4	9	1	0	0	1	1	1:	-1 1-1		413	12	
1:	1	1	5	7	0	1	0	0	1	5:	-1 1-1		413	8	
1:	1	1	6	7	1	1	0	1	1	4:	-1 1-1		413	12	
1:	1	1	7	7	1	1	0	0	1	6:	-1 1-1		413	24	
1:	1	2	3	5	1	0	0	0	0	1:	-1 1-1		413	8	
1:	1	2	3	6	1	1	0	0	1	2:	-1 1-1		413	4	
1:	1	2	3	7	1	0	2	0	0	3:	-1 1-1		413	4	
1:	1	2	4	4	1	0	1	0	1	1:	-1 1-1		413	8	
1:	1	3	3	4	1	1	-1	1	0	0:	-1 1-1		413	8	
1:	1	3	3	4	0	0	1	-1	-2	2:	-1 1-1		413	8	
2:	1	1	1	52	1	1	0	0	1	0:	-1-1 1		413	48	73/48
2:	1	1	3	10	1	1	0	0	1	2:	-1-1 1		413	8	
2:	1	1	3	13	1	1	0	0	0	1:	-1-1 1		413	12	
2:	1	1	4	10	1	1	0	0	0	3:	-1-1 1		413	12	
2:	1	1	5	6	0	1	1	1	0	0:	-1-1 1		413	8	
2:	1	1	5	8	1	1	0	0	1	2:	-1-1 1		413	12	
2:	1	2	3	5	0	0	1	-1	-1	1:	-1-1 1		413	4	
2:	1	3	3	4	1	0	1	0	0	3:	-1-1 1		413	4	
2:	2	2	3	3	1	2	0	0	1	-1:	-1-1 1		413	2	

D= 416

GENUS#:	F11	F22	F33	F44	F12	F13	F23	F14	F24	F34:	HASSE	SYM	LEVEL	AUTOS	MASS
1:	1	1	4	10	1	0	0	1	0	4:	1 1		416	24	25/24
1:	1	2	2	7	0	0	0	1	1	1:	1 1		416	8	
1:	1	2	2	8	1	0	1	0	0	0:	1 1		416	8	
1:	1	2	2	9	1	1	0	1	0	0:	1 1		416	8	
1:	1	2	4	4	0	1	1	1	1	0:	1 1		416	8	
1:	1	3	3	4	1	0	2	0	2	0:	1 1		416	4	

GENUS#	F11	F22	F33	F44	F12	F13	F23	F14	F24	F34:	HASSE	SYM	LEVEL	AUTOS	MASS
1:	2	3	3	3	1	1	-2	0	2	2:	1 1		416	4	
2:	1	1	1	26	0	0	0	0	0	0:	1 1		104	96	25/96
2:	1	1	3	9	0	0	0	0	0	2:	1 1		104	16	
2:	1	2	2	7	0	0	0	0	2	0:	1 1		104	16	
2:	2	2	3	3	0	0	0	0	2	2:	1 1		104	8	
3:	1	1	1	35	1	0	0	1	0	0:	1 1		416	24	25/24
3:	1	1	3	9	0	0	0	1	0	1:	1 1		416	8	
3:	1	1	4	7	0	0	0	1	0	2:	1 1		416	8	
3:	1	2	4	4	1	0	1	0	1	0:	1 1		416	8	
3:	1	3	4	4	1	0	3	0	1	4:	1 1		416	4	
3:	1	3	4	4	1	0	3	0	3	0:	1 1		416	8	
3:	2	2	3	3	1	1	0	-1	0	2:	1 1		416	4	
4:	1	1	1	52	1	1	0	0	0	0:	-1-1		208	96	25/96
4:	1	1	5	6	0	1	1	0	0	2:	-1-1		208	16	
4:	1	2	4	5	0	0	0	1	2	4:	-1-1		208	16	
4:	1	3	3	5	1	1	-1	1	3	2:	-1-1		208	8	
5:	1	1	2	13	0	0	0	0	0	0:	-1-1		104	32	25/96
5:	1	1	5	6	0	0	0	0	0	4:	-1-1		104	16	
5:	1	2	4	4	0	0	0	0	2	2:	-1-1		104	8	
5:	2	2	2	5	2	0	0	2	0	0:	-1-1		104	24	
6:	1	1	2	14	0	0	0	1	1	2:	-1-1		208	32	25/32
6:	1	1	2	18	0	1	1	1	1	0:	-1-1		208	16	
6:	1	2	4	4	0	0	0	1	2	0:	-1-1		208	16	
6:	1	2	4	4	1	0	0	1	0	0:	-1-1		208	8	
6:	2	2	2	4	0	0	0	2	1	1:	-1-1		208	8	
6:	2	2	2	4	1	1	0	0	1	1:	-1-1		208	4	
6:	2	2	2	5	2	0	0	1	1	2:	-1-1		208	8	
7:	1	1	2	15	0	1	0	0	0	1:	-1-1		416	8	25/24
7:	1	1	3	13	1	1	0	0	0	0:	-1-1		416	24	
7:	1	1	4	7	0	1	0	0	0	1:	-1-1		416	8	
7:	1	1	4	8	0	1	0	0	0	4:	-1-1		416	8	
7:	1	2	4	4	0	0	2	0	1	-1:	-1-1		416	4	
7:	1	2	4	5	1	0	2	0	2	0:	-1-1		416	4	
7:	1	3	3	4	0	0	2	0	1	3:	-1-1		416	8	
8:	1	1	3	11	0	1	1	1	1	0:	1 1		208	16	25/96
8:	1	1	4	7	0	0	0	1	1	0:	1 1		208	32	
8:	1	1	4	9	1	0	0	1	0	0:	1 1		208	24	
8:	1	2	3	5	0	1	0	1	0	0:	1 1		208	8	
9:	1	1	3	14	1	1	0	1	1	3:	-1-1		416	24	25/24
9:	1	2	4	5	1	0	2	1	0	2:	-1-1		416	4	
9:	1	2	4	5	1	0	0	1	0	4:	-1-1		416	8	
9:	1	3	3	4	1	1	-1	1	1	0:	-1-1		416	8	
9:	1	3	3	4	0	0	2	1	2	2:	-1-1		416	8	
9:	2	2	2	4	1	1	0	1	1	1:	-1-1		416	4	
9:	2	2	3	4	0	2	2	-1	1	-2:	-1-1		416	8	
10:	1	1	4	8	0	0	0	1	1	4:	1 1		208	32	25/32
10:	1	1	4	8	0	1	1	1	1	0:	1 1		208	16	
10:	1	2	2	8	1	0	0	0	2	0:	1 1		208	8	
10:	1	2	2	9	1	1	0	0	1	1:	1 1		208	8	
10:	2	2	2	4	1	0	0	0	2	0:	1 1		208	4	
10:	2	2	2	7	2	2	0	-1	-1	1:	1 1		208	16	
10:	2	2	3	4	0	2	0	0	2	3:	1 1		208	8	

D= 417

GENUS#	F11	F22	F33	F44	F12	F13	F23	F14	F24	F34:	HASSE	SYM	LEVEL	AUTOS	MASS
1:	1	1	1	35	1	0	0	0	0	0:	-1-1	1	417	48	47/16
1:	1	1	5	6	0	1	0	1	1	0:	-1-1	1	417	8	
1:	1	1	5	7	1	0	0	0	0	1:	-1-1	1	417	24	
1:	1	2	2	11	1	0	2	0	2	1:	-1-1	1	417	4	
1:	1	2	2	8	0	0	1	1	2	2:	-1-1	1	417	8	
1:	1	2	2	9	1	1	0	1	1	0:	-1-1	1	417	8	
1:	1	2	3	6	1	1	0	1	1	2:	-1-1	1	417	4	
1:	1	2	3	6	1	0	0	1	0	3:	-1-1	1	417	8	
1:	1	3	3	4	1	0	0	0	1	3:	-1-1	1	417	8	
1:	2	2	2	4	1	1	1	1	0	0:	-1-1	1	417	4	
1:	2	2	2	5	2	1	0	1	1	1:	-1-1	1	417	2	
1:	2	2	2	6	2	1	-1	1	0	1:	-1-1	1	417	4	
1:	2	2	3	3	1	1	0	1	-1	2:	-1-1	1	417	2	
1:	2	2	3	4	1	2	-1	2	2	1:	-1-1	1	417	4	
2:	1	1	2	18	0	1	1	1	0	2:	-1 1	1-1	417	16	47/16
2:	1	1	2	21	1	1	0	0	0	1:	-1 1	1-1	417	12	
2:	1	1	3	10	0	1	0	1	1	0:	-1 1	1-1	417	8	

```
2:   1   1   3  12   1   0   0   1   0   1: -1 1-1      417   12
2:   1   1   4  10   1   1   0   1   1   2: -1 1-1      417   12
2:   1   2   3   6   1   0   1   1   2   2: -1 1-1      417    4
2:   1   2   3   6   1   0   1   0   0   3: -1 1-1      417    4
2:   1   2   4   4   0   1   1   1   0   2: -1 1-1      417    4
2:   1   2   4   5   1   0   2   0   0   3: -1 1-1      417    4
2:   1   3   3   4   1   0   2   1   0   1: -1 1-1      417    4
2:   1   3   3   4   0   0   1   1   3   0: -1 1-1      417    4
2:   1   3   4   4   0   1   2   1   3   4: -1 1-1      417    4
2:   2   2   3   3   0   1   1   2   1   1: -1 1-1      417    2
2:   2   2   3   4   2   1  -1   1   2   0: -1 1-1      417    4

D= 420
GENUS#:  F11 F22 F33 F44 F12 F13 F23 F14 F24 F34: HASSE SYM  LEVEL AUTOS  MASS
   1:   1   1   1  27   0   0   0   1   1   1:  1-1 1-1    210   96   3/32
   1:   1   3   3   5   1   1   0   1   3   3:  1-1 1-1    210   16
   1:   2   2   3   3   2   0   0   0   1   1:  1-1 1-1    210   48
   2:   1   1   1  35   1   0   0   0   0   0: -1 1-1 1    420   48   9/8
   2:   1   1   2  21   1   1   0   0   0   0: -1 1-1 1    420   24
   2:   1   1   3   9   0   0   0   0   1   0: -1 1-1 1    420   16
   2:   1   1   4   9   1   0   0   0   0   2: -1 1-1 1    420   24
   2:   1   1   6   7   1   1   0   0   1   3: -1 1-1 1    420   12
   2:   1   2   3   6   0   0   2   0   1   3: -1 1-1 1    420    8
   2:   1   2   3   7   1   0   2   1   0   2: -1 1-1 1    420    4
   2:   1   3   3   4   0   0   3   0   0   1: -1 1-1 1    420    8
   2:   1   3   3   4   1   0   0   0   3   0: -1 1-1 1    420    8
   2:   2   2   3   4   1   2  -1   2   2   0: -1 1-1 1    420    4
   3:   1   1   1  53   1   1   0   1   0   0:  1 1-1-1    210   96   3/32
   3:   1   1   6   6   1   0   0   0   0   2:  1 1-1-1    210   48
   3:   2   2   3   3   2   2   1   0   3   0:  1 1-1-1    210   16
   4:   1   1   2  15   0   1   0   0   0   0: -1-1-1-1    420   16   9/8
   4:   1   1   4   7   0   1   0   0   0   0: -1-1-1-1    420   16
   4:   1   2   4   5   0   0   2   1  -1   3: -1-1-1-1    420    8
   4:   1   3   3   4   1   1  -1   0   1   1: -1-1-1-1    420    8
   4:   2   2   2   5   1   1  -1   2   0   2: -1-1-1-1    420    4
   4:   2   2   3   3   1   1   0   2   0   0: -1-1-1-1    420    2
   5:   1   1   3  11   0   1   1   1   1   1:  1-1-1 1    210   32   3/32
   5:   1   2   2   9   0   0   2   1   0   0:  1-1-1 1    210   48
   5:   2   2   3   5   2   2   0   2   2   3:  1-1-1 1    210   24
   6:   1   1   3  10   0   1   0   1   0   2: -1 1 1-1    420    8   9/8
   6:   1   1   3  12   1   0   0   1   0   0: -1 1 1-1    420   24
   6:   1   1   3  12   1   0   0   0   0   0: -1 1 1-1    420   24
   6:   1   1   4  10   1   1   0   0   1   1: -1 1 1-1    420   12
   6:   1   1   5   7   1   0   0   0   0   0: -1 1 1 1    420   18
   6:   1   2   2   7   0   0   1   0   0   0: -1 1 1-1    420   16
   6:   1   2   3   5   1   0   0   0   0   0: -1 1 1-1    420   16
   6:   1   2   4   4   0   1   1   1   1   1: -1 1 1-1    420    8
   6:   1   3   3   3   0   0   0   0   1   0: -1 1 1-1    420   16
   6:   1   3   3   4   1   0   2   0   0   2: -1 1 1-1    420    4
   6:   1   3   3   4   0   0   0   0   3   2: -1 1 1-1    420    8
   6:   2   3   3   3   1   1  -2   1   3   1: -1 1 1-1    420    8
   7:   1   1   4   8   0   1   1   1   1   1: -1 1 1-1    210   32   27/32
   7:   1   2   2   8   1   0   0   0   0   2: -1 1 1-1    210   16
   7:   1   3   3   4   0   0   0   1   1   3: -1 1 1-1    210    8
   7:   1   3   4   4   0   1   3   1   1  -2: -1 1 1-1    210    4
   7:   1   3   4   4   0   1   3   1   3   1: -1 1 1-1    210   16
   7:   2   2   2   4   1   0   0   0   0   2: -1 1 1-1    210   16
   7:   2   2   3   3   0   1   1  -1   1  -2: -1 1 1-1    210    4
   8:   1   1   5   6   0   1   0   0   0   3: -1-1 1 1    420    8   9/8
   8:   1   2   3   5   0   1   1   0   0   0: -1-1 1 1    420    8
   8:   1   2   3   6   1   1   0   1   0   2: -1-1 1 1    420    4
   8:   1   2   4   4   0   0   1   0   1   3: -1-1 1 1    420    8
   8:   1   2   4   5   1   0   2   1   2   0: -1-1 1 1    420    4
   8:   2   2   3   3   1   1  -1   0   0   2: -1-1 1 1    420    4
   9:   1   1   2  18   1   0   0   0   0   2:  1 1 1 1    210   48   3/32
   9:   1   1   5   7   0   0   0   1   1   5:  1 1 1 1    210   32
   9:   1   1   5   9   1   1   0   1   1   5:  1 1 1 1    210   24
  10:   1   1   2  18   0   1   1   1   1   1: -1-1 1 1    210   32   27/32
  10:   1   2   2  11   1   0   2   0   1   2: -1-1 1 1    210   16
  10:   1   2   4   5   1   0   2   0   1  -2: -1-1 1 1    210    4
  10:   2   2   2   5   2   0   0   2   1   1: -1-1 1 1    210    8
  10:   2   2   3   3   1   1  -1  -1   1   0: -1-1 1 1    210    8
```

```
10:  2  2  3   4  2  1   0   2   2   2: -1-1 1 1    210    4
11:  1  1  3  10  0  0   0   1   1   3: -1 1-1 1    210   32    27/32
11:  1  2  2   9  1  1   0   0  -1   1: -1 1-1 1    210    8
11:  1  2  3   6  0  0   2   1   2   1: -1 1-1 1    210   16
11:  1  2  3   6  1  0   1   0   2   2: -1 1-1 1    210    4
11:  2  2  3   4  0  2   1   2   2   0: -1 1-1 1    210    4
11:  2  2  3   4  1  1   1   2   2  -2: -1 1-1 1    210    8
12:  1  1  6   6  0  1   1   1   1   5: -1-1-1-1    210   32    27/32
12:  1  2  4   4  0  0   2   1   0   0: -1-1-1-1    210   16
12:  1  2  4   4  1  0   0   0   0   2: -1-1-1-1    210   16
12:  2  2  2   4  1  1   0   0  -1   1: -1-1-1-1    210    4
12:  2  2  2   5  2  1   0   1   0   1: -1-1-1-1    210    4
12:  2  2  2   6  2  1  -1   0   0   1: -1-1-1-1    210    8
12:  2  2  3   4  2  1  -1  -1   1  -1: -1-1-1-1    210   16
```

D= 421

GENUS#:	F11	F22	F33	F44	F12	F13	F23	F14	F24	F34:	HASSE SYM	LEVEL	AUTOS	MASS
1:	1	1	1	53	1	1	0	0	1	0:	-1-1	421	48	209/48
1:	1	1	2	18	1	0	0	1	0	1:	-1-1	421	12	
1:	1	1	3	11	0	1	1	1	0	2:	-1-1	421	8	
1:	1	1	3	14	1	1	0	0	1	2:	-1-1	421	12	
1:	1	1	5	6	0	1	0	0	1	2:	-1-1	421	8	
1:	1	1	5	8	1	1	0	0	0	3:	-1-1	421	12	
1:	1	1	5	8	1	1	0	1	1	2:	-1-1	421	12	
1:	1	2	2	9	1	0	1	1	0	2:	-1-1	421	4	
1:	1	2	3	5	0	0	1	1	1	1:	-1-1	421	4	
1:	1	2	3	5	0	1	0	0	1	1:	-1-1	421	4	
1:	1	2	3	6	1	1	0	0	1	-2:	-1-1	421	4	
1:	1	2	3	6	0	1	2	0	1	1:	-1-1	421	4	
1:	1	2	4	5	0	1	1	1	2	4:	-1-1	421	4	
1:	1	3	3	4	1	0	2	0	2	1:	-1-1	421	4	
1:	1	3	3	4	1	0	1	1	2	2:	-1-1	421	4	
1:	1	3	3	4	1	1	0	0	2	1:	-1-1	421	4	
1:	1	3	3	4	1	1	0	1	1	2:	-1-1	421	4	
1:	1	3	3	5	1	0	3	1	2	2:	-1-1	421	4	
1:	2	2	3	3	1	2	0	1	0	1:	-1-1	421	2	
1:	2	3	3	3	2	1	0	2	3	2:	-1-1	421	2	

D= 424

GENUS#:	F11	F22	F33	F44	F12	F13	F23	F14	F24	F34:	HASSE SYM	LEVEL	AUTOS	MASS
1:	1	1	1	53	1	1	0	0	0	0:	-1-1	424	96	87/32
1:	1	1	2	14	0	0	0	0	1	2:	-1-1	424	16	
1:	1	1	5	6	0	0	0	0	1	3:	-1-1	424	8	
1:	1	1	5	6	0	0	0	1	1	2:	-1-1	424	16	
1:	1	1	5	8	1	1	0	0	1	1:	-1-1	424	12	
1:	1	2	2	9	0	0	2	0	0	1:	-1-1	424	8	
1:	1	2	3	5	0	0	1	0	1	2:	-1-1	424	4	
1:	1	2	3	6	0	1	2	0	0	1:	-1-1	424	8	
1:	1	2	4	4	0	0	1	0	0	3:	-1-1	424	4	
1:	1	2	4	5	1	1	0	0	2	2:	-1-1	424	4	
1:	1	2	4	5	0	1	2	1	2	0:	-1-1	424	8	
1:	1	3	3	4	1	1	0	1	0	2:	-1-1	424	4	
1:	1	3	3	4	1	0	0	1	2	2:	-1-1	424	4	
1:	2	2	2	5	2	1	0	1	0	0:	-1-1	424	4	
1:	2	2	3	4	1	2	-1	1	0	3:	-1-1	424	2	
2:	1	1	1	27	0	0	0	1	1	1:	1 1	424	32	87/32
2:	1	1	2	18	1	0	0	1	0	0:	1 1	424	24	
2:	1	1	3	10	0	1	0	0	0	2:	1 1	424	8	
2:	1	1	3	11	0	1	1	0	0	2:	1 1	424	16	
2:	1	1	4	10	1	1	0	1	0	0:	1 1	424	12	
2:	1	2	2	7	0	0	0	1	1	0:	1 1	424	8	
2:	1	2	2	9	1	0	1	0	2	1:	1 1	424	4	
2:	1	2	3	5	0	0	0	1	0	2:	1 1	424	8	
2:	1	2	3	5	0	0	0	0	2	1:	1 1	424	8	
2:	1	2	4	5	0	1	1	0	2	4:	1 1	424	4	
2:	1	3	3	4	1	0	2	0	1	2:	1 1	424	4	
2:	2	2	3	3	0	1	0	2	0	2:	1 1	424	4	
2:	2	2	3	3	1	2	0	0	1	0:	1 1	424	2	
2:	2	2	3	4	2	2	0	1	1	1:	1 1	424	4	
2:	2	3	3	3	2	0	3	0	0	2:	1 1	424	4	

D= 425

GENUS#:	F11	F22	F33	F44	F12	F13	F23	F14	F24	F34:	HASSE	SYM	LEVEL	AUTOS	MASS
1:	1	1	1	36	1	0	0	1	0	1:	-1	1-1	425	24	65/24
1:	1	1	2	16	0	1	0	0	1	2:	-1	1-1	425	8	
1:	1	1	6	6	0	1	1	0	1	5:	-1	1-1	425	8	
1:	1	1	6	8	1	1	0	1	1	6:	-1	1-1	425	24	
1:	1	2	2	9	1	1	0	0	0	1:	-1	1-1	425	4	
1:	1	2	3	5	0	0	1	1	1	0:	-1	1-1	425	4	
1:	1	2	3	7	1	0	2	0	1	3:	-1	1-1	425	8	
1:	1	2	4	4	1	0	0	0	1	1:	-1	1-1	425	4	
1:	2	2	2	4	1	1	0	1	1	0:	-1	1-1	425	2	
1:	2	2	2	5	2	1	0	1	1	0:	-1	1-1	425	2	
1:	2	2	3	4	1	2	0	0	1	3:	-1	1-1	425	2	
2:	1	1	2	22	1	1	0	1	1	2:	-1	1-1	85	24	13/24
2:	1	2	2	11	1	0	2	1	1	1:	-1	1-1	85	8	
2:	1	2	3	6	1	0	1	1	0	2:	-1	1-1	85	4	
2:	2	2	3	3	1	1	-1	1	-1	1:	-1	1-1	85	8	
3:	1	1	2	18	0	1	1	0	1	0:	-1	1-1	425	8	65/24
3:	1	1	3	10	0	1	0	0	1	1:	-1	1-1	425	8	
3:	1	1	4	10	1	1	0	0	1	0:	-1	1-1	425	12	
3:	1	1	4	8	0	1	1	0	1	2:	-1	1-1	425	8	
3:	1	2	2	8	1	0	0	1	0	1:	-1	1-1	425	4	
3:	1	2	4	4	0	0	1	1	1	2:	-1	1-1	425	4	
3:	1	2	4	5	1	0	2	0	1	3:	-1	1-1	425	4	
3:	1	2	4	5	1	0	2	0	2	1:	-1	1-1	425	4	
3:	1	3	3	4	1	0	2	1	1	1:	-1	1-1	425	4	
3:	1	3	4	4	0	1	2	0	3	4:	-1	1-1	425	2	
3:	2	2	3	3	0	1	1	0	1	-2:	-1	1-1	425	2	
3:	2	2	3	4	2	1	0	1	1	3:	-1	1-1	425	4	
4:	1	1	4	8	0	1	0	1	1	3:	-1-1	1	85	8	13/24
4:	1	3	4	4	1	0	3	1	3	0:	-1-1	1	85	8	
4:	2	2	2	4	1	0	0	1	-1	-1:	-1-1	1	85	4	
4:	2	2	2	5	1	1	-1	1	-1	-1:	-1-1	1	85	24	

D= 428

GENUS#:	F11	F22	F33	F44	F12	F13	F23	F14	F24	F34:	HASSE	SYM	LEVEL	AUTOS	MASS
1:	1	1	1	27	0	0	0	0	0	1:	-1-1		428	32	197/96
1:	1	1	1	36	1	0	0	1	0	0:	-1-1		428	24	
1:	1	1	3	10	0	1	0	1	0	0:	-1-1		428	8	
1:	1	1	3	9	0	0	0	0	0	1:	-1-1		428	16	
1:	1	1	4	10	1	1	0	0	0	2:	-1-1		428	12	
1:	1	1	4	7	0	0	0	0	1	1:	-1-1		428	8	
1:	1	1	5	6	0	1	0	1	0	2:	-1-1		428	8	
1:	1	1	5	8	1	1	0	1	0	0:	-1-1		428	12	
1:	1	2	2	11	1	0	2	1	0	0:	-1-1		428	8	
1:	1	2	3	5	0	0	1	0	0	2:	-1-1		428	4	
1:	1	2	3	6	1	1	0	0	0	2:	-1-1		428	4	
1:	1	2	4	5	1	0	1	0	0	4:	-1-1		428	4	
1:	1	3	3	4	0	0	1	0	2	3:	-1-1		428	4	
1:	2	2	3	4	2	1	0	1	2	2:	-1-1		428	4	
2:	1	1	1	54	1	1	0	1	0	0:	1 1		428	96	197/96
2:	1	1	2	14	0	0	0	1	1	1:	1 1		428	16	
2:	1	1	2	18	0	1	1	0	0	1:	1 1		428	16	
2:	1	1	2	22	1	1	0	0	1	1:	1 1		428	12	
2:	1	1	5	6	0	1	1	0	0	1:	1 1		428	16	
2:	1	1	6	6	0	1	1	0	0	5:	1 1		428	16	
2:	1	1	6	7	1	1	0	0	0	4:	1 1		428	12	
2:	1	2	3	6	1	0	0	1	2	2:	1 1		428	4	
2:	1	2	3	6	0	0	2	1	0	2:	1 1		428	8	
2:	1	2	3	7	1	0	2	0	2	0:	1 1		428	4	
2:	2	2	2	5	2	1	0	0	0	1:	1 1		428	4	
2:	2	2	3	3	0	1	0	1	2	1:	1 1		428	4	
2:	2	2	3	4	2	2	0	1	0	1:	1 1		428	4	
2:	2	2	3	4	0	2	2	2	1	2:	1 1		428	4	

D= 429

GENUS#:	F11	F22	F33	F44	F12	F13	F23	F14	F24	F34:	HASSE	SYM	LEVEL	AUTOS	MASS
1:	1	1	1	36	1	0	0	0	0	1:	-1	1-1 1	429	48	1/1
1:	1	1	3	12	1	0	0	0	0	1:	-1	1-1 1	429	24	
1:	1	1	3	13	1	0	0	1	0	3:	-1	1-1 1	429	24	
1:	1	1	4	9	1	0	0	0	0	1:	-1	1-1 1	429	24	
1:	1	1	5	8	1	1	0	0	1	0:	-1	1-1 1	429	12	

	F11	F22	F33	F44	F12	F13	F23	F14	F24	HASSE: SYM	LEVEL	AUTOS	MASS
1:	1	1	5	9	1	1	0	0	1	4: -1 1-1 1	429	12	
1:	1	1	6	6	0	1	0	0	1	5: -1 1-1 1	429	16	
1:	1	2	3	5	0	0	1	1	0	1: -1 1-1 1	429	4	
1:	1	3	3	5	0	0	3	1	-2	1: -1 1-1 1	429	8	
1:	2	3	3	3	2	1	-1	0	3	0: -1 1-1 1	429	4	
2:	1	1	1	54	1	1	0	0	1	0: -1 1 1-1	429	48	1/1
2:	1	1	2	18	1	0	0	0	0	1: -1 1 1-1	429	24	
2:	1	1	3	10	0	1	0	0	1	0: -1 1 1-1	429	16	
2:	1	1	3	14	1	1	0	1	1	2: -1 1 1-1	429	12	
2:	1	1	4	11	1	1	0	1	1	4: -1 1 1-1	429	24	
2:	1	1	6	6	1	0	0	0	0	1: -1 1 1-1	429	48	
2:	1	1	6	7	1	0	0	0	0	5: -1 1 1-1	429	24	
2:	1	2	3	6	1	0	0	0	1	3: -1 1 1-1	429	8	
2:	1	3	3	4	1	0	1	1	0	2: -1 1 1-1	429	4	
2:	1	3	3	4	1	0	0	0	0	3: -1 1 1-1	429	16	
2:	1	3	3	5	1	0	3	1	2	0: -1 1 1-1	429	8	
2:	2	2	3	4	1	2	-1	1	1	-2: -1 1 1-1	429	8	
3:	1	1	4	8	0	1	0	0	1	3: -1-1-1-1	429	8	1/1
3:	1	2	4	5	0	0	1	1	2	4: -1-1-1-1	429	4	
3:	1	2	5	5	1	1	0	1	2	5: -1-1-1-1	429	8	
3:	2	2	3	3	1	2	0	0	0	1: -1-1-1-1	429	2	
4:	1	1	3	11	0	1	1	1	0	0: -1-1 1 1	429	8	1/1
4:	1	2	2	9	1	0	1	1	1	2: -1-1 1 1	429	8	
4:	1	2	3	5	0	1	0	0	1	0: -1-1 1 1	429	8	
4:	1	3	3	4	1	1	0	0	1	-2: -1-1 1 1	429	4	
4:	1	3	4	4	1	1	-1	0	2	3: -1-1 1 1	429	8	
4:	2	2	3	3	1	1	0	0	1	2: -1-1 1 1	429	4	

D= 432

GENUS#:	F11	F22	F33	F44	F12	F13	F23	F14	F24	F34	HASSE: SYM	LEVEL	AUTOS	MASS
1:	1	1	1	27	0	0	0	0	0	0:	-1-1	108	96	9/32
1:	1	1	4	7	0	0	0	0	0	2:	-1-1	108	16	
1:	1	2	4	5	0	0	2	0	0	4:	-1-1	108	8	
1:	2	2	3	4	2	0	0	2	2	2:	-1-1	108	12	
2:	1	1	1	36	1	0	0	0	0	0:	-1-1	144	48	1/4
2:	1	1	3	10	0	0	0	1	0	3:	-1-1	144	16	
2:	1	1	4	9	1	0	0	0	0	0:	-1-1	144	48	
2:	1	2	2	9	1	1	0	0	0	0:	-1-1	144	16	
2:	1	3	3	4	0	0	3	0	0	0:	-1-1	144	48	
2:	1	3	4	4	0	0	0	0	3	4:	-1-1	144	16	
3:	1	1	1	54	1	1	0	0	0	0:	1 1	216	96	9/32
3:	1	1	5	7	0	1	1	1	1	4:	1 1	216	16	
3:	1	1	6	7	1	0	0	1	1	4:	1 1	216	12	
3:	1	2	3	7	0	1	2	1	2	0:	1 1	216	8	
4:	1	1	2	16	0	1	0	0	0	2:	-1-1	432	8	9/4
4:	1	1	4	7	0	0	0	1	0	0:	-1-1	432	16	
4:	1	1	4	8	0	0	0	1	0	4:	-1-1	432	16	
4:	1	1	5	7	0	1	0	0	0	5:	-1-1	432	8	
4:	1	2	2	7	0	0	0	0	1	-1:	-1-1	432	8	
4:	1	2	3	6	1	1	0	1	0	0:	-1-1	432	4	
4:	1	2	4	4	0	0	1	0	2	0:	-1-1	432	4	
4:	1	2	4	4	1	0	0	0	1	0:	-1-1	432	8	
4:	1	2	4	5	1	0	1	1	2	3:	-1-1	432	4	
4:	1	3	3	4	1	1	-1	0	1	-1:	-1-1	432	8	
4:	2	2	2	5	1	1	0	0	2	2:	-1-1	432	4	
4:	2	2	3	3	1	1	-1	1	0	-1:	-1-1	432	2	
5:	1	1	2	22	1	1	0	1	0	0:	1 1	432	12	9/4
5:	1	1	3	14	1	1	0	0	1	1:	1 1	432	12	
5:	1	1	6	7	1	1	0	1	1	3:	1 1	432	12	
5:	1	2	3	6	1	0	1	0	-1	2:	1 1	432	4	
5:	1	2	3	6	0	0	2	1	1	2:	1 1	432	4	
5:	1	2	3	7	1	0	2	1	2	0:	1 1	432	4	
5:	1	2	4	4	0	0	0	1	-1	2:	1 1	432	4	
5:	1	3	3	4	0	0	2	1	2	0:	1 1	432	4	
5:	1	3	3	4	1	0	2	1	0	0:	1 1	432	4	
5:	2	3	3	3	1	1	-2	1	-3	1:	1 1	432	2	
6:	1	1	3	12	1	0	0	0	0	0:	1 1	48	48	1/12
6:	1	3	3	4	0	0	0	1	3	0:	1 1	48	16	
7:	1	1	3	9	0	0	0	0	0	0:	-1-1	36	32	1/32
8:	1	1	4	10	1	0	0	0	0	4:	-1-1	72	48	1/32
8:	2	2	3	3	0	0	0	0	0	3:	-1-1	72	96	
9:	1	1	5	6	0	1	1	0	0	0:	1 1	72	32	1/32

GENUS#	F11	F22	F33	F44	F12	F13	F23	F14	F24	F34	HASSE	SYM	LEVEL	AUTOS	MASS
10:	1	1	5	8	1	0	0	0	0	4	-1	-1	144	24	1/2
10:	1	2	2	8	0	0	1	0	2	2	-1	-1	144	8	
10:	1	3	3	5	0	0	3	0	3	0	-1	-1	144	24	
10:	1	3	3	5	1	0	3	0	2	0	-1	-1	144	8	
10:	2	2	2	4	1	1	1	0	0	0	-1	-1	144	24	
10:	2	3	3	3	1	1	-1	1	3	2	-1	-1	144	8	
11:	1	2	2	10	1	1	0	1	2	2	-1	-1	72	16	3/32
11:	2	2	3	4	0	0	0	2	2	3	-1	-1	72	32	
12:	1	2	2	11	1	0	2	0	1	0	-1	-1	216	8	27/32
12:	1	2	2	8	0	0	0	1	2	2	-1	-1	216	32	
12:	1	2	4	5	1	0	0	0	1	4	-1	-1	216	8	
12:	2	2	2	4	1	1	0	1	0	0	-1	-1	216	4	
12:	2	2	2	7	2	2	0	1	1	1	-1	-1	216	16	
12:	2	2	3	4	2	1	0	0	2	1	-1	-1	216	4	
13:	1	3	3	3	0	0	0	0	0	0	1	1	12	96	1/96
14:	2	2	2	6	2	1	-1	0	0	0	-1	-1	24	32	1/32
15:	1	1	2	14	0	0	0	0	0	2	1	1	108	32	9/32
15:	1	2	3	6	0	0	2	0	2	0	1	1	108	8	
15:	2	2	3	4	0	2	2	2	0	0	1	1	108	8	
16:	1	1	2	14	0	0	0	1	1	0	1	1	216	32	27/32
16:	1	1	4	8	0	1	1	0	0	0	1	1	216	16	
16:	1	2	2	8	1	0	0	1	0	0	1	1	216	8	
16:	1	2	3	6	1	0	1	0	2	2	1	1	216	4	
16:	2	2	2	4	0	0	0	0	1	2	1	1	216	8	
16:	2	2	3	4	0	2	1	0	2	3	1	1	216	4	
17:	1	1	2	18	0	1	1	0	0	0	1	1	72	32	3/32
17:	2	2	2	5	2	0	0	2	1	0	1	1	72	16	
18:	1	1	2	18	1	0	0	0	0	0	-1	-1	72	48	1/16
18:	3	3	3	3	3	1	0	1	3	3	-1	-1	72	24	
19:	1	1	6	6	0	0	0	0	0	6	1	1	36	96	1/32
19:	2	2	2	5	2	0	0	0	0	2	1	1	36	48	
20:	1	1	6	6	0	0	0	0	0	0	1	1	24	96	1/96
21:	1	2	2	7	0	0	0	1	0	0	-1	-1	216	32	9/32
21:	1	2	3	5	0	1	0	0	0	1	-1	-1	216	8	
21:	1	3	3	5	1	1	-1	0	3	1	-1	-1	216	8	
22:	1	2	2	9	0	0	2	0	0	0	1	1	36	48	1/16
22:	2	3	3	3	2	2	0	2	0	0	1	1	36	24	
23:	1	2	3	5	0	0	0	0	2	0	-1	-1	36	16	1/16
24:	1	2	3	6	0	0	0	1	2	3	-1	-1	72	16	3/16
24:	2	2	3	3	1	1	1	1	1	2	-1	-1	72	8	
25:	1	2	3	6	0	1	2	0	0	0	1	1	72	16	1/16
26:	1	2	3	6	0	1	1	0	0	3	1	1	144	4	1/2
26:	2	3	3	3	2	1	0	2	0	3	1	1	144	4	
27:	1	2	4	4	0	1	0	1	0	2	1	1	72	16	3/16
27:	2	2	3	4	2	1	-1	0	0	2	1	1	72	8	
28:	1	2	4	5	1	0	2	1	1	2	1	1	144	4	1/4
29:	1	3	4	4	0	1	3	1	3	2	1	1	24	32	1/32
30:	2	2	2	5	1	1	1	2	2	2	-1	-1	48	12	1/12
31:	2	2	3	3	2	0	0	0	0	0	-1	-1	12	96	1/96
32:	2	3	3	3	0	0	0	0	3	3	-1	-1	24	96	1/96

D= 433

GENUS#:	F11	F22	F33	F44	F12	F13	F23	F14	F24	F34	HASSE	SYM	LEVEL	AUTOS	MASS
1:	1	1	2	16	0	1	0	1	1	0	-1	-1	433	8	163/24
1:	1	1	2	22	1	1	0	0	1	0	-1	-1	433	12	
1:	1	1	4	8	0	1	1	0	1	0	-1	-1	433	8	
1:	1	1	5	6	0	1	0	0	1	1	-1	-1	433	8	
1:	1	1	5	8	1	0	0	1	1	3	-1	-1	433	12	
1:	1	2	2	11	1	0	2	0	1	1	-1	-1	433	4	
1:	1	2	2	8	1	0	0	0	-1	1	-1	-1	433	4	
1:	1	2	2	8	0	0	1	1	2	0	-1	-1	433	4	
1:	1	2	3	6	1	1	0	1	1	0	-1	-1	433	4	
1:	1	2	3	6	0	1	1	1	0	3	-1	-1	433	4	
1:	1	2	3	6	1	0	1	1	2	0	-1	-1	433	4	
1:	1	2	3	6	0	0	1	1	2	3	-1	-1	433	4	
1:	1	2	3	6	0	1	1	1	2	0	-1	-1	433	4	
1:	1	2	3	6	1	0	1	1	1	2	-1	-1	433	4	
1:	1	2	4	4	0	1	1	1	0	0	-1	-1	433	4	
1:	1	2	4	4	0	1	0	0	1	2	-1	-1	433	4	
1:	1	2	4	5	1	0	0	1	2	3	-1	-1	433	4	
1:	1	3	3	4	1	0	2	0	-1	1	-1	-1	433	4	
1:	2	2	2	4	1	1	0	0	1	0	-1	-1	433	2	

GENUS#	F11	F22	F33	F44	F12	F13	F23	F14	F24	F34:	HASSE SYM	LEVEL	AUTOS	MASS
1:	2	2	3	3	0	1	0	2	1	0:	-1-1	433	2	
1:	2	2	3	3	1	1	1	-1	1	-1:	-1-1	433	2	
1:	2	2	3	3	1	1	1	0	1	2:	-1-1	433	2	
1:	2	2	3	4	2	1	-1	0	1	1:	-1-1	433	2	
1:	2	3	3	3	1	1	-3	-1	2	-2:	-1-1	433	2	

D= 436

GENUS#	F11	F22	F33	F44	F12	F13	F23	F14	F24	F34:	HASSE SYM	LEVEL	AUTOS	MASS
1:	1	1	1	28	0	0	0	1	1	1:	1 1	218	96	45/32
1:	1	1	5	6	0	0	0	1	1	1:	1 1	218	16	
1:	1	2	2	10	0	0	2	1	2	0:	1 1	218	24	
1:	1	2	4	5	1	1	0	0	1	3:	1 1	218	4	
1:	1	2	4	5	0	1	2	0	2	3:	1 1	218	8	
1:	2	2	2	6	2	1	0	1	2	2:	1 1	218	4	
1:	2	2	2	7	2	2	0	1	1	0:	1 1	218	12	
1:	2	2	3	3	0	1	0	1	2	0:	1 1	218	4	
1:	2	2	3	4	2	2	0	0	0	1:	1 1	218	12	
1:	2	2	3	4	2	1	0	0	0	3:	1 1	218	4	
2:	1	1	1	55	1	1	0	1	0	0:	-1-1	218	96	45/32
2:	1	1	2	19	1	0	0	1	0	2:	-1-1	218	24	
2:	1	1	3	11	0	1	1	0	0	1:	-1-1	218	16	
2:	1	1	3	14	1	1	0	0	0	2:	-1-1	218	12	
2:	1	1	5	8	1	1	0	0	0	2:	-1-1	218	12	
2:	1	2	3	7	0	1	2	1	0	3:	-1-1	218	8	
2:	1	3	3	4	1	0	1	0	2	2:	-1-1	218	4	
2:	1	3	3	4	1	1	0	0	2	0:	-1-1	218	4	
2:	1	3	3	5	1	0	3	1	2	1:	-1-1	218	4	
2:	2	3	3	3	2	0	3	0	-1	1:	-1-1	218	4	
3:	1	1	2	14	0	0	0	0	1	1:	-1-1	436	8	45/8
3:	1	1	3	10	0	1	0	0	0	1:	-1-1	436	8	
3:	1	1	3	14	1	1	0	1	0	0:	-1-1	436	12	
3:	1	1	4	11	1	1	0	0	1	3:	-1-1	436	12	
3:	1	1	5	6	0	1	0	1	0	0:	-1-1	436	8	
3:	1	1	6	8	1	1	0	0	1	5:	-1-1	436	12	
3:	1	2	2	9	1	0	1	0	0	0:	-1-1	436	4	
3:	1	2	3	5	0	0	0	1	1	1:	-1-1	436	4	
3:	1	2	3	5	0	0	1	0	1	-1:	-1-1	436	4	
3:	1	2	3	5	0	0	0	0	1	-2:	-1-1	436	4	
3:	1	2	3	6	1	1	0	0	1	1:	-1-1	436	4	
3:	1	2	3	7	1	0	2	1	1	2:	-1-1	436	4	
3:	1	3	3	4	1	0	0	0	2	2:	-1-1	436	4	
3:	1	3	3	4	0	0	2	0	1	-2:	-1-1	436	4	
3:	1	3	3	4	1	0	1	-1	1	-1:	-1-1	436	4	
3:	1	3	3	4	0	0	2	1	2	1:	-1-1	436	4	
3:	1	3	3	5	1	0	3	0	0	2:	-1-1	436	4	
3:	1	3	4	4	1	0	1	2		4:	-1-1	436	4	
3:	2	2	3	3	0	1	1	0	2	0:	-1-1	436	2	
3:	2	2	3	4	1	2	0	0	2	2:	-1-1	436	2	
3:	2	2	3	4	1	2	-1	2	0	0:	-1-1	436	2	
3:	2	3	3	3	2	1	-1	1	1	3:	-1-1	436	2	

D= 437

GENUS#	F11	F22	F33	F44	F12	F13	F23	F14	F24	F34:	HASSE SYM	LEVEL	AUTOS	MASS
1:	1	1	1	37	1	0	0	1	0	1:	-1 1-1	437	24	77/48
1:	1	1	1	55	1	1	0	0	1	0:	-1 1-1	437	48	
1:	1	1	3	14	1	1	0	0	1	0:	-1 1-1	437	12	
1:	1	1	4	10	1	0	0	1	1	3:	-1 1-1	437	12	
1:	1	1	5	7	0	1	1	1	0	4:	-1 1-1	437	8	
1:	1	2	2	9	1	0	1	1	2	0:	-1 1-1	437	4	
1:	1	2	4	4	0	1	1	0	-1	1:	-1 1-1	437	4	
1:	1	2	4	5	1	0	1	0	2	3:	-1 1-1	437	4	
1:	1	3	3	4	1	0	1	1	1	2:	-1 1-1	437	4	
1:	1	3	3	5	1	0	2	1	0	3:	-1 1-1	437	4	
2:	1	1	2	16	0	1	0	0	1	1:	-1-1 1	437	8	77/48
2:	1	1	2	22	1	1	0	0	0	1:	-1-1 1	437	12	
2:	1	1	3	11	0	1	0	0	1	3:	-1-1 1	437	8	
2:	1	1	4	10	1	1	0	0	0	1:	-1-1 1	437	12	
2:	1	1	5	6	0	1	0	0	1	0:	-1-1 1	437	16	
2:	1	1	6	7	1	1	0	0	1	2:	-1-1 1	437	12	
2:	1	1	7	7	1	1	0	1	1	6:	-1-1 1	437	24	
2:	1	2	3	5	0	0	1	1	0	0:	-1-1 1	437	8	
2:	1	2	3	7	1	0	2	0	2	1:	-1-1 1	437	4	

#	F11	F22	F33	F44	F12	F13	F23	F14	F24	F34:	HASSE SYM	LEVEL	AUTOS	MASS
2:	1	2	3	7	1	1	0	1	2	3:	-1-1 1	437	8	
2:	1	3	3	4	1	1	-1	0	0	1:	-1-1 1	437	4	
2:	2	2	3	3	1	1	0	0	1	-2:	-1-1 1	437	4	

D= 440

GENUS#:	F11	F22	F33	F44	F12	F13	F23	F14	F24	F34:	HASSE SYM	LEVEL	AUTOS	MASS
1:	1	1	1	37	1	0	0	1	0	0:	1-1-1	440	24	103/96
1:	1	1	3	10	0	0	0	1	1	2:	1-1-1	440	16	
1:	1	1	5	6	0	0	0	1	1	0:	1-1-1	440	32	
1:	1	1	5	7	0	0	0	0	1	5:	1-1-1	440	16	
1:	1	2	4	5	1	0	1	1	0	3:	1-1-1	440	4	
1:	2	2	2	5	2	0	0	0	0	0:	1-1-1	440	8	
1:	2	2	2	5	1	1	-1	2	0	0:	1-1-1	440	4	
1:	2	2	3	3	0	2	1	0	0	1:	1-1-1	440	4	
2:	1	1	1	55	1	1	0	0	0	0:	-1-1 1	440	96	103/96
2:	1	1	2	14	0	0	0	0	1	0:	-1-1 1	440	16	
2:	1	1	2	16	0	1	0	1	0	0:	-1-1 1	440	8	
2:	1	1	5	7	0	1	1	0	0	4:	-1-1 1	440	16	
2:	1	2	2	11	1	0	2	0	0	0:	-1-1 1	440	16	
2:	1	2	2	9	1	0	1	0	0	2:	-1-1 1	440	4	
2:	1	2	3	5	0	1	0	0	0	0:	-1-1 1	440	16	
2:	1	2	4	4	0	0	0	0	2	1:	-1-1 1	440	8	
2:	1	2	4	4	0	0	0	0	0	3:	-1-1 1	440	16	
2:	2	2	4	4	1	0	0	1	-1	0:	-1-1 1	440	8	
2:	2	3	3	3	2	2	1	0	-2	2:	-1-1 1	440	8	
3:	1	1	2	19	0	1	1	1	1	0:	1 1 1	440	16	103/96
3:	1	1	3	10	0	1	0	0	0	0:	1 1 1	440	16	
3:	1	1	3	11	0	1	1	0	0	0:	1 1 1	440	32	
3:	1	1	4	10	1	1	0	0	0	0:	1 1 1	440	24	
3:	1	1	5	6	0	1	0	0	0	2:	1 1 1	440	8	
3:	1	2	3	6	1	1	0	0	-1	1:	1 1 1	440	4	
3:	1	3	4	4	0	0	2	1	2	-3:	1 1 1	440	8	
3:	1	3	4	4	1	1	0	1	3	3:	1 1 1	440	4	
3:	2	2	3	4	0	2	0	1	2	3:	1 1 1	440	8	
4:	1	1	1	28	0	0	0	1	1	0:	-1 1-1	440	32	103/96
4:	1	1	2	22	1	1	0	0	0	0:	-1 1-1	440	24	
4:	1	1	4	8	0	1	0	1	0	3:	-1 1-1	440	8	
4:	1	1	4	9	0	1	1	0	0	4:	-1 1-1	440	16	
4:	1	2	2	7	0	0	0	0	0	1:	-1 1-1	440	8	
4:	1	2	2	8	1	0	0	0	1	0:	-1 1-1	440	8	
4:	1	2	3	6	0	0	2	0	2	1:	-1 1-1	440	16	
4:	1	2	4	4	0	0	1	0	-1	2:	-1 1-1	440	8	
4:	1	2	4	5	0	1	2	1	2	1:	-1 1-1	440	8	
4:	1	2	4	5	1	1	0	1	2	2:	-1 1-1	440	4	

D= 441

GENUS#:	F11	F22	F33	F44	F12	F13	F23	F14	F24	F34:	HASSE SYM	LEVEL	AUTOS	MASS
1:	1	1	1	37	1	0	0	0	0	1:	-1-1 1	147	48	7/12
1:	1	1	2	19	1	0	1	1	0	2:	-1-1 1	147	16	
1:	1	2	2	10	1	1	0	0	2	1:	-1-1 1	147	8	
1:	1	2	4	5	1	0	2	1	0	1:	-1-1 1	147	4	
1:	2	2	3	4	1	0	0	2	2	3:	-1-1 1	147	8	
2:	1	1	2	16	0	1	0	0	1	0:	-1 1-1	63	16	3/8
2:	1	2	4	4	1	0	0	0	0	1:	-1 1-1	63	16	
2:	2	2	2	5	1	1	0	-1	2	1:	-1 1-1	63	4	
3:	1	1	3	13	1	0	0	0	0	3:	-1-1 1	147	48	7/12
3:	1	1	4	8	0	1	0	1	1	2:	-1-1 1	147	8	
3:	1	3	4	4	0	1	0	3	0	0:	-1-1 1	147	8	
3:	2	2	2	4	1	0	0	1	1	1:	-1-1 1	147	4	
3:	2	2	2	7	2	1	-1	1	2	1:	-1-1 1	147	16	
4:	1	1	4	9	0	1	1	1	0	4:	-1 1-1	63	16	3/8
4:	1	2	2	8	1	0	0	0	0	1:	-1 1-1	63	8	
4:	1	3	4	4	1	0	3	0	3	1:	-1 1-1	63	8	
4:	2	3	3	3	1	1	-1	1	-2	-2:	-1 1-1	63	16	
5:	1	1	7	7	1	0	0	0	0	7:	-1-1 1	21	144	1/9
5:	1	2	3	6	1	0	0	0	0	3:	-1-1 1	21	16	
5:	2	2	4	4	2	1	-1	1	2	3:	-1-1 1	21	24	
6:	1	2	3	6	0	1	1	1	2	1:	-1 1-1	21	8	1/4
6:	2	2	3	3	1	1	1	1	1	-1:	-1 1-1	21	8	

D= 444

GENUS#:	F11	F22	F33	F44	F12	F13	F23	F14	F24	F34:	HASSE SYM	LEVEL	AUTOS	MASS

GENUS#	F11	F22	F33	F44	F12	F13	F23	F14	F24	F34:	HASSE SYM	LEVEL	AUTOS	MASS
1:	1	1	1	28	0	0	0	1	0	0:	-1-1 1	444	32	61/48
1:	1	1	1	37	1	0	0	0	0	0:	-1-1 1	444	48	
1:	1	1	2	16	0	1	0	0	0	1:	-1-1 1	444	8	
1:	1	1	3	10	0	0	0	0	0	3:	-1-1 1	444	32	
1:	1	1	3	12	0	1	1	0	0	3:	-1-1 1	444	16	
1:	1	1	4	7	0	0	0	0	0	1:	-1-1 1	444	16	
1:	1	1	4	8	0	0	0	1	1	3:	-1-1 1	444	16	
1:	1	2	4	5	0	0	2	1	2	3:	-1-1 1	444	8	
1:	2	2	2	5	1	0	0	2	2	2:	-1-1 1	444	8	
1:	2	2	2	5	1	1	1	0	2	2:	-1-1 1	444	4	
1:	2	2	3	3	0	1	0	0	2	1:	-1-1 1	444	4	
1:	2	2	3	4	2	0	0	1	1	3:	-1-1 1	444	8	
2:	1	1	1	56	1	1	0	1	0	0:	-1 1-1	444	96	61/48
2:	1	1	2	14	0	0	0	0	0	1:	-1 1-1	444	16	
2:	1	1	2	19	0	1	1	1	1	1:	-1 1-1	444	32	
2:	1	1	4	8	0	1	0	0	0	3:	-1 1-1	444	8	
2:	1	1	4	8	0	1	1	0	0	1:	-1 1-1	444	16	
2:	1	1	5	6	0	0	0	1	0	2:	-1 1-1	444	8	
2:	1	1	5	6	0	0	0	0	0	3:	-1 1-1	444	16	
2:	1	1	6	8	1	0	0	1	0	6:	-1 1-1	444	24	
2:	1	2	2	9	0	0	0	0	2	2:	-1 1-1	444	8	
2:	1	2	3	5	0	0	1	0	1	1:	-1 1-1	444	4	
2:	1	3	4	4	1	1	-1	1	3	2:	-1 1-1	444	4	
2:	1	3	4	4	1	0	3	1	3	1:	-1 1-1	444	8	
3:	1	1	2	19	1	0	0	0	0	2:	1-1-1	444	48	61/48
3:	1	2	2	9	1	0	1	1	0	1:	1-1-1	444	4	
3:	1	2	3	5	0	0	0	1	1	0:	1-1-1	444	8	
3:	1	2	3	6	0	1	0	1	2	2:	1-1-1	444	8	
3:	2	2	2	7	2	1	-1	2	0	0:	1-1-1	444	8	
3:	2	2	3	3	1	0	0	-1	-1	2:	1-1-1	444	4	
3:	2	2	3	3	1	0	0	2	0	0:	1-1-1	444	4	
3:	2	3	3	3	0	0	3	2	2	2:	1-1-1	444	8	
4:	1	1	3	13	1	0	0	1	0	2:	1 1 1	444	12	61/48
4:	1	2	2	10	0	0	2	-1	1	-1:	1 1 1	444	16	
4:	1	2	3	6	0	1	1	0	2	2:	1 1 1	444	4	
4:	1	2	4	5	1	1	0	1	1	3:	1 1 1	444	4	
4:	1	2	4	5	0	1	2	1	0	3:	1 1 1	444	8	
4:	1	3	3	4	1	1	0	1	0	0:	1 1 1	444	8	
4:	1	3	3	4	0	0	0	1	2	2:	1 1 1	444	8	
4:	2	3	3	3	2	2	0	1	2	2:	1 1 1	444	4	

D= 445

GENUS#:	F11	F22	F33	F44	F12	F13	F23	F14	F24	F34:	HASSE SYM	LEVEL	AUTOS	MASS
1:	1	1	1	56	1	1	0	0	1	0:	-1-1 1	445	48	55/24
1:	1	1	2	23	1	1	0	1	1	2:	-1-1 1	445	24	
1:	1	1	3	12	0	1	1	1	0	3:	-1-1 1	445	16	
1:	1	1	3	14	1	1	0	0	0	1:	-1-1 1	445	12	
1:	1	1	6	7	1	1	0	1	1	2:	-1-1 1	445	12	
1:	1	2	3	6	0	1	0	1	1	3:	-1-1 1	445	4	
1:	1	2	3	7	0	1	2	0	1	3:	-1-1 1	445	4	
1:	1	2	3	7	1	0	2	1	0	1:	-1-1 1	445	4	
1:	1	3	3	4	1	0	1	0	2	-1:	-1-1 1	445	4	
1:	2	2	2	5	1	1	-1	1	0	2:	-1-1 1	445	4	
1:	2	3	3	3	2	1	-1	2	3	0:	-1-1 1	445	2	
1:	2	3	3	3	2	2	1	1	-2	0:	-1-1 1	445	4	
2:	1	1	2	19	1	0	0	1	0	1:	-1 1-1	445	12	55/24
2:	1	1	5	7	0	1	0	1	1	4:	-1 1-1	445	8	
2:	1	1	5	8	1	1	0	0	0	1:	-1 1-1	445	12	
2:	1	1	5	9	1	1	0	1	1	4:	-1 1-1	445	12	
2:	1	1	5	9	1	0	0	1	0	5:	-1 1-1	445	24	
2:	1	2	2	8	0	0	1	1	1	-1:	-1 1-1	445	8	
2:	1	2	3	6	1	1	0	0	1	0:	-1 1-1	445	4	
2:	1	2	4	4	0	1	1	0	1	1:	-1 1-1	445	4	
2:	1	2	5	5	1	0	0	0	2	5:	-1 1-1	445	8	
2:	1	3	3	4	1	1	0	1	1	0:	-1 1-1	445	8	
2:	1	3	3	4	0	0	1	1	2	2:	-1 1-1	445	8	
2:	1	3	3	5	1	0	2	0	3	3:	-1 1-1	445	4	
2:	1	3	3	5	1	0	3	0	2	1:	-1 1-1	445	8	
2:	1	3	3	5	1	1	0	0	3	2:	-1 1-1	445	8	
2:	2	3	3	3	1	0	3	0	2	-1:	-1 1-1	445	4	

D= 448

GENUS#	F11	F22	F33	F44	F12	F13	F23	F14	F24	F34	HASSE	SYM	LEVEL	AUTOS	MASS
1:	1	1	1	28	0	0	0	0	0	0:	-1-1		112	96	1/12
1:	1	2	4	5	0	0	0	0	2	4:	-1-1		112	16	
1:	2	2	2	7	2	2	0	0	0	0:	-1-1		112	96	
2:	1	1	1	56	1	1	0	0	0	0:	-1-1		224	96	1/3
2:	1	1	2	15	0	0	0	1	1	2:	-1-1		224	32	
2:	1	1	3	14	1	1	0	0	0	0:	-1-1		224	24	
2:	1	2	3	7	0	1	2	1	2	1:	-1-1		224	8	
2:	2	3	3	3	2	0	3	0	1	2:	-1-1		224	8	
3:	1	1	2	14	0	0	0	0	0	0:	-1-1		56	32	1/8
3:	1	2	2	7	0	0	0	0	0	0:	-1-1		56	32	
3:	2	2	3	4	0	2	2	0	0	2:	-1-1		56	16	
4:	1	1	2	16	0	1	0	0	0	0:	-1-1		448	16	4/3
4:	1	1	5	8	1	0	0	1	1	2:	-1-1		448	12	
4:	1	1	6	6	0	1	0	1	0	5:	-1-1		448	16	
4:	1	2	2	8	0	0	1	0	2	0:	-1-1		448	4	
4:	1	2	3	5	0	0	1	0	1	0:	-1-1		448	4	
4:	1	2	4	5	0	0	2	0	1	4:	-1-1		448	8	
4:	1	3	3	4	1	0	1	1	2	1:	-1-1		448	4	
4:	1	3	4	4	1	0	0	1	2	4:	-1-1		448	8	
4:	2	2	3	3	0	1	1	1	1	-1:	-1-1		448	8	
5:	1	1	2	19	1	0	0	1	0	0:	1 1		224	24	1/3
5:	1	1	5	8	1	1	0	0	0	0:	1 1		224	24	
5:	1	2	3	5	0	0	0	1	0	1:	1 1		224	8	
5:	1	3	3	5	1	0	3	0	1	2:	1 1		224	8	
6:	1	1	2	23	1	1	0	0	1	1:	1 1		448	12	4/3
6:	1	2	2	9	1	0	1	1	1	1:	1 1		448	4	
6:	1	2	3	6	0	0	2	1	1	0:	1 1		448	4	
6:	1	2	3	6	0	1	1	1	1	-1:	1 1		448	4	
6:	2	3	3	3	1	0	2	2	3	2:	1 1		448	2	
7:	1	1	3	11	0	1	0	0	0	3:	-1-1		448	8	4/3
7:	1	1	4	11	1	1	0	1	1	3:	-1-1		448	12	
7:	1	2	4	5	1	0	0	0	0	4:	-1-1		448	16	
7:	1	3	3	4	0	0	1	0	2	-2:	-1-1		448	8	
7:	1	3	3	4	1	0	1	0	1	-2:	-1-1		448	4	
7:	1	3	3	5	1	1	-1	0	2	2:	-1-1		448	16	
7:	2	2	2	5	1	1	0	1	2	2:	-1-1		448	8	
7:	2	2	3	3	1	1	0	1	0	2:	-1-1		448	4	
7:	2	3	3	3	1	0	3	0	-1	2:	-1-1		448	4	
8:	1	1	3	15	1	1	0	1	1	3:	-1-1		112	24	1/6
8:	1	3	3	4	1	1	-1	0	0	0:	-1-1		112	16	
8:	3	3	3	3	2	2	-2	1	1	3:	-1-1		112	16	
9:	1	1	4	7	0	0	0	0	0	0:	-1-1		112	32	1/4
9:	1	2	2	0	0	0	0	0	2	2:	-1-1		112	32	
9:	1	2	4	4	0	0	0	0	2	0:	-1-1		112	16	
9:	1	3	3	4	0	0	2	0	2	2:	-1-1		112	8	
10:	1	1	6	7	1	1	0	0	1	1:	1 1		448	12	4/3
10:	1	2	3	6	1	0	1	0	1	2:	1 1		448	4	
10:	1	2	3	7	1	0	2	0	0	2:	1 1		448	4	
10:	1	3	3	4	1	0	0	1	1	2:	1 1		448	4	
10:	2	2	3	4	1	2	-1	0	2	0:	1 1		448	2	
11:	1	1	7	7	1	1	0	0	1	5:	1 1		112	24	1/6
11:	1	3	3	5	1	0	3	1	1	1:	1 1		112	8	
12:	1	2	2	10	0	0	2	0	2	0:	1 1		112	24	1/12
12:	2	2	3	4	2	2	0	0	0	0:	1 1		112	24	
13:	1	2	2	12	1	0	2	0	2	0:	1 1		224	8	1/1
13:	1	2	4	5	0	1	2	0	0	3:	1 1		224	8	
13:	1	2	4	5	1	1	0	0	1	-3:	1 1		224	4	
13:	2	2	3	3	1	1	-1	1	0	0:	1 1		224	4	
13:	2	2	3	4	2	1	0	2	0	0:	1 1		224	4	
14:	1	2	4	4	1	0	0	0	0	0:	-1-1		112	32	1/2
14:	1	3	3	4	0	0	2	-1	-1	1:	-1-1		112	8	
14:	1	4	4	4	1	0	4	0	4	0:	-1-1		112	32	
14:	2	2	2	4	1	1	0	0	0	0:	-1-1		112	8	
14:	2	2	3	3	0	1	1	1	1	2:	-1-1		112	8	
14:	2	3	3	3	1	1	2	1	-2	2:	-1-1		112	16	
15:	1	1	4	8	0	1	1	0	0	0:	-1-1		224	32	1/1
15:	1	1	6	6	0	1	1	1	1	4:	-1-1		224	32	
15:	1	2	2	8	1	0	0	0	0	0:	-1-1		224	16	
15:	1	2	4	4	0	1	0	0	0	2:	-1-1		224	8	
15:	1	2	4	4	0	1	0	1	0	0:	-1-1		224	16	

GENUS#	F11	F22	F33	F44	F12	F13	F23	F14	F24	F34:	HASSE SYM	LEVEL	AUTOS	MASS
15:	2	2	2	6	2	1	0	0	2	1:	-1 -1	224	4	
15:	2	2	3	3	0	1	0	1	0	-2:	-1 -1	224	8	
15:	2	2	3	4	2	1	-1	1	1	-1:	-1 -1	224	4	
15:	2	2	3	4	0	2	2	2	0	1:	-1 -1	224	16	
16:	1	1	4	8	0	0	0	0	0	4:	-1 -1	28	32	1/24
16:	2	2	2	5	0	0	0	2	2	2:	-1 -1	28	96	
17:	1	2	3	5	0	0	0	0	0	2:	1 1	56	8	1/8
18:	1	2	3	6	0	0	2	0	0	2:	1 1	112	8	1/4
18:	2	3	3	3	2	2	0	0	2	-2:	1 1	112	8	
19:	1	2	3	6	1	0	1	1	0	1:	1 1	112	4	1/2
19:	2	2	3	4	1	1	-1	0	2	2:	1 1	112	4	
20:	2	3	3	3	2	2	0	2	2	0:	1 1	28	24	1/24

D= 449

GENUS#:	F11	F22	F33	F44	F12	F13	F23	F14	F24	F34:	HASSE SYM	LEVEL	AUTOS	MASS
1:	1	1	1	38	1	0	0	1	0	1:	-1 -1	449	24	51/8
1:	1	1	2	19	0	1	1	1	0	0:	-1 -1	449	8	
1:	1	1	4	8	0	1	0	1	1	0:	-1 -1	449	8	
1:	1	1	4	8	0	1	0	0	1	2:	-1 -1	449	8	
1:	1	1	5	7	0	1	0	0	1	4:	-1 -1	449	8	
1:	1	1	6	7	1	0	0	0	0	3:	-1 -1	449	12	
1:	1	2	2	10	1	1	0	1	2	1:	-1 -1	449	4	
1:	1	2	2	9	1	0	1	0	1	-1:	-1 -1	449	4	
1:	1	2	3	6	1	0	0	0	2	1:	-1 -1	449	4	
1:	1	2	3	6	1	1	0	0	0	1:	-1 -1	449	4	
1:	1	2	3	7	1	0	2	0	-1	1:	-1 -1	449	4	
1:	1	2	4	4	0	0	1	1	0	0:	-1 -1	449	4	
1:	1	2	4	5	1	0	2	0	-1	1:	-1 -1	449	4	
1:	1	2	4	5	1	0	2	1	1	1:	-1 -1	449	4	
1:	1	2	4	5	1	0	1	0	-1	3:	-1 -1	449	4	
1:	1	2	4	5	1	1	0	0	2	1:	-1 -1	449	4	
1:	1	2	4	5	1	1	0	1	0	3:	-1 -1	449	4	
1:	1	2	4	5	0	1	2	-1	-1	1:	-1 -1	449	4	
1:	1	3	3	4	1	0	2	0	1	1:	-1 -1	449	4	
1:	2	2	2	4	1	0	0	0	-1	1:	-1 -1	449	2	
1:	2	2	3	4	1	1	0	0	2	3:	-1 -1	449	2	
1:	2	2	3	4	0	2	1	1	2	3:	-1 -1	449	2	
1:	2	2	3	4	2	1	0	-1	1	1:	-1 -1	449	2	
1:	2	2	3	4	1	2	0	2	1	3:	-1 -1	449	2	

D= 452

GENUS#:	F11	F22	F33	F44	F12	F13	F23	F14	F24	F34:	HASSE SYM	LEVEL	AUTOS	MASS
1:	1	1	1	29	0	0	0	1	1	1:	1 1	226	96	3/8
1:	1	1	1	57	1	1	0	1	0	0:	1 1	226	96	
1:	1	1	5	7	0	1	1	1	1	3:	1 1	226	16	
1:	1	1	5	9	1	1	0	0	1	3:	1 1	226	12	
1:	1	2	3	6	0	1	0	0	2	2:	1 1	226	8	
1:	2	2	3	5	2	2	0	0	2	1:	1 1	226	12	
2:	1	1	1	38	1	0	0	1	0	0:	-1 -1	452	24	9/2
2:	1	1	2	23	1	1	0	1	0	0:	-1 -1	452	12	
2:	1	1	3	10	0	0	0	0	1	2:	-1 -1	452	8	
2:	1	1	4	10	1	0	0	1	1	2:	-1 -1	452	12	
2:	1	1	5	6	0	1	0	0	0	1:	-1 -1	452	8	
2:	1	1	6	6	0	0	0	0	1	5:	-1 -1	452	8	
2:	1	1	6	7	1	1	0	1	0	0:	-1 -1	452	12	
2:	1	1	7	7	1	0	0	1	1	6:	-1 -1	452	12	
2:	1	2	3	5	0	0	1	0	0	1:	-1 -1	452	4	
2:	1	2	3	6	1	0	0	1	1	2:	-1 -1	452	4	
2:	1	2	3	6	0	0	2	0	1	2:	-1 -1	452	4	
2:	1	2	3	7	1	0	2	0	1	2:	-1 -1	452	4	
2:	1	2	3	7	1	1	0	0	2	2:	-1 -1	452	4	
2:	1	2	4	5	0	1	1	0	0	4:	-1 -1	452	4	
2:	1	2	4	5	1	0	2	0	0	2:	-1 -1	452	4	
2:	1	2	4	5	1	0	1	1	1	3:	-1 -1	452	4	
2:	1	2	4	5	1	0	2	1	0	0:	-1 -1	452	4	
2:	1	3	3	4	1	0	2	0	1	0:	-1 -1	452	4	
2:	1	3	3	4	0	0	1	0	3	0:	-1 -1	452	4	
2:	2	2	3	3	1	1	-1	1	0	1:	-1 -1	452	2	
2:	2	2	3	4	1	2	0	2	2	2:	-1 -1	452	2	
3:	1	1	2	19	0	1	1	0	0	1:	-1 -1	226	16	27/8
3:	1	1	3	10	0	0	0	1	1	1:	-1 -1	226	16	
3:	1	2	2	10	1	1	0	0	2	0:	-1 -1	226	4	

GENUS#	F11	F22	F33	F44	F12	F13	F23	F14	F24	F34:	HASSE	SYM	LEVEL	AUTOS	MASS
3:	1	2	2	12	1	0	2	0	0	2:	-1-1		226	8	
3:	1	2	3	6	1	0	1	0	0	2:	-1-1		226	4	
3:	1	2	3	6	0	0	2	1	0	1:	-1-1		226	8	
3:	1	2	3	6	1	0	1	1	1	1:	-1-1		226	4	
3:	1	2	4	5	1	0	2	0	1	2:	-1-1		226	4	
3:	1	2	4	6	1	0	2	1	1	4:	-1-1		226	4	
3:	2	2	2	5	2	0	0	1	1	1:	-1-1		226	4	
3:	2	2	3	4	2	1	0	0	2	0:	-1-1		226	4	
3:	2	2	3	4	0	2	1	2	2	1:	-1-1		226	4	
3:	2	2	3	4	0	1	0	2	2	3:	-1-1		226	4	
3:	2	2	3	4	1	1	-1	-1	2	1:	-1-1		226	2	
3:	2	2	4	4	2	1	-1	1	1	4:	-1-1		226	4	

D= 453

GENUS#:	F11	F22	F33	F44	F12	F13	F23	F14	F24	F34:	HASSE	SYM	LEVEL	AUTOS	MASS
1:	1	1	1	38	1	0	0	0	0	1:	-1-1	1	453	48	91/48
1:	1	1	2	17	0	1	0	0	1	2:	-1-1	1	453	8	
1:	1	1	2	19	1	0	0	0	0	1:	-1-1	1	453	24	
1:	1	1	4	10	1	0	0	0	0	3:	-1-1	1	453	24	
1:	1	1	5	8	1	0	0	0	0	3:	-1-1	1	453	24	
1:	1	2	2	8	0	0	1	1	1	1:	-1-1	1	453	8	
1:	1	2	2	9	1	0	1	1	0	0:	-1-1	1	453	4	
1:	1	3	3	5	1	1	-1	1	3	1:	-1-1	1	453	4	
1:	1	3	3	5	1	0	3	1	0	0:	-1-1	1	453	8	
1:	1	3	3	5	1	0	0	0	3	3:	-1-1	1	453	8	
1:	2	2	2	5	1	1	1	2	1	2:	-1-1	1	453	4	
1:	2	2	3	4	1	2	-1	1	1	2:	-1-1	1	453	4	
1:	2	3	3	3	1	0	2	0	3	3:	-1-1	1	453	4	
2:	1	1	1	57	1	1	0	0	1	0:	-1	1-1	453	48	91/48
2:	1	1	2	23	1	1	0	0	1	0:	-1	1-1	453	12	
2:	1	1	3	11	0	1	0	1	1	2:	-1	1-1	453	8	
2:	1	1	3	13	1	0	0	1	1	1:	-1	1-1	453	12	
2:	1	1	3	15	1	1	0	0	1	2:	-1	1-1	453	12	
2:	1	1	4	11	1	1	0	0	1	2:	-1	1-1	453	12	
2:	1	1	6	7	1	1	0	0	1	0:	-1	1-1	453	12	
2:	1	1	6	7	1	0	0	1	1	3:	-1	1-1	453	12	
2:	1	2	3	6	0	1	1	0	-1	2:	-1	1-1	453	4	
2:	1	2	3	7	1	0	2	1	1	1:	-1	1-1	453	4	
2:	1	2	4	5	0	1	1	1	0	4:	-1	1-1	453	4	
2:	1	3	3	4	1	0	1	1	2	0:	-1	1-1	453	4	
2:	1	3	3	4	1	0	2	0	0	1:	-1	1-1	453	4	

D= 456

GENUS#:	F11	F22	F33	F44	F12	F13	F23	F14	F24	F34:	HASSE	SYM	LEVEL	AUTOS	MASS
1:	1	1	1	38	1	0	0	0	0	0:	1 1	1	456	48	11/0
1:	1	1	1	57	1	1	0	0	0	0:	1 1	1	456	96	
1:	1	1	2	15	0	0	0	0	1	2:	1 1	1	456	16	
1:	1	1	3	10	0	0	0	1	1	0:	1 1	1	456	32	
1:	1	1	5	9	1	1	0	0	0	4:	1 1	1	456	12	
1:	1	1	6	7	1	0	0	0	0	4:	1 1	1	456	24	
1:	1	2	3	6	1	0	0	0	2	0:	1 1	1	456	8	
1:	1	2	4	4	0	1	1	0	0	1:	1 1	1	456	4	
1:	1	3	3	5	0	0	3	0	3	1:	1 1	1	456	8	
1:	1	3	3	5	1	0	2	-1	1	-2:	1 1	1	456	4	
1:	2	2	3	3	0	0	2	1	0	0:	1 1	1	456	8	
1:	2	3	3	3	2	1	-1	0	0	3:	1 1	1	456	4	
2:	1	1	3	12	0	1	1	1	1	0:	-1-1	1	456	16	11/8
2:	1	1	4	8	0	1	0	1	0	2:	-1-1	1	456	8	
2:	1	2	2	10	0	0	2	0	2	1:	-1-1	1	456	16	
2:	1	2	3	6	0	1	0	0	0	3:	-1-1	1	456	8	
2:	1	2	4	4	0	0	1	0	1	2:	-1-1	1	456	8	
2:	1	2	5	5	1	0	2	1	0	4:	-1-1	1	456	4	
2:	2	2	2	7	2	1	-1	2	0	1:	-1-1	1	456	8	
2:	2	2	3	3	0	2	0	1	0	1:	-1-1	1	456	4	
2:	2	2	3	4	2	1	0	-1	0	2:	-1-1	1	456	4	11/8
3:	1	1	2	19	1	0	0	0	0	0:	-1	1-1	456	48	11/8
3:	1	1	3	13	1	0	0	0	0	2:	-1	1-1	456	24	
3:	1	1	3	13	1	0	0	1	0	0:	-1	1-1	456	24	
3:	1	1	5	6	0	0	0	0	1	1:	-1	1-1	456	8	
3:	1	1	5	7	0	0	0	1	1	4:	-1	1-1	456	16	
3:	1	1	6	8	1	1	0	1	1	5:	-1	1-1	456	12	
3:	1	2	3	5	0	0	0	1	0	0:	-1	1-1	456	16	

GENUS#:	F11	F22	F33	F44	F12	F13	F23	F14	F24	F34:	HASSE SYM	LEVEL	AUTOS	MASS
3:	1	2	3	6	0	0	0	0	2	3:	-1 1-1	456	16	
3:	1	2	3	6	0	0	1	0	2	3:	-1 1-1	456	4	
3:	1	2	3	7	1	0	2	1	0	0:	-1 1-1	456	4	
3:	1	3	3	4	0	0	0	0	1	3:	-1 1-1	456	8	
3:	2	2	2	4	1	0	0	1	1	0:	-1 1-1	456	8	
3:	2	3	3	4	2	0	3	0	3	0:	-1 1-1	456	8	
4:	1	1	1	29	0	0	0	0	1	1:	1-1-1	456	32	11/8
4:	1	1	2	19	0	1	1	0	0	0:	1-1-1	456	32	
4:	1	1	5	6	0	1	0	0	0	0:	1-1-1	456	16	
4:	1	2	2	8	0	0	0	1	1	2:	1-1-1	456	8	
4:	1	2	3	6	1	1	0	0	0	0:	1-1-1	456	8	
4:	1	3	3	4	1	1	0	0	1	1:	1-1-1	456	8	
4:	2	2	3	3	1	1	0	0	0	2:	1-1-1	456	2	
4:	2	2	3	4	2	0	0	2	1	2:	1-1-1	456	8	
4:	2	2	3	5	2	2	0	2	1	3:	1-1-1	456	4	

D= 457

GENUS#:	F11	F22	F33	F44	F12	F13	F23	F14	F24	F34:	HASSE SYM	LEVEL	AUTOS	MASS
1:	1	1	2	23	1	1	0	0	0	1:	-1-1	457	12	15/2
1:	1	1	3	11	0	1	0	0	1	2:	-1-1	457	8	
1:	1	1	4	11	1	1	0	0	0	3:	-1-1	457	12	
1:	1	1	5	8	1	0	0	1	1	1:	-1-1	457	12	
1:	1	1	6	6	0	1	1	0	1	4:	-1-1	457	8	
1:	1	2	2	12	1	0	2	1	0	1:	-1-1	457	4	
1:	1	2	2	8	0	0	1	1	0	1:	-1-1	457	4	
1:	1	2	2	9	1	0	1	0	1	1:	-1-1	457	4	
1:	1	2	3	6	0	0	1	1	0	3:	-1-1	457	4	
1:	1	2	3	6	1	0	1	1	0	0:	-1-1	457	4	
1:	1	2	3	6	1	0	1	0	1	-1:	-1-1	457	4	
1:	1	2	4	4	0	1	0	0	1	1:	-1-1	457	4	
1:	1	2	4	5	0	1	0	1	1	4:	-1-1	457	4	
1:	1	2	4	5	0	1	2	0	1	3:	-1-1	457	4	
1:	1	2	4	5	1	1	0	1	2	1:	-1-1	457	4	
1:	1	2	4	5	1	1	0	0	0	3:	-1-1	457	4	
1:	1	3	3	4	0	0	1	-1	-2	1:	-1-1	457	4	
1:	1	3	3	4	1	0	0	-1	1	-1:	-1-1	457	4	
1:	1	3	4	4	1	1	0	0	1	4:	-1-1	457	4	
1:	2	2	2	6	2	1	0	2	1	2:	-1-1	457	2	
1:	2	2	3	3	0	1	0	-1	1	-2:	-1-1	457	2	
1:	2	2	3	3	1	1	-1	0	1	0:	-1-1	457	2	
1:	2	2	3	4	1	2	-1	2	0	1:	-1-1	457	2	
1:	2	2	3	4	1	2	0	0	0	3:	-1-1	457	2	
1:	2	2	3	4	2	1	-1	0	1	0:	-1-1	457	2	
1:	2	3	3	3	2	1	-1	-1	2	0:	-1-1	457	2	

D= 460

GENUS#:	F11	F22	F33	F44	F12	F13	F23	F14	F24	F34:	HASSE SYM	LEVEL	AUTOS	MASS
1:	1	1	1	58	1	1	0	1	0	0:	1-1-1	460	96	139/96
1:	1	1	2	15	0	0	0	1	1	1:	1-1-1	460	16	
1:	1	2	2	10	0	0	2	1	1	1:	1-1-1	460	8	
1:	1	2	2	9	1	0	1	0	1	0:	1-1-1	460	4	
1:	1	2	3	7	0	1	2	0	2	0:	1-1-1	460	8	
1:	2	2	2	5	1	1	-1	0	1	1:	1-1-1	460	4	
1:	2	2	3	4	0	2	2	-1	0	1:	1-1-1	460	4	
1:	2	3	3	3	2	2	1	2	0	1:	1-1-1	460	8	
1:	2	3	3	3	2	2	0	0	2	1:	1-1-1	460	4	
2:	1	1	2	17	0	1	0	0	0	2:	-1-1 1	460	8	139/96
2:	1	1	5	6	0	0	0	0	1	0:	-1-1 1	460	16	
2:	1	1	5	7	0	0	0	0	0	5:	-1-1 1	460	32	
2:	1	1	5	8	1	0	0	1	0	0:	-1-1 1	460	24	
2:	1	2	2	12	1	0	2	0	1	2:	-1-1 1	460	16	
2:	1	2	2	8	0	0	1	0	-1	1:	-1-1 1	460	8	
2:	1	2	3	5	0	0	1	0	0	0:	-1-1 1	460	8	
2:	1	2	3	6	0	1	0	1	2	0:	-1-1 1	460	8	
2:	1	2	4	5	0	0	2	1	0	3:	-1-1 1	460	8	
2:	1	3	3	4	1	1	0	0	1	-1:	-1-1 1	460	8	
2:	2	2	3	3	1	1	0	1	1	-1:	-1-1 1	460	2	
3:	1	1	1	29	0	0	0	0	0	1:	-1 1-1	460	32	139/96
3:	1	1	2	20	1	0	0	1	0	2:	-1 1-1	460	24	
3:	1	1	4	8	0	0	0	0	1	3:	-1 1-1	460	8	
3:	1	2	3	5	0	0	0	0	1	1:	-1 1-1	460	4	
3:	1	2	4	5	0	1	2	0	2	0:	-1 1-1	460	8	

GENUS#	F11	F22	F33	F44	F12	F13	F23	F14	F24	F34:	HASSE SYM	LEVEL	AUTOS	MASS
3:	1	2	4	5	0	0	1	0	2	4:	-1 1-1	460	4	
3:	1	2	4	5	1	1	0	0	2	0:	-1 1-1	460	4	
3:	1	3	4	4	1	1	-1	0	3	1:	-1 1-1	460	4	
3:	1	3	4	4	1	1	-1	1	2	3:	-1 1-1	460	8	
4:	1	1	2	23	1	1	0	0	0	0:	1 1 1	460	24	139/96
4:	1	1	3	12	0	1	1	1	1	1:	1 1 1	460	32	
4:	1	1	4	9	0	1	1	1	1	3:	1 1 1	460	16	
4:	1	2	2	9	1	0	0	1	0	2:	1 1 1	460	8	
4:	1	2	3	6	0	0	2	1	0	0:	1 1 1	460	16	
4:	1	2	3	6	0	1	1	0	2	0:	1 1 1	460	4	
4:	1	2	3	6	0	1	1	1	1	2:	1 1 1	460	4	
4:	1	2	4	5	0	1	0	1	2	3:	1 1 1	460	8	
4:	2	2	3	3	0	1	0	0	2	0:	1 1 1	460	8	
4:	2	2	3	4	2	1	-1	1	1	0:	1 1 1	460	8	
4:	2	3	3	3	1	0	2	2	3	0:	1 1 1	460	4	

D= 461

GENUS#	F11	F22	F33	F44	F12	F13	F23	F14	F24	F34:	HASSE SYM	LEVEL	AUTOS	MASS
1:	1	1	1	39	1	0	0	1	0	1:	-1-1	461	24	61/16
1:	1	1	1	58	1	1	0	0	1	0:	-1-1	461	48	
1:	1	1	2	17	0	1	0	1	1	0:	-1-1	461	8	
1:	1	1	3	11	0	1	0	1	1	0:	-1-1	461	8	
1:	1	1	3	12	0	1	1	0	1	2:	-1-1	461	8	
1:	1	1	3	15	1	1	0	1	1	2:	-1-1	461	12	
1:	1	1	4	10	1	0	0	1	1	1:	-1-1	461	12	
1:	1	1	4	11	1	0	1	1	2	2:	-1-1	461	12	
1:	1	1	4	8	0	1	0	0	1	1:	-1-1	461	8	
1:	1	1	4	9	0	1	0	0	1	4:	-1-1	461	8	
1:	1	1	5	7	0	1	1	1	0	3:	-1-1	461	8	
1:	1	2	2	9	1	0	1	0	0	1:	-1-1	461	4	
1:	1	2	4	5	1	0	1	1	2	2:	-1-1	461	4	
1:	1	2	4	5	0	0	1	1	0	4:	-1-1	461	4	
1:	1	2	4	5	0	1	1	1	2	3:	-1-1	461	4	
1:	1	3	3	4	1	0	1	0	2	1:	-1-1	461	4	
1:	1	3	3	5	1	0	2	1	1	3:	-1-1	461	4	
1:	1	3	3	5	1	1	-1	0	3	0:	-1-1	461	4	
1:	1	3	4	4	1	0	3	0	2	-1:	-1-1	461	4	
1:	1	3	4	4	1	0	0	1	3	3:	-1-1	461	4	
1:	2	2	3	4	1	2	0	1	1	3:	-1-1	461	2	

D= 464

GENUS#	F11	F22	F33	F44	F12	F13	F23	F14	F24	F34:	HASSE SYM	LEVEL	AUTOS	MASS
1:	1	1	1	29	0	0	0	0	0	0:	1 1	116	96	5/32
1:	1	1	5	6	0	0	0	0	0	2:	1 1	116	16	
1:	2	2	3	5	2	2	0	2	0	0:	1 1	116	12	
2:	1	1	1	39	1	0	0	1	0	0:	-1-1	464	24	5/2
2:	1	1	3	10	0	0	0	1	0	1:	-1-1	464	8	
2:	1	1	3	11	0	1	0	1	0	2:	-1-1	464	8	
2:	1	1	4	11	1	0	0	1	1	1:	-1-1	464	12	
2:	1	1	4	8	0	1	0	0	0	2:	-1-1	464	8	
2:	1	1	4	8	0	1	0	1	0	0:	-1-1	464	8	
2:	1	1	5	7	0	1	0	1	0	4:	-1-1	464	8	
2:	1	2	3	7	1	1	0	1	1	3:	-1-1	464	4	
2:	1	2	4	4	0	0	0	0	1	2:	-1-1	464	4	
2:	1	2	4	5	1	0	1	0	2	2:	-1-1	464	4	
2:	1	3	3	4	1	0	1	0	1	2:	-1-1	464	4	
2:	1	3	4	4	0	0	2	0	3	4:	-1-1	464	4	
2:	2	2	3	4	1	1	0	1	-2	-2:	-1-1	464	2	
3:	1	1	1	58	1	1	0	0	0	0:	-1-1	232	96	5/8
3:	1	1	2	15	0	0	0	1	1	0:	-1-1	232	32	
3:	1	1	3	12	0	1	1	0	0	2:	-1-1	232	16	
3:	1	1	5	7	0	1	1	1	1	0:	-1-1	232	16	
3:	1	1	5	9	1	1	0	1	1	3:	-1-1	232	12	
3:	1	3	3	5	1	0	1	1	3	3:	-1-1	232	4	
3:	2	2	3	3	0	0	0	2	0	1:	-1-1	232	8	
4:	1	1	2	20	0	1	1	1	1	0:	1 1	232	16	15/8
4:	1	1	4	8	0	0	0	1	1	2:	1 1	232	16	
4:	1	1	6	6	0	1	1	0	0	4:	1 1	232	16	
4:	1	2	2	8	0	0	0	1	2	0:	1 1	232	16	
4:	1	2	4	5	1	0	2	0	1	0:	1 1	232	4	
4:	1	2	4	6	1	0	2	0	0	4:	1 1	232	4	
4:	2	2	2	4	1	0	0	1	0	0:	1 1	232	4	

```
4:  2 2 2  5 2 0 0 1 1  0:  1 1      232   8
4:  2 2 2  6 2 1 0 2 0  0:  1 1      232   4
4:  2 2 3  4 0 2 1 0 0  3:  1 1      232   4
4:  2 2 3  4 2 1 0 1 2  1:  1 1      232   4
5:  1 1 3 15 1 1 0 0 1  1: -1-1      116  12    5/8
5:  1 1 4 11 1 0 0 1 0  4: -1-1      116  24
5:  1 3 3  4 1 0 1 0 2  0: -1-1      116   4
5:  1 3 3  5 0 0 2 1 3  3: -1-1      116   8
5:  1 3 4  4 1 0 2 0 0  4: -1-1      116   8
6:  1 1 4 10 1 0 0 1 0  0: -1-1      464  24    5/2
6:  1 1 6  7 1 1 0 0 0  2: -1-1      464  12
6:  1 2 2 10 1 1 0 0 1  1: -1-1      464   8
6:  1 2 3  6 1 0 0 0 1 -2: -1-1      464   4
6:  1 2 3  7 1 0 2 0 1  0: -1-1      464   4
6:  1 2 4  4 0 0 0 1 1  0: -1-1      464   8
6:  1 2 4  5 0 0 0 1 1  4: -1-1      464   8
6:  1 3 3  4 1 0 2 0 0  0: -1-1      464   8
6:  1 3 4  4 1 0 1 1 0  1: -1-1      464   4
6:  1 3 4  4 0 1 2 0 0  4: -1-1      464   8
6:  2 2 3  3 0 0 0 -1 1 2: -1-1      464   4
6:  2 2 3  4 0 2 2 1 1  2: -1-1      464   4
6:  2 2 3  4 1 2 0 2 2  0: -1-1      464   2
7:  1 1 2 15 0 0 0 0 0  2: -1-1      116  32    15/32
7:  1 1 3 10 0 0 0 0 0  2: -1-1      116  16
7:  1 3 3  4 0 0 2 0 2  0: -1-1      116   4
7:  2 2 3  4 0 2 0 2 2  0: -1-1      116   8
8:  1 2 2 10 1 1 0 1 0  0: -1-1      116   8    5/8
8:  1 3 3  4 0 0 2 1 1  1: -1-1      116   8
8:  1 4 4  4 0 0 4 1 4  0: -1-1      116  24
8:  2 2 2  4 0 0 0 1 1  1: -1-1      116  12
8:  2 2 3  4 0 1 1 2 2  3: -1-1      116   4
```

D= 465

GENUS#:	F11	F22	F33	F44	F12	F13	F23	F14	F24	F34:	HASSE SYM	LEVEL	AUTOS	MASS
1:	1	1	1	39	1	0	0	0	0	1:	-1 1-1 1	465	48	7/4
1:	1	1	5	9	1	0	0	0	0	5:	-1 1-1 1	465	48	
1:	1	1	6	6	0	1	0	0	1	4:	-1 1-1 1	465	16	
1:	1	1	6	8	1	1	0	0	1	4:	-1 1-1 1	465	12	
1:	1	2	2	10	1	1	0	1	1	0:	-1 1-1 1	465	8	
1:	1	2	2	8	0	0	1	1	0	0:	-1 1-1 1	465	16	
1:	1	2	3	6	0	0	1	1	2	2:	-1 1-1 1	465	4	
1:	1	2	3	7	1	0	2	0	1	1:	-1 1-1 1	465	4	
1:	1	3	3	5	0	0	3	1	2	2:	-1 1-1 1	465	8	
1:	2	2	2	4	1	0	0	0	0	1:	-1 1-1 1	465	8	
1:	2	2	3	3	1	1	1	1	1	1:	-1 1-1 1	465	8	
1:	2	2	3	4	1	1	1	2	2	3:	-1 1-1 1	465	4	
1:	2	3	3	3	2	1	-1	1	2	2:	-1 1-1 1	465	4	
2:	1	1	2	17	0	1	0	0	1	1:	-1-1-1-1	465	8	7/4
2:	1	2	2	12	1	0	2	1	1	1:	-1-1-1-1	465	8	
2:	1	2	4	5	1	0	0	1	1	3:	-1-1-1-1	465	4	
2:	2	2	2	5	1	1	-1	1	1	1:	-1-1-1-1	465	2	
2:	2	2	2	5	1	1	0	2	2	1:	-1-1-1-1	465	2	
2:	2	2	2	7	2	1	-1	1	0	2:	-1-1-1-1	465	4	
2:	2	2	3	3	1	0	0	-1	1	-1:	-1-1-1-1	465	4	
3:	1	1	2	20	0	1	1	1	0	2:	-1-1 1 1	465	16	7/4
3:	1	1	4	8	0	1	0	0	1	0:	-1-1 1 1	465	16	
3:	1	2	3	6	0	1	1	1	0	2:	-1-1 1 1	465	4	
3:	1	2	4	4	0	1	0	0	1	0:	-1-1 1 1	465	8	
3:	1	2	4	5	1	0	2	0	1	1:	-1-1 1 1	465	4	
3:	1	2	5	5	1	0	2	0	1	5:	-1-1 1 1	465	8	
3:	2	2	3	3	0	1	0	0	1	-2:	-1-1 1 1	465	4	
3:	2	2	3	4	2	1	0	1	1	-2:	-1-1 1 1	465	2	
3:	2	2	3	5	2	1	-1	1	2	2:	-1-1 1 1	465	8	
4:	1	1	2	24	1	1	0	1	1	1:	-1 1 1-1	465	24	7/4
4:	1	1	3	13	1	0	0	0	0	1:	-1 1 1-1	465	24	
4:	1	1	3	14	1	0	0	1	0	3:	-1 1 1-1	465	24	
4:	1	1	4	9	0	1	1	1	0	1:	-1 1 1-1	465	8	
4:	1	2	2	9	1	0	0	1	2	1:	-1 1 1-1	465	4	
4:	1	2	3	6	1	0	1	0	1	1:	-1 1 1-1	465	4	
4:	1	3	4	4	0	1	3	0	3	1:	-1 1 1-1	465	8	
4:	1	3	4	4	1	0	3	1	0	2:	-1 1 1-1	465	4	
4:	1	3	4	4	0	1	2	1	3	0:	-1 1 1-1	465	4	

```
    4:   1   3   4   4   0   1   3   1   0   3: -1 1 1-1    465    8
    4:   2   3   3   3   1   0   0   1  -2  -3: -1 1 1-1    465    4

D= 468
GENUS#: F11 F22 F33 F44 F12 F13 F23 F14 F24 F34: HASSE  SYM  LEVEL AUTOS  MASS
    1:   1   1   1  30   0   0   0   1   1   1:  1 1 1     234    96   5/8
    1:   1   1   6   6   0   1   1   1   1   3:  1 1 1     234    32
    1:   1   2   2  12   1   0   2   1   0   0:  1 1 1     234     8
    1:   2   2   2   6   2   1   0   0   2   0:  1 1 1     234     4
    1:   2   2   2   8   2   0   0   2   0   1:  1 1 1     234    12
    1:   2   2   3   4   2   0   0   2   0   1:  1 1 1     234    12
    1:   2   2   4   4   2   0   0   2   2   3:  1 1 1     234    24
    2:   1   1   1  39   1   0   0   0   0   0: -1-1 1     156    48   5/12
    2:   1   1   3  10   0   0   0   0   1   0: -1-1 1     156    16
    2:   1   1   3  13   0   0   0   0   0   0: -1-1 1     156    48
    2:   1   2   2  10   1   1   0   0   1  -1: -1-1 1     156     8
    2:   1   3   3   4   0   0   0   0   3   0: -1-1 1     156    16
    2:   1   3   4   4   0   1   3   0   0   3: -1-1 1     156     8
    3:   1   1   1  59   1   1   0   1   0   0: -1 1-1     234    96   5/8
    3:   1   1   3  15   1   1   0   1   0   0: -1 1-1     234    12
    3:   1   1   5   7   0   1   1   1   1   1: -1 1-1     234    32
    3:   1   1   6   7   1   0   0   1   0   2: -1 1-1     234    12
    3:   1   1   7   7   1   1   0   1   1   5: -1 1-1     234    24
    3:   1   2   3   7   0   0   2   1   0   3: -1 1-1     234     8
    3:   1   3   3   4   1   0   1   0   0   2: -1 1-1     234     4
    4:   1   1   2  17   0   1   0   1   0   0: -1 1-1     468     8   5/2
    4:   1   1   5   7   0   1   0   0   0   4: -1 1-1     468     8
    4:   1   2   2   9   1   0   1   0   0   0: -1 1-1     468     8
    4:   1   2   3   7   1   1   0   1   2   2: -1 1-1     468     4
    4:   1   2   4   4   0   0   1   0  -1   1: -1 1-1     468     8
    4:   1   2   4   5   1   0   0   1   2   2: -1 1-1     468     4
    4:   1   2   4   5   0   0   2  -1   1  -2: -1 1-1     468     4
    4:   2   2   2   5   1   1   0   2   0   2: -1 1-1     468     2
    4:   2   2   3   4   1   2   0   1   2   2: -1 1-1     468     2
    4:   2   3   3   3   1   0   1   2   3   2: -1 1-1     468     4
    5:   1   1   2  15   0   0   0   0   1   1: -1 1-1     468     8   5/2
    5:   1   1   2  24   1   1   0   0   1   1: -1 1-1     468    12
    5:   1   1   3  11   0   1   0   0   0   2: -1 1-1     468     8
    5:   1   1   3  15   1   1   0   0   0   2: -1 1-1     468    12
    5:   1   1   4  11   1   1   0   1   0   0: -1 1-1     468    12
    5:   1   2   3   6   1   0   1   0   1   0: -1 1-1     468     4
    5:   1   2   3   6   0   0   2   0   1   1: -1 1-1     468     4
    5:   1   2   3   7   1   0   1   1   0   3: -1 1-1     468     4
    5:   1   3   3   4   0   0   2   0   1   2: -1 1-1     468     4
    5:   1   3   3   4   1   0   1   1   1   1: -1 1-1     468     4
    5:   1   3   3   4   0   0   2   1   1   0: -1 1-1     468     4
    5:   1   3   4   4   0   1   0   0   2   4: -1 1-1     468     4
    5:   2   2   3   3   0   1   1  -1   1  -1: -1 1-1     468     4
    6:   1   1   3  11   0   0   0   1   1   3: -1-1 1      78    32   5/48
    6:   1   1   4  10   1   0   0   0   0   2: -1-1 1      78    24
    6:   1   3   3   5   0   0   0   1   3   3: -1-1 1      78    32
    7:   1   1   4   9   0   1   1   0   0   3:  1 1 1     234    16   5/8
    7:   1   2   2   9   1   0   0   0   1   2:  1 1 1     234     8
    7:   1   2   4   5   0   1   2   0   2   1:  1 1 1     234    16
    7:   1   2   4   5   1   1   0   1   2   4:  1 1 1     234     8
    7:   2   2   3   4   0   2   1   0   1   3:  1 1 1     234     4
    8:   1   1   5   8   1   0   0   0   0   2: -1 1-1     156    24   5/3
    8:   1   2   2   8   0   0   1   0   1   0: -1 1-1     156     8
    8:   1   2   3   5   0   0   0   0   1   0: -1 1-1     156     8
    8:   1   2   3   6   0   1   1   1   1   0: -1 1-1     156     4
    8:   1   2   3   6   0   0   0   1   1   3: -1 1-1     156     8
    8:   1   3   3   4   1   0   0   1   2   0: -1 1-1     156     8
    8:   1   3   3   6   1   0   3   0   3   0: -1 1-1     156     8
    8:   2   2   3   4   1   2  -1   0   0   2: -1 1-1     156     4
    8:   2   3   3   3   1   1  -1   2   1   3: -1 1-1     156     2
    9:   1   1   2  20   1   0   0   0   0   2: -1 1-1      78    48   5/12
    9:   1   1   6   8   1   0   0   0   0   6: -1 1-1      78    48
    9:   1   2   3   7   0   1   2   1   0   2: -1 1-1      78     8
    9:   1   3   3   5   1   0   3   0   1   0: -1 1-1      78     8
    9:   3   3   3   3   1   1   0   3   3   3: -1 1-1      78     8
   10:   1   1   2  20   0   1   1   1   1   1:  1-1-1      78    32   5/48
   10:   2   2   2   5   2   0   0   0   0   1:  1-1-1      78    24
```

```
10:  2  2  2  7  2  1 -1  1 -1   1:  1-1-1    78  32
11:  1  1  5  7  0  1  1  0  0   3: -1 1-1   234  16   5/8
11:  1  2  3  7  0  1  2  0  2   1: -1 1-1   234  16
11:  1  2  4  5  0  0  2  1  2   0: -1 1-1   234   8
11:  1  3  3  5  1  1 -1  1  3   0: -1 1-1   234   4
11:  2  3  3  3  0  0  3  2  2   1: -1 1-1   234   8
12:  1  2  2 10  0  0  2  1  0   0:  1 1 1    78  48   5/12
12:  1  2  4  5  0  1  0  0  2   3:  1 1 1    78   8
12:  2  2  3  3  1  1  1  1  0   0:  1 1 1    78   8
12:  2  2  3  4  2  1 -1  0  0   1:  1 1 1    78   8
12:  2  2  3  4  2  0  0  0  0   3:  1 1 1    78  48
```

D= 469

GENUS#	F11	F22	F33	F44	F12	F13	F23	F14	F24	F34:	HASSE	SYM	LEVEL	AUTOS	MASS
1:	1	1	1	59	1	1	0	0	1	0:	-1-1	1	469	48	5/2
1:	1	1	2	17	0	1	0	0	1	0:	-1-1	1	469	16	
1:	1	1	3	11	0	1	0	0	1	1:	-1-1	1	469	8	
1:	1	1	3	15	1	1	0	0	1	0:	-1-1	1	469	12	
1:	1	1	4	11	1	1	0	0	1	0:	-1-1	1	469	12	
1:	1	2	2	7	0	1	2	1	1	0:	-1-1	1	469	4	
1:	1	3	3	4	1	1	0	0	1	0:	-1-1	1	469	4	
1:	1	3	3	5	1	1	-1	1	2	2:	-1-1	1	469	8	
1:	1	3	3	5	1	0	1	0	3	3:	-1-1	1	469	4	
1:	1	3	3	5	1	1	0	1	3	2:	-1-1	1	469	4	
1:	2	2	3	3	1	1	0	1	0	-1:	-1-1	1	469	4	
1:	2	3	3	3	1	0	3	2	2	1:	-1-1	1	469	2	
1:	2	3	3	3	2	1	0	2	3	1:	-1-1	1	469	4	
2:	1	1	2	20	1	0	0	1	0	1:	-1	1-1	469	12	5/2
2:	1	1	3	12	0	1	1	1	0	0:	-1	1-1	469	8	
2:	1	1	5	7	0	1	0	1	1	3:	-1	1-1	469	8	
2:	1	1	5	9	1	1	0	0	1	2:	-1	1-1	469	12	
2:	1	1	6	8	1	1	0	0	0	5:	-1	1-1	469	12	
2:	1	2	3	6	0	1	1	0	1	2:	-1	1-1	469	4	
2:	1	2	3	7	1	1	0	0	0	3:	-1	1-1	469	4	
2:	1	2	3	7	1	0	2	0	0	1:	-1	1-1	469	4	
2:	1	3	3	4	1	0	0	0	2	1:	-1	1-1	469	4	
2:	1	3	3	5	1	0	3	0	0	1:	-1	1-1	469	4	
2:	1	3	4	4	1	0	3	0	2	3:	-1	1-1	469	4	
2:	2	2	3	4	1	2	-1	1	-1	0:	-1	1-1	469	2	

D= 472

GENUS#	F11	F22	F33	F44	F12	F13	F23	F14	F24	F34:	HASSE	SYM	LEVEL	AUTOS	MASS
1:	1	1	1	59	1	1	0	0	0	0:	1	1	472	96	277/96
1:	1	1	2	15	0	0	0	1	0	0:	1	1	472	16	
1:	1	1	2	20	1	0	0	1	0	0:	1	1	472	24	
1:	1	1	3	11	0	1	0	1	0	0:	1	1	472	8	
1:	1	1	4	11	1	1	0	0	0	2:	1	1	472	12	
1:	1	1	4	9	0	1	1	1	1	0:	1	1	472	16	
1:	1	2	2	12	1	0	2	0	1	0:	1	1	472	8	
1:	1	2	2	9	1	0	0	0	2	0:	1	1	472	8	
1:	1	2	3	5	0	0	0	0	0	1:	1	1	472	8	
1:	1	2	3	6	0	0	0	1	2	2:	1	1	472	8	
1:	1	2	3	7	0	1	2	0	0	2:	1	1	472	8	
1:	1	2	4	4	0	1	0	0	0	1:	1	1	472	8	
1:	1	2	4	5	0	1	0	0	0	4:	1	1	472	8	
1:	1	2	4	5	0	0	0	1	2	3:	1	1	472	8	
1:	1	3	3	4	1	0	0	0	-1	2:	1	1	472	4	
1:	2	2	3	3	1	1	1	0	1	1:	1	1	472	2	
1:	2	2	3	4	1	2	-1	0	-1	2:	1	1	472	2	
1:	2	2	3	5	2	1	-1	0	2	1:	1	1	472	4	
2:	1	1	1	30	0	0	0	1	1	0:	-1-1		472	32	277/96
2:	1	1	2	17	0	1	0	0	0	1:	-1-1		472	8	
2:	1	1	2	24	1	1	0	1	0	0:	-1-1		472	12	
2:	1	1	5	9	1	0	0	1	0	4:	-1-1		472	12	
2:	1	1	6	7	0	1	1	0	0	6:	-1-1		472	16	
2:	1	2	2	10	0	0	2	0	0	1:	-1-1		472	8	
2:	1	2	2	8	0	0	0	0	1	2:	-1-1		472	8	
2:	1	2	2	8	0	0	1	0	0	1:	-1-1		472	4	
2:	1	2	3	6	0	1	1	0	0	2:	-1-1		472	4	
2:	1	2	3	6	0	1	1	1	1	1:	-1-1		472	4	
2:	1	2	3	6	0	0	2	0	0	1:	-1-1		472	8	
2:	1	2	4	5	0	1	1	0	2	3:	-1-1		472	4	

	F11	F22	F33	F44	F12	F13	F23	F14	F24	F34:	HASSE SYM	LEVEL	AUTOS	MASS
2:	1	2	4	5	0	0	2	0	2	3:	-1-1	472	8	
2:	1	3	3	5	1	0	2	0	3	0:	-1-1	472	4	
2:	2	2	2	6	2	1	0	1	2	1:	-1-1	472	4	
2:	2	3	3	3	1	1	-3	-1	1	0:	-1-1	472	2	

D= 473

GENUS#:	F11	F22	F33	F44	F12	F13	F23	F14	F24	F34:	HASSE SYM	LEVEL	AUTOS	MASS
1:	1	1	1	40	1	0	0	1	0	1:	-1-1 1	473	24	51/16
1:	1	1	2	24	1	1	0	0	1	0:	-1-1 1	473	12	
1:	1	1	6	7	0	1	1	1	0	6:	-1-1 1	473	16	
1:	1	2	2	10	1	1	0	0	1	0:	-1-1 1	473	4	
1:	1	2	3	6	1	0	1	0	0	1:	-1-1 1	473	4	
1:	1	2	3	7	1	0	1	0	2	3:	-1-1 1	473	4	
1:	1	2	4	5	1	1	0	0	1	2:	-1-1 1	473	4	
1:	1	2	4	5	1	0	1	0	1	3:	-1-1 1	473	4	
1:	1	2	4	5	1	1	0	1	1	2:	-1-1 1	473	4	
1:	1	2	4	5	0	1	2	1	1	0:	-1-1 1	473	4	
1:	2	2	2	6	2	1	0	1	1	2:	-1-1 1	473	4	
1:	2	2	3	4	2	1	0	-1	1	0:	-1-1 1	473	2	
1:	2	2	3	4	1	2	0	2	1	1:	-1-1 1	473	2	
2:	1	1	2	20	0	1	1	1	0	0:	-1 1-1	473	8	51/16
2:	1	1	3	11	0	1	0	0	1	0:	-1 1-1	473	16	
2:	1	1	4	12	1	1	0	1	1	4:	-1 1-1	473	24	
2:	1	1	6	7	1	1	0	0	0	1:	-1 1-1	473	12	
2:	1	2	2	10	1	0	1	1	0	2:	-1 1-1	473	4	
2:	1	2	2	12	1	0	2	0	1	1:	-1 1-1	473	4	
2:	1	2	3	6	1	0	0	1	1	1:	-1 1-1	473	4	
2:	1	2	3	8	1	0	2	1	1	3:	-1 1-1	473	4	
2:	1	2	4	5	1	0	2	0	0	1:	-1 1-1	473	4	
2:	1	2	4	6	1	0	2	1	0	3:	-1 1-1	473	4	
2:	1	3	3	4	0	0	1	1	2	1:	-1 1-1	473	4	
2:	1	3	4	4	1	0	1	1	1	4:	-1 1-1	473	8	
2:	2	2	3	4	0	1	1	-2	-1	2:	-1 1-1	473	4	
2:	2	2	3	4	1	1	-1	1	2	2:	-1 1-1	473	2	
2:	2	2	3	4	1	1	0	1	2	3:	-1 1-1	473	4	

D= 476

GENUS#:	F11	F22	F33	F44	F12	F13	F23	F14	F24	F34:	HASSE SYM	LEVEL	AUTOS	MASS
1:	1	1	1	40	1	0	0	1	0	0:	-1 1-1	476	24	31/24
1:	1	1	2	17	0	1	0	0	0	0:	-1 1-1	476	16	
1:	1	1	2	20	0	1	1	0	0	1:	-1 1-1	476	16	
1:	1	1	3	10	0	0	0	0	0	1:	-1 1-1	476	16	
1:	1	1	3	12	0	1	1	0	0	1:	-1 1-1	476	16	
1:	1	1	4	9	0	1	1	1	1	1:	-1 1-1	476	32	
1:	1	1	5	6	0	0	0	0	0	1:	-1 1-1	476	16	
1:	1	1	5	7	0	0	0	0	1	4:	-1 1-1	476	8	
1:	1	1	6	6	0	0	0	0	0	5:	-1 1-1	476	32	
1:	1	2	2	9	1	0	0	0	0	2:	-1 1-1	476	16	
1:	1	2	4	4	0	0	1	0	1	1:	-1 1-1	476	8	
1:	1	2	4	5	0	0	2	1	2	1:	-1 1-1	476	16	
1:	1	2	4	5	1	0	1	0	2	0:	-1 1-1	476	4	
1:	2	2	3	4	0	2	0	2	1	3:	-1 1-1	476	4	
2:	1	1	1	30	0	0	0	0	0	1:	-1-1 1	476	32	31/24
2:	1	1	1	60	1	1	0	1	0	0:	-1-1 1	476	96	
2:	1	1	2	15	0	0	0	0	0	1:	-1-1 1	476	16	
2:	1	1	4	8	0	0	0	0	0	3:	-1-1 1	476	16	
2:	1	1	4	8	0	0	0	1	1	1:	-1-1 1	476	16	
2:	1	1	4	8	0	1	0	0	0	1:	-1-1 1	476	8	
2:	1	1	4	9	0	1	0	0	0	4:	-1-1 1	476	8	
2:	1	1	5	8	0	1	1	0	0	5:	-1-1 1	476	16	
2:	1	3	4	4	1	0	3	0	1	-2:	-1-1 1	476	4	
2:	2	2	2	5	1	0	0	0	2	2:	-1-1 1	476	4	
2:	2	2	3	3	1	1	0	0	1	1:	-1-1 1	476	4	
3:	1	1	5	10	1	1	0	1	1	5:	1-1-1	476	24	31/24
3:	1	2	3	6	0	1	0	0	2	1:	1-1-1	476	8	
3:	1	2	4	5	0	1	1	-1	-1	2:	1-1-1	476	4	
3:	1	3	3	5	1	0	2	1	3	0:	1-1-1	476	4	
3:	2	2	3	4	1	2	0	-1	0	2:	1-1-1	476	2	
3:	2	2	3	5	2	2	0	1	1	-2:	1-1-1	476	8	
4:	1	1	6	7	1	1	0	0	0	0:	1 1 1	476	24	31/24
4:	1	2	2	10	1	0	1	0	2	2:	1 1 1	476	4	
4:	1	2	3	6	1	0	0	0	0	2:	1 1 1	476	8	

											HASSE SYM		
4:	1	2	3	7	1	0	2	0	0	0:	1 1 1	476	8	
4:	1	2	4	5	1	1	0	1	0	2:	1 1 1	476	4	
4:	1	2	4	5	0	1	2	1	0	2:	1 1 1	476	8	
4:	1	2	5	5	1	0	1	1	1	5:	1 1 1	476	8	
4:	2	2	3	4	2	1	0	1	2	0:	1 1 1	476	8	
4:	2	3	3	3	2	0	2	0	-2	1:	1 1 1	476	8	

D= 477

GENUS#:	F11	F22	F33	F44	F12	F13	F23	F14	F24	F34:	HASSE SYM	LEVEL	AUTOS	MASS
1:	1	1	1	40	1	0	0	0	0	1:	-1 1-1	159	48	35/48
1:	1	1	4	10	1	0	0	0	0	1:	-1 1-1	159	24	
1:	1	1	5	7	0	1	1	0	1	1:	-1 1-1	159	8	
1:	1	1	6	7	1	0	0	0	0	3:	-1 1-1	159	24	
1:	1	2	3	7	1	0	0	0	2	3:	-1 1-1	159	8	
1:	1	2	4	5	1	0	1	1	0	2:	-1 1-1	159	4	
1:	1	3	4	4	0	1	0	0	3	3:	-1 1-1	159	8	
2:	1	1	1	60	1	1	0	0	1	0:	-1 1-1	477	48	35/16
2:	1	1	2	24	1	1	0	0	0	1:	-1 1-1	477	12	
2:	1	1	3	15	1	1	0	0	0	1:	-1 1-1	477	12	
2:	1	1	5	9	1	1	0	1	1	2:	-1 1-1	477	12	
2:	1	1	5	9	1	1	0	0	0	3:	-1 1-1	477	12	
2:	1	1	6	7	1	0	0	1	1	1:	-1 1-1	477	12	
2:	1	1	6	7	0	1	0	0	1	6:	-1 1-1	477	8	
2:	1	1	6	8	1	0	0	1	1	5:	-1 1-1	477	12	
2:	1	1	7	7	1	1	0	0	1	4:	-1 1-1	477	24	
2:	1	2	3	6	0	0	1	1	1	-2:	-1 1-1	477	4	
2:	1	2	3	7	0	1	2	1	1	1:	-1 1-1	477	8	
2:	1	2	4	5	1	0	1	0	0	3:	-1 1-1	477	4	
2:	1	3	3	4	0	0	1	1	0	2:	-1 1-1	477	4	
2:	1	3	3	4	1	0	1	1	0	0:	-1 1-1	477	4	
2:	1	3	3	5	0	0	3	1	-1	1:	-1 1-1	477	8	
2:	2	3	3	3	1	1	1	2	3	2:	-1 1-1	477	4	
3:	1	1	2	20	1	0	0	0	0	1:	-1-1 1	159	24	35/48
3:	1	1	3	14	1	0	0	0	0	3:	-1-1 1	159	48	
3:	1	1	5	8	1	0	0	0	0	1:	-1-1 1	159	24	
3:	1	2	2	9	0	0	1	1	2	2:	-1-1 1	159	8	
3:	1	3	3	6	1	0	3	1	3	0:	-1-1 1	159	8	
3:	2	3	3	3	2	0	3	1	1	0:	-1-1 1	159	4	
3:	3	3	3	3	2	1	0	0	3	3:	-1-1 1	159	8	
4:	1	1	5	7	0	1	0	0	1	3:	-1 1-1	477	8	35/16
4:	1	1	5	8	0	1	1	1	0	5:	-1 1-1	477	16	
4:	1	2	3	6	0	1	0	1	1	-1:	-1 1-1	477	4	
4:	1	2	3	7	1	1	0	0	2	1:	-1 1-1	477	4	
4:	1	2	3	7	0	1	2	0	1	-1:	-1 1-1	477	4	
4:	1	2	4	5	1	0	1	1	2	0:	-1 1-1	477	4	
4:	1	2	4	5	0	0	1	1	2	3:	-1 1-1	477	4	
4:	2	2	2	5	1	1	1	2	1	0:	-1 1-1	477	4	
4:	2	3	3	3	2	0	1	1	-2	1:	-1 1-1	477	2	

D= 480

GENUS#:	F11	F22	F33	F44	F12	F13	F23	F14	F24	F34:	HASSE SYM	LEVEL	AUTOS	MASS
1:	1	1	1	30	0	0	0	0	0	0:	-1-1 1	120	96	17/96
1:	1	2	2	8	0	0	0	0	2	0:	-1-1 1	120	16	
1:	1	2	3	5	0	0	0	0	0	0:	-1-1 1	120	16	
1:	2	2	3	4	2	0	0	2	0	0:	-1-1 1	120	24	
2:	1	1	1	40	1	0	0	0	0	0:	1-1-1	480	48	17/24
2:	1	1	3	11	0	0	0	1	0	3:	1-1-1	480	16	
2:	1	3	4	4	1	0	2	0	1	4:	1-1-1	480	8	
2:	2	2	2	5	1	1	1	0	0	2:	1-1-1	480	4	
2:	2	2	3	3	1	1	0	0	-1	1:	1-1-1	480	4	
3:	1	1	1	60	1	1	0	0	0	0:	-1 1-1	240	96	17/96
3:	1	1	6	7	1	0	0	1	0	0:	-1 1-1	240	24	
3:	1	2	4	5	0	0	0	1	0	4:	-1 1-1	240	16	
3:	2	3	3	3	2	2	1	2	0	1:	-1 1-1	240	16	
4:	1	1	2	15	0	0	0	0	0	0:	-1 1-1	120	32	17/96
4:	1	2	2	10	0	0	2	0	0	0:	-1 1-1	120	48	
4:	1	2	3	6	0	0	2	0	0	0:	-1 1-1	120	16	
4:	1	2	4	4	0	0	0	0	0	2:	-1 1-1	120	16	
5:	1	1	2	16	0	0	0	1	1	2:	-1 1-1	240	32	17/32
5:	1	2	3	6	1	0	1	0	0	0:	-1 1-1	240	8	
5:	1	2	4	4	0	0	0	1	0	0:	-1 1-1	240	16	
5:	2	2	2	4	0	0	0	1	-1	0:	-1 1-1	240	8	

```
5:    2   2   3    4    2   1  -1   0   0   0:  -1 1-1    240   16
5:    2   2   4    4    2   1  -1   0   0   4:  -1 1-1    240   16
5:    2   3   3    4    2   2   0   2   3   3:  -1 1-1    240   16
6:    1   1   2   24    1   1   0   0   0   0:  -1 1-1    480   24    17/24
6:    1   1   3   15    1   1   0   0   0   0:  -1 1-1    480   24
6:    1   2   3    7    0   0   2   1   1   3:  -1 1-1    480    8
6:    1   3   3    4    0   0   2   1   0   0:  -1 1-1    480   16
6:    1   3   3    5    0   0   2  -1  -2   2:  -1 1-1    480   16
6:    2   2   3    4    0   2   2   1  -1   0:  -1 1-1    480    8
6:    2   3   3    3    2   1  -1   2   0   2:  -1 1-1    480    4
7:    1   1   3   10    0   0   0   0   0   0:   1-1-1    120   32    17/96
7:    2   2   3    3    0   0   0   2   0   0:   1-1-1    120   16
7:    2   2   3    5    2   2   0   0   2   0:   1-1-1    120   12
8:    1   1   3   12    0   1   1   0   0   0:   1-1-1    240   32    17/96
8:    1   1   4   10    1   0   0   0   0   0:   1-1-1    240   48
8:    2   3   3    3    0   0   0   2   3   1:   1-1-1    240    8
9:    1   1   3   16    1   1   0   1   1   3:  -1 1-1    480   24    17/24
9:    1   1   4    8    0   1   0   0   0   0:  -1 1-1    480   16
9:    1   1   6    6    0   1   0   1   0   4:  -1 1-1    480   16
9:    1   1   7    8    1   1   0   1   1   7:  -1 1-1    480   24
9:    1   2   3    6    0   0   1   0   0   3:  -1 1-1    480    4
9:    1   2   3    7    1   0   1   1   1   3:  -1 1-1    480    8
9:    1   3   3    4    0   0   2   0  -1   1:  -1 1-1    480    8
10:   1   1   4   11    1   0   0   0   0   4:   1-1-1    480   48    17/24
10:   1   2   2   11    1   1   0   1   2   2:   1-1-1    480   16
10:   1   3   3    4    1   0   0   0   2   0:   1-1-1    480    8
10:   2   2   2    5    1   1  -1   1   0   1:   1-1-1    480    4
10:   2   2   3    4    1   2  -1  -1  -1   1:   1-1-1    480    4
11:   1   1   3   11    0   1   0   0   0   1:   1 1 1    480    8    17/24
11:   1   1   4   12    1   1   0   0   1   3:   1 1 1    480   12
11:   1   2   4    5    1   0   2   0   0   0:   1 1 1    480    8
11:   1   3   3    4    0   0   1   0   2   2:   1 1 1    480    8
11:   1   3   4    4    0   1   3   0  -2   1:   1 1 1    480    4
12:   1   1   2   20    1   0   0   0   0   0:  -1-1 1    240   48    17/96
12:   1   1   4    9    0   0   0   1   1   4:  -1-1 1    240   32
12:   1   2   5    5    0   0   0   1   2   5:  -1-1 1    240   16
12:   2   3   3    3    0   0   3   0   2  -1:  -1-1 1    240   16
13:   1   1   5    6    0   0   0   0   0   0:   1 1 1    120   32    17/96
13:   1   3   3    4    0   0   0   0   2  -2:   1 1 1    120    8
13:   2   2   2    5    2   0   0   0   0   0:   1 1 1    120   48
14:   1   1   4    8    0   0   0   0   1   2:  -1-1 1    480    8    17/24
14:   1   1   5    8    1   0   0   0   0   0:  -1-1 1    480   48
14:   1   2   2    8    0   0   1   0   0   0:  -1-1 1    480   16
14:   1   2   4    4    0   0   1   0   1   0:  -1-1 1    480    8
14:   1   2   5    5    1   0   2   1   2   4:  -1-1 1    480    8
14:   1   3   3    5    1   0   3   0   0   0:  -1-1 1    480   16
14:   1   3   4    4    0   0   3   0   3   0:  -1-1 1    480   16
14:   2   2   3    4    1   1  -1  -1   1   2:  -1-1 1    480    8
15:   1   1   5    7    1   1   0   0   1   1:   1 1 1    240   12    17/96
15:   1   1   6    7    0   0   0   1   1   6:   1 1 1    240   32
15:   1   3   3    4    1   1   0   0   0   0:   1 1 1    240   16
16:   1   1   7    7    1   0   0   0   0   6:  -1-1 1    480   48    17/24
16:   1   2   2    8    0   0   0   1   1   1:  -1-1 1    480    8
16:   1   2   3    6    1   0   0   1   0   0:  -1-1 1    480    8
16:   1   3   3    6    1   0   3   1   2   3:  -1-1 1    480   16
16:   2   2   3    3    1   0   0  -1   1   0:  -1-1 1    480    8
16:   2   3   3    3    1   0   0   0   0   2:  -1-1 1    480   16
16:   2   3   3    3    1   0   2   0   2  -2:  -1-1 1    480    8
16:   3   3   3    3    2   0   0   2  -2  -3:  -1-1 1    480   16
17:   1   1   4    8    0   0   0   1   1   0:  -1-1 1    240   32    17/32
17:   1   2   2   12    1   0   2   0   0   0:  -1-1 1    240   16
17:   1   2   3    6    0   0   0   1   0   3:  -1-1 1    240   16
17:   2   2   2    4    1   0   0   0   0   0:  -1-1 1    240   16
17:   2   2   2    8    2   2   0   1  -1   1:  -1-1 1    240   16
17:   2   2   3    3    0   1   0   1   0   2:  -1-1 1    240    8
17:   2   2   4    4    2   1   0   0   2   3:  -1-1 1    480   12    17/24
18:   1   1   3   14    1   0   0   1   1   2:   1 1 1    480    8
18:   1   2   4    6    1   0   2   0   1   4:   1 1 1    480    4
18:   1   3   3    4    0   0   0   1   2  -1:   1 1 1    480    8
18:   1   3   4    4    1   1   0   0   0   4:   1 1 1    480    8
18:   2   2   3    3    0   1   1   1   1   1:   1 1 1    480    8
19:   1   1   6    6    0   1   1   1   1   0:   1-1-1    240   32    17/32
```

```
19:   1  2  2  10  1  1  0  0  0  0:  1-1-1    240  16
19:   2  2  2   6  2  1  0  0  0  2:  1-1-1    240   4
19:   2  2  2   7  2  1 -1  1  1  1:  1-1-1    240   8
19:   2  2  3   4  0  2  0  2  2  1:  1-1-1    240  16
20:   1  1  2  20  0  1  1  0  0  0:  1 1 1    240  32   17/32
20:   1  3  4   4  0  1  3  1  1 -1:  1 1 1    240   4
20:   1  4  4   4  1  0  4  1  2  4:  1 1 1    240  16
20:   2  2  2   6  2  0  0  2  1  2:  1 1 1    240  16
20:   2  2  3   3  0  1  1  1  1  0:  1 1 1    240   8
```

D= 481

GENUS#:	F11	F22	F33	F44	F12	F13	F23	F14	F24	F34:	HASSE SYM	LEVEL	AUTOS	MASS
1:	1	1	2	18	0	1	0	0	1	2:	-1-1 1	481	8	101/24
1:	1	1	3	12	0	1	0	0	1	3:	-1-1 1	481	8	
1:	1	1	4	11	1	1	0	0	0	1:	-1-1 1	481	12	
1:	1	2	3	6	0	1	1	0	1	-1:	-1-1 1	481	4	
1:	1	2	3	6	0	1	1	1	0	0:	-1-1 1	481	4	
1:	1	2	4	5	0	1	2	1	1	1:	-1-1 1	481	8	
1:	1	2	4	5	1	1	0	0	-1	2:	-1-1 1	481	4	
1:	1	2	4	5	1	0	0	0	1	3:	-1-1 1	481	4	
1:	1	3	3	4	0	0	1	1	1	1:	-1-1 1	481	4	
1:	1	3	4	4	0	0	1	1	2	4:	-1-1 1	481	4	
1:	2	2	2	5	1	1	0	0	1	2:	-1-1 1	481	2	
1:	2	2	3	3	0	1	0	1	-1	-1:	-1-1 1	481	2	
1:	2	2	3	4	1	2	0	1	0	3:	-1-1 1	481	4	
1:	2	2	3	4	2	1	0	0	1	-2:	-1-1 1	481	2	
1:	2	2	3	4	1	2	-1	0	1	1:	-1-1 1	481	2	
2:	1	1	4	9	0	1	1	1	0	2:	-1 1-1	481	8	101/24
2:	1	1	6	6	0	1	0	1	1	3:	-1 1-1	481	8	
2:	1	1	6	8	1	1	0	1	1	4:	-1 1-1	481	12	
2:	1	2	2	10	1	0	1	1	1	2:	-1 1-1	481	8	
2:	1	2	2	9	1	0	0	1	1	1:	-1 1-1	481	4	
2:	1	2	3	6	0	0	1	1	2	0:	-1 1-1	481	4	
2:	1	2	3	8	1	0	2	0	0	3:	-1 1-1	481	4	
2:	1	2	4	5	0	1	1	0	1	-3:	-1 1-1	481	4	
2:	1	3	3	5	1	0	2	0	0	3:	-1 1-1	481	4	
2:	1	3	4	4	1	1	-1	1	3	1:	-1 1-1	481	4	
2:	1	3	4	4	1	0	3	0	1	3:	-1 1-1	481	4	
2:	2	2	2	6	2	1	0	2	1	0:	-1 1-1	481	2	
2:	2	2	3	3	1	1	1	0	1	0:	-1 1-1	481	2	
2:	2	2	3	4	1	1	1	2	0	3:	-1 1-1	481	2	
2:	2	3	3	3	1	0	3	1	-2	-2:	-1 1-1	481	2	

D= 484

GENUS#:	F11	F22	F33	F44	F12	F13	F23	F14	F24	F34:	HASSE SYM	LEVEL	AUTOS	MASS
1:	1	1	1	31	0	0	0	1	1	1:	1 1	242	96	55/96
1:	1	1	5	7	0	0	0	1	1	3:	1 1	242	16	
1:	1	1	5	9	1	1	0	1	0	0:	1 1	242	12	
1:	1	2	2	11	0	0	2	1	2	0:	1 1	242	24	
1:	1	2	5	5	0	1	0	0	2	5:	1 1	242	8	
1:	2	3	3	3	0	0	1	2	0	3:	1 1	242	4	
2:	1	1	1	61	1	1	0	1	0	0:	1 1	242	96	55/96
2:	1	1	2	21	1	0	0	1	0	2:	1 1	242	24	
2:	1	1	3	13	0	1	1	0	0	3:	1 1	242	16	
2:	1	2	3	7	0	1	2	1	0	0:	1 1	242	8	
2:	1	3	3	5	1	1	0	0	3	1:	1 1	242	4	
2:	2	2	3	5	2	2	0	2	0	1:	1 1	242	12	
3:	1	1	3	11	0	1	0	0	0	0:	-1-1	44	16	25/48
3:	1	1	4	11	1	1	0	0	0	0:	-1-1	44	24	
3:	1	3	3	4	1	0	0	0	0	2:	-1-1	44	8	
3:	1	3	4	4	0	0	2	0	-2	3:	-1-1	44	24	
3:	2	2	3	4	1	2	-1	-1	1	-1:	-1-1	44	4	
4:	1	1	6	6	0	1	1	1	1	1:	-1-1	22	64	25/64
4:	1	3	4	4	0	0	2	1	1	4:	-1-1	22	8	
4:	2	2	2	6	2	1	0	1	2	0:	-1-1	22	8	
4:	2	2	3	3	0	1	1	-1	1	0:	-1-1	22	8	
5:	1	2	3	6	0	1	0	0	2	0:	1 1	22	16	1/16

D= 485

GENUS#:	F11	F22	F33	F44	F12	F13	F23	F14	F24	F34:	HASSE SYM	LEVEL	AUTOS	MASS
1:	1	1	1	41	1	0	0	1	0	1:	-1-1 1	485	24	101/48
1:	1	1	1	61	1	1	0	0	1	0:	-1-1 1	485	48	

#	F11	F22	F33	F44	F12	F13	F23	F14	F24	F34:	HASSE	SYM	LEVEL	AUTOS	MASS
1:	1	1	3	16	1	1	0	0	1	2:	-1-1	1	485	12	
1:	1	1	4	11	1	0	0	1	1	3:	-1-1	1	485	12	
1:	1	1	5	7	0	1	1	0	1	0:	-1-1	1	485	8	
1:	1	2	3	6	0	1	0	0	1	2:	-1-1	1	485	4	
1:	1	2	4	5	1	0	1	1	1	2:	-1-1	1	485	4	
1:	1	2	5	5	1	1	0	0	0	5:	-1-1	1	485	8	
1:	1	3	3	4	1	0	1	0	-1	1:	-1-1	1	485	4	
1:	1	3	3	5	0	0	1	1	3	3:	-1-1	1	485	8	
1:	2	2	2	5	1	1	-1	1	0	0:	-1-1	1	485	4	
1:	2	2	3	3	1	1	0	1	0	1:	-1-1	1	485	4	
1:	2	3	3	3	2	1	-1	2	1	2:	-1-1	1	485	4	
2:	1	1	2	25	1	1	0	1	1	2:	-1	1-1	485	24	101/48
2:	1	1	3	13	1	0	1	1	0	3:	-1	1-1	485	16	
2:	1	1	4	9	0	1	0	1	1	3:	-1	1-1	485	8	
2:	1	1	5	10	1	1	0	0	1	4:	-1	1-1	485	12	
2:	1	1	5	7	0	1	0	1	1	2:	-1	1-1	485	8	
2:	1	1	5	9	1	1	0	0	1	0:	-1	1-1	485	12	
2:	1	1	7	7	1	1	0	0	0	5:	-1	1-1	485	12	
2:	1	2	3	6	1	0	0	0	-1	1:	-1	1-1	485	4	
2:	1	2	3	7	1	1	0	1	2	1:	-1	1-1	485	4	
2:	1	2	4	5	0	1	1	1	2	0:	-1	1-1	485	4	
2:	1	3	3	5	1	0	2	0	3	1:	-1	1-1	485	4	
2:	1	3	4	4	0	0	3	1	-2	1:	-1	1-1	485	4	
2:	1	3	4	4	1	0	2	1	0	3:	-1	1-1	485	4	

D= 488

GENUS#:	F11	F22	F33	F44	F12	F13	F23	F14	F24	F34:	HASSE	SYM	LEVEL	AUTOS	MASS
1:	1	1	1	31	0	0	0	1	1	0:	1	1	488	32	77/32
1:	1	1	1	41	1	0	0	1	0	0:	1	1	488	24	
1:	1	1	2	25	1	1	0	0	1	1:	1	1	488	12	
1:	1	1	3	11	0	0	0	1	1	2:	1	1	488	16	
1:	1	1	4	9	0	1	1	0	0	2:	1	1	488	16	
1:	1	1	5	7	0	1	0	1	0	3:	1	1	488	8	
1:	1	2	2	10	1	0	1	0	2	0:	1	1	488	4	
1:	1	2	2	8	0	0	0	1	0	1:	1	1	488	8	
1:	1	2	2	9	1	0	0	1	0	0:	1	1	488	8	
1:	1	2	3	6	0	1	0	1	0	2:	1	1	488	8	
1:	1	2	3	7	1	1	0	0	2	0:	1	1	488	4	
1:	1	2	3	7	0	0	2	0	0	3:	1	1	488	8	
1:	1	2	4	5	1	0	1	0	-1	2:	1	1	488	4	
1:	2	2	3	3	0	0	0	0	0	2:	1	1	488	4	
1:	2	2	3	4	2	1	0	1	0	2:	1	1	488	4	
1:	2	3	3	3	0	0	2	-2	2	-1:	1	1	488	4	
?:	1	1	1	61	1	1	0	0	0	0:	-1-1		488	96	77/32
2:	1	1	2	16	0	0	0	0	1	2:	-1-1		488	16	
2:	1	1	2	18	0	1	0	0	0	2:	-1-1		488	8	
2:	1	1	2	21	0	1	1	1	1	0:	-1-1		488	16	
2:	1	1	5	7	0	1	1	0	0	2:	-1-1		488	16	
2:	1	1	6	6	0	0	0	0	1	4:	-1-1		488	8	
2:	1	1	6	8	1	1	0	0	1	3:	-1-1		488	12	
2:	1	2	2	13	1	0	2	0	2	0:	-1-1		488	8	
2:	1	2	3	6	0	0	1	0	2	2:	-1-1		488	4	
2:	1	2	3	8	1	0	2	1	0	2:	-1-1		488	4	
2:	1	2	4	4	0	0	1	0	0	1:	-1-1		488	4	
2:	1	2	4	5	0	1	2	0	0	2:	-1-1		488	8	
2:	1	2	4	5	1	1	0	1	0	0:	-1-1		488	4	
2:	1	2	4	5	0	0	2	0	0	3:	-1-1		488	8	
2:	1	2	5	5	1	1	0	1	2	4:	-1-1		488	4	
2:	2	2	3	4	0	2	1	2	1	0:	-1-1		488	4	

D= 489

GENUS#:	F11	F22	F33	F44	F12	F13	F23	F14	F24	F34:	HASSE	SYM	LEVEL	AUTOS	MASS
1:	1	1	1	41	1	0	0	0	0	1:	-1-1	1	489	48	187/48
1:	1	1	2	18	0	1	0	1	1	0:	-1-1	1	489	8	
1:	1	1	5	8	0	1	0	0	1	5:	-1-1	1	489	8	
1:	1	2	2	10	1	0	1	1	2	0:	-1-1	1	489	4	
1:	1	2	2	11	0	1	0	0	2	1:	-1-1	1	489	8	
1:	1	2	3	7	1	1	0	0	1	2:	-1-1	1	489	4	
1:	1	2	4	5	1	0	0	1	2	1:	-1-1	1	489	4	
1:	1	2	4	5	0	0	1	-1	-1	3:	-1-1	1	489	4	
1:	2	2	2	5	1	1	1	-1	1	1:	-1-1	1	489	4	
1:	2	2	2	5	1	1	0	1	2	1:	-1-1	1	489	2	

GENUS#	F11	F22	F33	F44	F12	F13	F23	F14	F24	F34	HASSE SYM	LEVEL	AUTOS	MASS
1:	2	2	2	5	1	0	0	2	2	1:	-1-1 1	489	4	
1:	2	2	2	7	2	1	-1	1	0	1:	-1-1 1	489	4	
1:	2	2	3	3	1	1	1	0	0	1:	-1-1 1	489	4	
1:	2	2	3	4	1	2	0	0	2	1:	-1-1 1	489	2	
1:	2	2	3	4	1	1	1	2	2	-1:	-1-1 1	489	4	
1:	2	2	3	4	1	0	0	2	0	3:	-1-1 1	489	4	
2:	1	1	2	21	0	1	1	1	0	2:	-1 1-1	489	16	187/48
2:	1	1	3	14	1	0	0	1	0	1:	-1 1-1	489	12	
2:	1	1	4	9	0	1	0	0	1	3:	-1 1-1	489	8	
2:	1	1	4	9	0	1	1	1	0	0:	-1 1-1	489	8	
2:	1	2	2	9	1	0	0	0	1	-1:	-1 1-1	489	4	
2:	1	2	3	6	0	1	1	0	1	1:	-1 1-1	489	4	
2:	1	2	4	5	0	1	2	0	-1	1:	-1 1-1	489	4	
2:	1	2	4	5	1	1	0	1	1	0:	-1 1-1	489	4	
2:	1	2	4	6	1	0	2	1	2	3:	-1 1-1	489	4	
2:	1	2	4	6	1	1	0	0	2	3:	-1 1-1	489	4	
2:	1	3	4	4	1	0	3	1	1	2:	-1 1-1	489	4	
2:	1	3	4	4	0	1	3	0	1	3:	-1 1-1	489	4	
2:	1	3	4	4	1	0	3	1	2	0:	-1 1-1	489	4	
2:	1	3	4	4	0	1	2	1	3	1:	-1 1-1	489	4	
2:	2	2	3	4	0	2	1	1	0	-2:	-1 1-1	489	2	
2:	2	3	3	3	2	1	0	1	-1	-2:	-1 1-1	489	2	

D= 492

GENUS#	F11	F22	F33	F44	F12	F13	F23	F14	F24	F34	HASSE SYM	LEVEL	AUTOS	MASS
1:	1	1	1	41	1	0	0	0	0	0:	-1 1-1	492	48	45/32
1:	1	1	2	21	1	0	0	0	0	2:	-1 1-1	492	48	
1:	1	1	3	11	0	0	0	0	0	3:	-1 1-1	492	32	
1:	1	1	3	12	0	1	0	0	0	3:	-1 1-1	492	8	
1:	1	1	4	12	1	1	0	1	1	3:	-1 1-1	492	12	
1:	1	1	5	9	1	1	0	0	0	2:	-1 1-1	492	12	
1:	1	1	5	9	1	0	0	0	0	4:	-1 1-1	492	24	
1:	1	2	2	9	0	0	1	0	2	2:	-1 1-1	492	8	
1:	1	2	3	6	0	0	0	0	1	3:	-1 1-1	492	8	
1:	1	2	4	5	1	1	0	0	0	2:	-1 1-1	492	4	
1:	1	2	4	5	0	1	2	1	0	0:	-1 1-1	492	8	
1:	1	3	3	4	0	0	1	0	2	-1:	-1 1-1	492	4	
1:	1	3	3	5	0	0	3	0	2	0:	-1 1-1	492	8	
2:	1	1	1	62	1	1	0	1	0	0:	1 1 1	492	96	45/32
2:	1	1	2	16	0	0	0	1	0	1:	1 1 1	492	16	
2:	1	1	2	25	1	1	0	1	0	0:	1 1 1	492	12	
2:	1	1	3	14	1	0	0	1	0	0:	1 1 1	492	24	
2:	1	1	3	14	1	0	0	0	0	2:	1 1 1	492	24	
2:	1	1	6	7	1	0	0	0	0	2:	1 1 1	492	24	
2:	1	2	3	6	1	0	0	0	1	0:	1 1 1	492	8	
2:	1	2	3	7	0	0	2	1	2	0:	1 1 1	492	8	
2:	1	3	3	4	1	0	0	1	0	0:	1 1 1	492	8	
2:	1	3	3	4	0	0	0	1	2	0:	1 1 1	492	8	
2:	2	2	3	4	0	2	2	1	0	0:	1 1 1	492	4	
2:	2	2	3	4	1	2	-1	1	1	1:	1 1 1	492	4	
2:	2	3	3	4	2	0	3	0	0	3:	1 1 1	492	8	
3:	1	1	2	21	0	1	1	1	1	1:	1-1-1	492	32	45/32
3:	1	1	6	6	0	1	1	0	0	3:	1-1-1	492	16	
3:	1	2	2	10	1	0	1	0	0	2:	1-1-1	492	4	
3:	1	2	2	11	0	0	2	1	1	2:	1-1-1	492	16	
3:	1	2	4	5	0	1	0	1	2	2:	1-1-1	492	8	
3:	2	2	2	6	2	1	0	-1	0	1:	1-1-1	492	4	
3:	2	2	3	5	2	1	-1	0	0	3:	1-1-1	492	8	
3:	2	2	3	5	2	2	0	2	1	0:	1-1-1	492	4	
3:	2	3	3	3	2	0	2	2	2	1:	1-1-1	492	4	
4:	1	1	1	31	0	0	0	0	1	0:	-1-1 1	492	32	45/32
4:	1	1	4	8	0	0	0	0	1	1:	-1-1 1	492	8	
4:	1	2	2	13	1	0	2	0	0	2:	-1-1 1	492	8	
4:	1	2	4	5	0	0	1	0	0	4:	-1-1 1	492	4	
4:	1	2	5	5	1	0	2	0	2	4:	-1-1 1	492	8	
4:	1	3	4	4	1	0	1	-1	1	-3:	-1-1 1	492	4	
4:	2	2	2	7	2	1	-1	1	1	0:	-1-1 1	492	8	
4:	2	2	3	3	1	1	0	1	0	0:	-1-1 1	492	4	
4:	2	2	3	4	2	0	0	2	1	1:	-1-1 1	492	8	

D= 493

GENUS#: F11 F22 F33 F44 F12 F13 F23 F14 F24 F34: HASSE SYM LEVEL AUTOS MASS

GENUS#	F11	F22	F33	F44	F12	F13	F23	F14	F24	F34	HASSE	SYM	LEVEL	AUTOS	MASS
1:	1	1	1	62	1	1	0	0	1	0	-1 1	-1	493	48	119/48
1:	1	1	3	16	1	1	0	1	1	2	-1 1	-1	493	12	
1:	1	1	5	7	0	1	0	1	1	0	-1 1	-1	493	8	
1:	1	1	5	9	1	0	0	1	1	3	-1 1	-1	493	12	
1:	1	1	6	9	1	1	0	1	1	6	-1 1	-1	493	24	
1:	1	2	2	9	0	0	1	1	2	0	-1 1	-1	493	4	
1:	1	2	3	6	0	1	1	0	1	0	-1 1	-1	493	4	
1:	1	2	3	7	0	1	2	0	1	1	-1 1	-1	493	4	
1:	1	2	3	7	0	1	1	1	2	3	-1 1	-1	493	4	
1:	1	2	3	7	1	1	0	1	1	2	-1 1	-1	493	4	
1:	1	2	3	8	1	0	2	0	1	3	-1 1	-1	493	8	
1:	1	3	3	4	1	0	1	0	1	1	-1 1	-1	493	4	
1:	2	2	3	3	1	1	0	0	1	0	-1 1	-1	493	4	
1:	3	3	3	3	3	2	0	1	2	3	-1 1	-1	493	8	
2:	1	1	2	18	0	1	0	0	1	1	-1 -1	1	493	12	119/48
2:	1	1	2	21	1	0	0	1	0	1	-1 -1	1	493	12	
2:	1	1	2	25	1	1	0	0	0	0	-1 -1	1	493	16	
2:	1	1	6	6	0	1	0	0	1	3	-1 -1	1	493	16	
2:	1	1	7	7	1	1	0	1	1	4	-1 -1	1	493	24	
2:	1	1	7	8	1	1	0	0	1	6	-1 -1	1	493	12	
2:	1	2	3	6	0	1	0	1	1	1	-1 -1	1	493	4	
2:	1	2	3	6	0	0	1	1	1	2	-1 -1	1	493	4	
2:	1	2	4	5	0	1	1	1	2	1	-1 -1	1	493	8	
2:	1	3	3	5	1	1	-1	0	2	1	-1 -1	1	493	4	
2:	1	3	3	5	1	0	2	0	1	3	-1 -1	1	493	8	
2:	1	3	3	6	1	0	3	0	3	1	-1 -1	1	493	4	
2:	2	3	3	3	1	0	3	2	1	1	-1 -1	1	493	4	
2:	2	3	3	3	1	1	-1	2	3	0	-1 -1	1	493	2	

D= 496

GENUS#	F11	F22	F33	F44	F12	F13	F23	F14	F24	F34	HASSE	SYM	LEVEL	AUTOS	MASS
1:	1	1	1	31	0	0	0	0	0	0	-1	-1	124	96	5/12
1:	1	1	2	16	0	0	0	0	0	2	-1	-1	124	32	
1:	1	1	4	8	0	0	0	0	0	4	-1	-1	124	16	
1:	1	1	5	7	0	0	0	0	0	4	-1	-1	124	16	
1:	1	2	4	5	0	0	2	0	2	0	-1	-1	124	8	
1:	2	2	3	4	0	2	0	0	2	2	-1	-1	124	8	
2:	1	1	1	62	1	1	0	0	0	0	-1	-1	248	96	5/12
2:	1	1	3	13	0	1	1	1	1	0	-1	-1	248	16	
2:	1	2	2	9	0	0	0	1	2	2	-1	-1	248	32	
2:	1	2	3	7	0	1	2	0	0	1	-1	-1	248	8	
2:	1	3	3	5	1	1	-1	-1	-2	1	-1	-1	248	8	
2:	2	2	3	5	0	2	2	0	0	3	-1	-1	248	16	
3:	1	1	2	18	0	1	0	1	0	0	-1	-1	496	8	10/3
3:	1	1	3	16	1	1	0	0	1	1	-1	-1	496	12	
3:	1	1	4	8	0	0	0	1	0	0	-1	-1	496	16	
3:	1	1	4	9	0	0	0	1	0	4	-1	-1	496	16	
3:	1	1	5	7	0	1	0	0	0	3	-1	-1	496	8	
3:	1	2	2	8	0	0	0	0	-1	1	-1	-1	496	8	
3:	1	2	3	7	1	1	0	1	0	2	-1	-1	496	4	
3:	1	2	4	4	0	0	0	0	1	0	-1	-1	496	8	
3:	1	2	4	5	0	0	0	0	1	4	-1	-1	496	8	
3:	1	2	4	5	0	0	2	0	1	3	-1	-1	496	4	
3:	1	2	4	5	1	0	0	0	2	0	-1	-1	496	8	
3:	1	3	3	4	0	0	2	0	1	1	-1	-1	496	8	
3:	1	3	3	5	0	0	2	0	3	3	-1	-1	496	8	
3:	1	3	3	5	1	1	-1	0	2	-2	-1	-1	496	8	
3:	1	3	4	4	1	1	-1	0	2	2	-1	-1	496	4	
3:	1	3	4	4	1	0	3	0	2	0	-1	-1	496	4	
3:	1	3	4	4	1	1	-1	1	3	0	-1	-1	496	4	
3:	2	2	2	5	1	1	0	2	0	0	-1	-1	496	2	
3:	2	3	3	3	1	1	-1	0	1	3	-1	-1	496	2	
4:	1	1	2	21	1	0	0	0	1	0	1	1	248	24	5/12
4:	1	2	3	6	0	1	0	0	0	2	1	1	248	8	
4:	1	3	3	5	1	1	0	1	1	3	1	1	248	4	
5:	1	1	2	16	0	0	0	1	1	0	-1	-1	248	32	5/4
5:	1	1	4	10	0	1	1	0	0	4	-1	-1	248	16	
5:	1	2	2	8	0	0	0	1	0	0	-1	-1	248	32	
5:	1	2	2	9	1	0	0	0	1	0	-1	-1	248	8	
5:	1	2	4	6	1	0	0	0	2	4	-1	-1	248	16	
5:	2	2	2	4	0	0	0	1	0	0	-1	-1	248	16	
5:	2	2	2	5	1	1	0	1	-1	-1	-1	-1	248	4	

	F11	F22	F33	F44	F12	F13	F23	F14	F24	F34	HASSE SYM	LEVEL	AUTOS	MASS
5:	2	2	2	5	0	0	0	2	2	1:	-1-1	248	16	
5:	2	2	2	8	2	2	0	1	0	0:	-1-1	248	16	
5:	2	2	3	4	0	2	0	0	0	3:	-1-1	248	8	
5:	2	2	3	4	0	2	1	0	2	2:	-1-1	248	4	
5:	2	2	3	4	0	2	2	0	0	1:	-1-1	248	16	
6:	1	1	6	8	1	1	0	0	0	4:	1 1	496	12	10/3
6:	1	2	2	10	1	0	1	1	0	1:	1 1	496	4	
6:	1	2	3	6	0	1	1	0	0	1:	1 1	496	4	
6:	1	2	3	7	0	1	1	0	2	3:	1 1	496	4	
6:	1	2	3	8	1	0	2	0	2	0:	1 1	496	4	
6:	1	2	4	5	0	1	1	1	1	3:	1 1	496	4	
6:	1	3	3	5	1	0	2	1	0	2:	1 1	496	4	
6:	1	3	3	6	1	0	3	1	3	1:	1 1	496	4	
6:	2	2	3	4	1	2	0	2	0	0:	1 1	496	2	
6:	2	3	3	3	2	1	0	2	2	2:	1 1	496	2	
6:	2	3	3	3	2	0	2	1	0	2:	1 1	496	2	
7:	1	2	2	11	0	0	2	0	2	0:	1 1	124	24	5/12
7:	1	2	3	6	0	0	0	0	2	2:	1 1	124	8	
7:	2	3	3	3	2	0	2	2	0	0:	1 1	124	4	
8:	1	2	3	6	0	0	0	1	2	1:	1 1	248	8	5/4
8:	1	2	4	5	1	1	0	0	1	1:	1 1	248	4	
8:	1	2	4	6	0	1	2	0	0	4:	1 1	248	8	
8:	2	2	3	4	2	1	0	0	0	2:	1 1	248	4	
8:	2	2	3	4	1	1	1	1	2	3:	1 1	248	2	

D= 497

GENUS#:	F11	F22	F33	F44	F12	F13	F23	F14	F24	F34	HASSE SYM	LEVEL	AUTOS	MASS
1:	1	1	1	42	1	0	0	1	0	1:	-1 1-1	497	24	27/8
1:	1	1	2	21	0	1	1	1	0	0:	-1 1-1	497	8	
1:	1	1	2	25	1	1	0	0	0	1:	-1 1-1	497	12	
1:	1	1	5	7	0	1	0	0	1	2:	-1 1-1	497	8	
1:	1	2	2	11	1	1	0	1	1	2:	-1 1-1	497	4	
1:	1	2	2	13	1	0	2	1	0	1:	-1 1-1	497	4	
1:	1	2	3	7	1	0	1	0	0	3:	-1 1-1	497	4	
1:	1	2	3	7	1	0	1	1	2	2:	-1 1-1	497	4	
1:	1	2	3	7	1	1	0	0	1	-2:	-1 1-1	497	4	
1:	1	2	4	6	1	0	2	1	1	3:	-1 1-1	497	4	
1:	1	2	4	6	1	0	1	1	0	4:	-1 1-1	497	4	
1:	1	2	4	6	1	0	2	0	2	3:	-1 1-1	497	4	
1:	2	2	3	4	1	1	-1	-1	2	0:	-1 1-1	497	2	
1:	2	2	3	4	0	2	1	1	2	2:	-1 1-1	497	2	
2:	1	1	2	18	0	1	0	0	1	0:	-1-1 1	497	16	27/8
2:	1	1	3	12	0	1	0	1	1	2:	-1-1 1	497	8	
2:	1	1	4	10	0	1	1	1	0	4:	-1-1 1	497	16	
2:	1	1	4	12	1	1	0	0	1	2:	-1-1 1	497	12	
2:	1	1	6	6	0	1	1	1	0	2:	-1-1 1	497	8	
2:	1	1	7	8	1	0	0	1	0	7:	-1-1 1	497	24	
2:	1	2	2	9	1	0	0	0	0	1:	-1-1 1	497	8	
2:	1	2	3	6	0	0	1	1	0	2:	-1-1 1	497	4	
2:	1	2	3	6	1	0	0	0	0	1:	-1-1 1	497	8	
2:	1	2	4	5	1	0	0	0	0	3:	-1-1 1	497	8	
2:	1	3	3	4	0	0	1	-1	-1	1:	-1-1 1	497	8	
2:	1	3	4	4	0	0	3	1	1	3:	-1-1 1	497	4	
2:	1	4	4	4	1	0	4	0	4	1:	-1-1 1	497	8	
2:	2	2	2	5	1	1	0	2	1	0:	-1-1 1	497	4	
2:	2	2	2	6	2	1	0	0	1	1:	-1-1 1	497	2	
2:	2	2	3	3	0	1	1	1	0	1:	-1-1 1	497	2	
2:	2	2	4	4	2	1	-1	0	1	3:	-1-1 1	497	4	
2:	2	3	3	3	1	1	-2	2	2	-1:	-1-1 1	497	4	

D- 500

GENUS#:	F11	F22	F33	F44	F12	F13	F23	F14	F24	F34	HASSE SYM	LEVEL	AUTOS	MASS
1:	1	1	1	32	0	0	0	1	1	1:	1 1	250	96	125/96
1:	1	1	2	21	0	1	1	0	0	1:	1 1	250	16	
1:	1	1	4	9	0	1	1	0	0	1:	1 1	250	16	
1:	1	2	2	10	1	0	0	0	2	2:	1 1	250	8	
1:	1	2	4	5	1	1	0	0	-1	1:	1 1	250	4	
1:	1	2	4	6	0	1	2	1	0	4:	1 1	250	8	
1:	2	2	2	6	2	0	0	2	2	1:	1 1	250	12	
1:	2	2	2	8	2	2	0	1	0	1:	1 1	250	12	
1:	2	2	3	4	2	1	0	-1	0	1:	1 1	250	4	
1:	2	2	3	4	0	1	0	0	2	3:	1 1	250	4	

	F11	F22	F33	F44	F12	F13	F23	F14	F24	F34:	HASSE	SYM	LEVEL	AUTOS	MASS
2:	1	1	1	42	1	0	0	1	0	0:	-1-1		500	24	125/24
2:	1	1	2	16	0	0	0	1	0	1:	-1-1		500	8	
2:	1	1	2	18	0	1	0	0	0	1:	-1-1		500	8	
2:	1	1	3	11	0	0	0	0	1	2:	-1-1		500	8	
2:	1	1	3	16	1	1	0	1	0	0:	-1-1		500	12	
2:	1	1	6	7	0	1	0	0	0	6:	-1-1		500	8	
2:	1	1	6	8	1	1	0	1	1	3:	-1-1		500	12	
2:	1	2	2	10	1	0	1	1	1	1:	-1-1		500	4	
2:	1	2	2	11	1	1	0	0	2	0:	-1-1		500	4	
2:	1	2	3	6	0	0	1	0	-1	2:	-1-1		500	4	
2:	1	2	3	8	1	0	2	1	2	0:	-1-1		500	4	
2:	1	2	4	5	1	0	1	1	0	1:	-1-1		500	4	
2:	1	2	4	5	1	0	0	1	1	2:	-1-1		500	4	
2:	1	2	4	5	0	0	2	1	1	2:	-1-1		500	4	
2:	1	3	3	4	0	0	2	0	1	0:	-1-1		500	4	
2:	1	3	3	4	1	0	1	0	1	0:	-1-1		500	4	
2:	1	3	3	5	0	0	2	1	3	0:	-1-1		500	4	
2:	2	2	2	5	1	1	0	0	0	2:	-1-1		500	2	
2:	2	2	3	4	1	1	0	2	1	3:	-1-1		500	2	
2:	2	2	3	4	1	2	0	2	1	0:	-1-1		500	2	
2:	2	3	3	3	1	0	1	2	-1	2:	-1-1		500	2	
3:	1	1	1	63	1	1	0	1	0	0:	-1-1		250	96	125/96
3:	1	1	3	11	0	0	0	1	1	1:	-1-1		250	16	
3:	1	1	3	16	1	1	0	0	0	2:	-1-1		250	12	
3:	1	1	4	11	1	0	0	1	0	2:	-1-1		250	12	
3:	1	1	5	7	0	1	1	0	0	1:	-1-1		250	16	
3:	1	2	3	7	0	1	0	0	2	3:	-1-1		250	8	
3:	1	2	4	5	0	0	2	1	0	2:	-1-1		250	8	
3:	1	3	3	5	1	0	1	1	0	3:	-1-1		250	4	
3:	1	3	3	5	1	1	-1	1	1	2:	-1-1		250	4	
3:	2	3	3	3	2	0	1	2	1	2:	-1-1		250	4	
4:	1	1	2	25	1	1	0	0	0	0:	-1-1		100	24	5/12
4:	1	2	3	7	1	0	1	0	2	2:	-1-1		100	4	
4:	1	2	3	7	0	0	2	0	1	3:	-1-1		100	8	
5:	1	1	3	13	0	1	1	1	1	1:	-1-1		50	32	5/32
5:	1	3	4	4	1	0	2	0	-2	2:	-1-1		50	8	
6:	1	1	5	8	0	0	0	1	1	5:	1 1		50	32	5/32
6:	2	2	5	1	0	0	1	-1	-2:	1 1			50	8	
7:	1	1	7	7	1	1	0	0	1	3:	-1-1		50	24	5/48
7:	1	2	3	7	0	0	2	1	2	1:	-1-1		50	16	
8:	1	2	2	13	1	0	2	0	1	2:	1 1		50	16	5/48
8:	2	2	4	4	2	2	0	0	2	1:	1 1		50	24	
9:	1	4	4	4	1	1	-2	1	3	3:	1 1		10	96	1/96
10:	1	1	4	9	0	1	0	1	0	3:	-1-1		100	8	5/8
10:	1	3	4	4	1	0	3	0	2	2:	-1-1		100	4	
10:	2	2	3	4	0	2	0	1	1	3:	-1-1		100	4	
11:	2	2	2	5	1	1	-1	0	0	0:	-1-1		20	24	1/24
12:	2	3	3	3	2	2	1	2	1	1:	-1-1		10	96	1/96

D= 729
GENUS#:	F11	F22	F33	F44	F12	F13	F23	F14	F24	F34:	HASSE	SYM	LEVEL	AUTOS	MASS
1:	1	1	1	61	1	0	0	0	0	1:	-1-1		243	48	27/16
1:	1	1	7	8	0	1	0	1	1	5:	-1-1		243	8	
1:	1	1	7	9	1	0	0	0	0	3:	-1-1		243	24	
1:	1	2	2	16	1	1	0	0	2	1:	-1-1		243	8	
1:	1	2	3	10	1	0	0	0	2	3:	-1-1		243	8	
1:	1	2	4	7	1	0	1	0	1	-1:	-1-1		243	4	
1:	1	3	3	8	0	0	3	1	0	0:	-1-1		243	24	
1:	1	3	4	5	0	0	3	1	0	1:	-1-1		243	8	
1:	2	2	2	7	1	1	1	1	1	1:	-1-1		243	12	
1:	2	2	2	7	1	0	0	2	2	1:	-1-1		243	4	
1:	2	2	4	4	1	1	1	2	2	1:	-1-1		243	4	
1:	2	3	4	4	0	1	0	2	3	4:	-1-1		243	4	
2:	1	1	2	31	0	1	1	1	0	2:	-1-1		243	16	27/16
2:	1	1	4	13	0	1	0	0	1	3:	-1-1		243	8	
2:	1	2	4	8	1	0	2	1	0	1:	-1-1		243	4	
2:	1	2	5	6	1	1	0	0	0	3:	-1-1		243	4	
2:	1	4	4	4	1	1	-1	0	0	3:	-1-1		243	4	
2:	2	2	4	4	0	1	1	2	1	3:	-1-1		243	2	
2:	2	2	4	5	2	1	-1	2	1	2:	-1-1		243	4	
3:	1	1	3	21	1	0	0	0	0	3:	-1-1		81	48	3/16
3:	1	3	4	6	0	1	3	0	3	0:	-1-1		81	8	

```
     3:  3  3  3  4  3  0  0  3  0   3: -1-1       81   24
     4:  1  1  7  7  0  1  0  0  1   0: -1-1       27   32     9/32
     4:  1  2  4  8  1  0  1  1  1   4: -1-1       27    8
     4:  2  2  4  4  0  2  1  1  2   1: -1-1       27    8
     5:  1  1  9  9  1  0  0  0  0   9: -1-1       27  144     1/24
     5:  1  3  3  7  0  0  3  1  0   0: -1-1       27   48
     5:  3  3  4  4  3  3  0  3  3   4: -1-1       27   72
     6:  2  2  2 11  2  1 -1  1  2   1: -1-1       81   16     3/16
     6:  2  2  3  5  1  0  0  1  1   3: -1-1       81    8
     7:  2  2  2  8  1  1  1  2  2  -1: -1-1       27   24     1/24
     8:  2  2  5  5  2  1 -1  1  2   4: -1-1       27   24     1/24
     9:  1  4  4  4  1  1 -1  1  2   2: -1-1       27   24     1/24

D=1729
GENUS#: F11 F22 F33 F44 F12 F13 F23 F14 F24 F34: HASSE SYM  LEVEL AUTOS   MASS
     1:  1  1 10 14  0  1  1  1  0  10: -1 1 1-1   1729   16  703/48
     1:  1  1  6 26  1  1  0  0  1   2: -1 1 1-1   1729   12
     1:  1  1  9 14  0  1  0  0  1   7: -1 1 1-1   1729    8
     1:  1  2  3 26  1  0  2  0  2   1: -1 1 1-1   1729    4
     1:  1  2  4 17  0  1  2  1  1   1: -1 1 1-1   1729    8
     1:  1  2  4 17  1  1  0  0 -1   2: -1 1 1-1   1729    4
     1:  1  2  6 11  0  1  2  0 -1   3: -1 1 1-1   1729    4
     1:  1  2  6 12  1  0  2  1  1   3: -1 1 1-1   1729    4
     1:  1  2  7 10  1  0  0  1  0   5: -1 1 1-1   1729    4
     1:  1  2  7  9  1  0  0  0  1   1: -1 1 1-1   1729    4
     1:  1  2  8  8  0  1  0  1  1   5: -1 1 1-1   1729    4
     1:  1  2  9  9  1  1  0  0  1  -8: -1 1 1-1   1729    4
     1:  1  3  4 13  1  0  3  0  2   3: -1 1 1-1   1729    4
     1:  1  3  5  8  1  0  0  0  1   1: -1 1 1-1   1729    4
     1:  1  3  6  7  0  1  1  1  2   2: -1 1 1-1   1729    4
     1:  1  3  6  8  0  1  1  0  3   5: -1 1 1-1   1729    4
     1:  1  3  6  8  1  1 -1  0  1   4: -1 1 1-1   1729    4
     1:  1  4  4  9  0  0  3 -1  3  -1: -1 1 1-1   1729    4
     1:  1  4  5  7  1  0  3  0  1   3: -1 1 1-1   1729    4
     1:  1  4  5  7  1  0  3  0  2  -1: -1 1 1-1   1729    4
     1:  1  4  5  7  1  0  3  1  1   2: -1 1 1-1   1729    4
     1:  1  4  5  8  1  0  3  1  4   4: -1 1 1-1   1729    4
     1:  2  2  3 11  1  1  1  2  1   2: -1 1 1-1   1729    2
     1:  2  2  3 12  0  2  1  1  2   1: -1 1 1-1   1729    4
     1:  2  2  4  8  1  1 -1  0  1   0: -1 1 1-1   1729    2
     1:  2  2  4  8  0  1  0  2  1   3: -1 1 1-1   1729    2
     1:  2  2  6  6  1  1  1  1  2   4: -1 1 1-1   1729    2
     1:  2  2  6  6  1  2  0  2  1   3: -1 1 1-1   1729    4
     1:  2  2  6  7  2  1 -1  1  0   3: -1 1 1-1   1729    2
     1:  2  3  3  7  1  1  0  0  1   2: -1 1 1-1   1729    2
     1:  2  3  3  7  1  1  0  1  1   2: -1 1 1-1   1729    2
     1:  2  3  3  8  1  0  2  2 -1   1: -1 1 1-1   1729    2
     1:  2  3  4  6  2  1  0  0  1   2: -1 1 1-1   1729    2
     1:  2  3  4  6  0  2  1 -1 -1   2: -1 1 1-1   1729    2
     1:  2  3  4  7  1  1 -3  0  1   2: -1 1 1-1   1729    2
     1:  2  3  4  7  1  1 -3  0  2  -3: -1 1 1-1   1729    2
     1:  2  3  4  8  2  0  3  1  0   3: -1 1 1-1   1729    2
     1:  2  4  4  5  2  1 -1  2  3   1: -1 1 1-1   1729    2
     1:  3  3  4  5  2  3  0  0  1  -3: -1 1 1-1   1729    2
     1:  3  3  4  5  2  3  1  0  3   0: -1 1 1-1   1729    4
     1:  3  3  4  6  3  2 -1  0  1   3: -1 1 1-1   1729    2
     1:  3  4  4  4  2  2 -1  3  3   2: -1 1 1-1   1729    2
     1:  3  4  4  4  3  2  0 -1  0   3: -1 1 1-1   1729    2
     2:  1  1 10 17  1  1  0  0  0   9: -1-1 1 1   1729   12  703/48
     2:  1  1  4 32  0  1  1  1  0   4: -1-1 1 1   1729   16
     2:  1  1  6 20  0  1  1  0  1   2: -1-1 1 1   1729    8
     2:  1  2  2 31  1  0  0  0  0   1: -1-1 1 1   1729    8
     2:  1  2  4 16  0  1  1  1  0   4: -1-1 1 1   1729    4
     2:  1  2  7 10  0  1  1  1  0   7: -1-1 1 1   1729    4
     2:  1  2  8  8  1  0  0  0  0   3: -1-1 1 1   1729   16
     2:  1  3  3 15  1  0  2  0  0   1: -1-1 1 1   1729    4
     2:  1  3  5 10  1  0  2  1  1   5: -1-1 1 1   1729    4
     2:  1  3  5 10  0  0  3 -1  2  -3: -1-1 1 1   1729    4
     2:  1  3  5  9  1  0  2  0  2  -1: -1-1 1 1   1729    4
     2:  1  3  5  9  0  1  2  0  1   4: -1-1 1 1   1729    4
     2:  1  3  6  8  1  0  1  1  3   4: -1-1 1 1   1729    4
     2:  1  4  4  8  1  1  0  1  1   0: -1-1 1 1   1729    8
```

```
2:   1   4   4   8    1   0   2   0   2    1: -1-1 1 1    1729    4
2:   1   4   5   6    0   0   1   1   2    1: -1-1 1 1    1729    4
2:   1   4   5   8    1   1  -2   1   0    5: -1-1 1 1    1729    4
2:   1   4   6   6    0   1   2   0   1    5: -1-1 1 1    1729    4
2:   1   5   5   5    1   0   0   0   0    3: -1-1 1 1    1729   16
2:   1   5   5   6    1   0   4  -1   1   -1: -1-1 1 1    1729    4
2:   1   5   5   6    1   0   2   0   1    5: -1-1 1 1    1729    8
2:   2   2   2  20    2   1   0   1   1   -1: -1-1 1 1    1729    2
2:   2   2   6   6    1   1   0   1   0    5: -1-1 1 1    1729    4
2:   2   2   6   6    1   1   0   2   2    3: -1-1 1 1    1729    2
2:   2   3   3   9    1   0   3   1   2    2: -1-1 1 1    1729    2
2:   2   3   3   9    2   1  -1  -1   2    0: -1-1 1 1    1729    2
2:   2   3   3   9    1   1  -2  -1   3    0: -1-1 1 1    1729    4
2:   2   3   4   5    1   1   1   0   1    0: -1-1 1 1    1729    2
2:   2   3   4   6    0   1   1   2   3    1: -1-1 1 1    1729    4
2:   2   3   4   6    1   1   1   2   3    0: -1-1 1 1    1729    2
2:   2   3   4   6    2   0   1  -1  -2    1: -1-1 1 1    1729    2
2:   2   3   4   6    0   0   3   1   2    1: -1-1 1 1    1729    4
2:   2   3   4   6    1   1  -2   2   1    0: -1-1 1 1    1729    2
2:   2   3   4   7    2   1  -2   1  -1    1: -1-1 1 1    1729    4
2:   2   3   5   6    1   0   2   2   0    5: -1-1 1 1    1729    2
2:   2   4   4   4    0   0   1   1   2    2: -1-1 1 1    1729    4
2:   2   4   4   5    1   0   2   2   3    3: -1-1 1 1    1729    2
2:   3   3   3   5    1   0   0   2  -2    1: -1-1 1 1    1729    4
2:   3   3   4   5    1   3  -1   2  -2    3: -1-1 1 1    1729    2
2:   3   3   4   5    0   3   2  -2  -1    1: -1-1 1 1    1729    2
2:   3   3   4   5    1   3   0  -2   2    1: -1-1 1 1    1729    8
2:   3   3   4   5    1   1  -1  -2   2    3: -1-1 1 1    1729    2
2:   3   3   4   5    1   2   1   2  -2   -3: -1-1 1 1    1729    2
2:   3   3   4   5    0   3   1  -2  -1    2: -1-1 1 1    1729    2
2:   3   3   4   6    1   3  -2   1   3    2: -1-1 1 1    1729    4
2:   3   4   4   5    3   1  -3   1   4    0: -1-1 1 1    1729    2
2:   3   4   4   5    3   1  -3   2   3    2: -1-1 1 1    1729   16    703/48
3:   1   1   2  62    0   1   0   0   1    0: -1-1-1-1    1729    8
3:   1   1   3  40    0   1   0   1   1    2: -1-1-1-1    1729    8
3:   1   1   4  40    1   0   0   1   1    2: -1-1-1-1    1729   12
3:   1   2   4  16    1   0   0   0   0    3: -1-1-1-1    1729    8
3:   1   2   5  12    0   0   1   1  -1    3: -1-1-1-1    1729    4
3:   1   2   8   8    0   0   1   1   2    4: -1-1-1-1    1729    4
3:   1   3   4  10    0   1   1   0   1   -1: -1-1-1-1    1729    4
3:   1   3   4  11    0   1   1   0   3    3: -1-1-1-1    1729    4
3:   1   3   4  12    1   1  -1   1   3    1: -1-1-1-1    1729    4
3:   1   3   6   8    1   1   0   1   0    5: -1-1-1-1    1729    4
3:   1   3   6   8    1   0   3   1   1    2: -1-1-1-1    1729    4
3:   1   4   4  11    1   0   4   1   0    3: -1-1-1-1    1729    4
3:   1   5   5   6    0   0   1   1   4   -3: -1-1-1-1    1729    4
3:   1   5   6   6    0   0   1   1   4    6: -1-1-1-1    1729    8
3:   1   6   6   6    1   0   6  -1   4   -1: -1-1-1-1    1729    4
3:   2   2   2  16    1   1   0   2   1    0: -1-1-1-1    1729    2
3:   2   2   3  10    0   1   1   0   1    1: -1-1-1-1    1729    2
3:   2   2   4   8    1   1   0   2   1    0: -1-1-1-1    1729    2
3:   2   2   4   9    1   1   0  -1   2    3: -1-1-1-1    1729    2
3:   2   2   5   9    2   1   0   2   1    5: -1-1-1-1    1729    4
3:   2   3   3   7    1   1  -1   1   0    0: -1-1-1-1    1729    2
3:   2   3   3   8    2   1   0   0   1   -2: -1-1-1-1    1729    2
3:   2   3   4   6    0   1   2   2   1    3: -1-1-1-1    1729    2
3:   2   3   4   6    0   1   2   1   3    2: -1-1-1-1    1729    4
3:   2   3   5   5    1   1  -2   1  -2    3: -1-1-1-1    1729    2
3:   2   3   5   5    1   1   0  -1   2    3: -1-1-1-1    1729    4
3:   2   3   5   5    2   1  -1   1   2    0: -1-1-1-1    1729    2
3:   2   3   5   5    2   1   2  -1  -1    0: -1-1-1-1    1729    2
3:   2   3   5   5    1   1   2   2   2    3: -1-1-1-1    1729    2
3:   2   3   5   5    1   1   2  -1   2    3: -1-1-1-1    1729    2
3:   2   3   5   6    2   1  -1   0   3    2: -1-1-1-1    1729    2
3:   2   3   5   6    2   1  -2   0  -3    1: -1-1-1-1    1729    2
3:   2   4   4   5    1   0   3   1  -3    3: -1-1-1-1    1729    4
3:   2   4   4   5    1   1   1  -1   3    1: -1-1-1-1    1729    2
3:   2   4   5   5    2   1  -3   1   4    0: -1-1-1-1    1729    8
3:   3   3   3   5    1   1   0   2   1    0: -1-1-1-1    1729    2
3:   3   3   3   6    2   1   0   3   3    2: -1-1-1-1    1729    2
3:   3   3   4   4    1   2  -1   2  -1    1: -1-1-1-1    1729    4
```

```
3:   3  3   4   4   1  2  -1  -1   0    2: -1-1-1-1   1729    2
3:   3  3   4   5   3  1  -1   0   2    1: -1-1-1-1   1729    2
3:   3  4   4   4   2  0   3  -1   2   -2: -1-1-1-1   1729    2
4:   1  1  11  14   1  0   0   1   1    5: -1 1-1 1   1729   12   703/48
4:   1  1   5  23   0  1   0   0   1    0: -1 1-1 1   1729   16
4:   1  1   8  15   0  1   0   0   1    5: -1 1-1 1   1729    8
4:   1  2   2  34   1  0   1   1   1    2: -1 1-1 1   1729    8
4:   1  2   3  21   0  1   1   0   1    2: -1 1-1 1   1729    8
4:   1  2   3  24   1  1   0   1   2    3: -1 1-1 1   1729    8
4:   1  2   4  15   0  0   1  -1  -1    3: -1 1-1 1   1729    4
4:   1  2   5  13   1  0   1   1   0    0: -1 1-1 1   1729    4
4:   1  2   5  14   1  0   2   0   0    1: -1 1-1 1   1729    4
4:   1  2   5  14   1  0   1   0   0    5: -1 1-1 1   1729    4
4:   1  2   6  11   1  0   1   0   1    3: -1 1-1 1   1729    4
4:   1  2   8   9   1  0   2   0   2    3: -1 1-1 1   1729    4
4:   1  3   3  14   1  0   0   1   3    1: -1 1-1 1   1729    4
4:   1  3   4  11   1  1   0   0   2    1: -1 1-1 1   1729    4
4:   1  3   4  12   0  1   2   1   3    4: -1 1-1 1   1729    4
4:   1  3   5   8   0  0   1   1   2   -1: -1 1-1 1   1729    4
4:   1  4   4   8   1  0   1   1   3    0: -1 1-1 1   1729    4
4:   1  4   5   8   1  0   4   1   3    1: -1 1-1 1   1729    4
4:   1  4   6   6   0  1   1   1   4    1: -1 1-1 1   1729    8
4:   1  4   6   6   1  1  -1   0   1    4: -1 1-1 1   1729    4
4:   1  4   6   7   1  1  -1   1   4    4: -1 1-1 1   1729    4
4:   1  5   6   6   1  0   5   1   4    4: -1 1-1 1   1729    4
4:   2  2   3  13   2  1   0   0   1    1: -1 1-1 1   1729    2
4:   2  2   3  14   1  2  -1   2   0    3: -1 1-1 1   1729    2
4:   2  2   4   9   1  2   0   1   1    3: -1 1-1 1   1729    2
4:   2  2   5   7   1  1   1   1   0    5: -1 1-1 1   1729    4
4:   2  2   5   7   0  2   1  -1   0    3: -1 1-1 1   1729    2
4:   2  2   5   7   1  1  -1   2   0    3: -1 1-1 1   1729    2
4:   2  3   4   6   1  0   0   1        4: -1 1-1 1   1729    2
4:   2  3   4   6   1  0   2  -2   1   -1: -1 1-1 1   1729    2
4:   2  3   4   6   1  0   2   0  -1    3: -1 1-1 1   1729    2
4:   2  3   4   6   1  2  -1   1  -1    0: -1 1-1 1   1729    2
4:   2  3   5   6   1  1  -3   1  -1   -3: -1 1-1 1   1729    4
4:   2  3   5   6   2  1   0   2   1    5: -1 1-1 1   1729    2
4:   2  3   5   6   1  0   3   2  -1    3: -1 1-1 1   1729    2
4:   2  4   4   4   1  1  -2   0   1    0: -1 1-1 1   1729    2
4:   2  4   4   5   2  2   1   1   0    3: -1 1-1 1   1729    2
4:   2  4   4   5   1  1  -2   1   3    2: -1 1-1 1   1729    2
4:   2  4   5   5   1  0   1   1   3    4: -1 1-1 1   1729    2
4:   2  4   5   5   2  1  -2   1   1    5: -1 1-1 1   1729    2
4:   3  3   4   4   2  2   1   1   1    2: -1 1-1 1   1729    2
4:   3  4   4   4   2  2   3   1   4    1: -1 1-1 1   1729    2
4:   3  4   4   4   1  1  -3  -2  -4    1: -1 1-1 1   1729    2
```

Appendix to Tables Through Discriminant 500, 728, and 1729

Containing, for each genus in the tables
and for each prime p|2d, the p-adic
density and a p-adic Jordan splitting

```
D=    4
 GENUS#1;P=2; 576; [A]+[2A]

D=    5
 GENUS#1;P=2; 15; [A+H]
 GENUS#1;P=5; 48/5; [(1)+(3/4)+(1)]+[5/12]

D=    8
 GENUS#1;P=2; 192; [A]+[1]+[2/3]

D=    9
 GENUS#1;P=2; 9; [A+A]
 GENUS#1;P=3; 96; [(1)+(1)]+[(3/4)+(3/4)]

D=   12
 GENUS#1;P=2; 96; [A]+[(5/3)+(3/5)]
 GENUS#1;P=3; 16/3; [(1)+(7/4)+(5/7)]+[3/5]
 GENUS#2;P=2; 96; [A]+[(1)+(1)]
 GENUS#2;P=3; 16/3; [(1)+(1)+(1)]+[3/4]

D=   13
 GENUS#1;P=2; 15; [A+H]
 GENUS#1;P=13; 336/13; [(1)+(3/4)+(2/3)]+[13/8]

D=   16
 GENUS#1;P=2; 6144; [(1)+(1)+(1)+(1)]
 GENUS#2;P=2; 1536; [A]+[(2/3)+(2)]

D=   17
 GENUS#1;P=2; 9; [A+A]
 GENUS#1;P=17; 576/17; [(1)+(3/4)+(1)]+[17/12]

D=   20
 GENUS#1;P=2; 192; [A]+[2H]
 GENUS#1;P=5; 48/5; [(1)+(3/4)+(2/3)]+[5/2]
 GENUS#2;P=2; 192; [H]+[2A]
 GENUS#2;P=5; 48/5; [(1)+(1)+(1)]+[5/4]
 GENUS#3;P=2; 48; [A]+[(1)+(5/3)]
 GENUS#3;P=5; 48/5; [(1)+(3/4)+(1)]+[5/3]

D=   21
 GENUS#1;P=2; 15; [A+H]
 GENUS#1;P=3; 16/3; [(1)+(1)+(7/4)]+[3/4]
 GENUS#1;P=7; 96/7; [(1)+(3/4)+(1)]+[7/4]
 GENUS#2;P=2; 15; [A+H]
 GENUS#2;P=3; 16/3; [(1)+(1)+(1/2)]+[21/8]
 GENUS#2;P=7; 96/7; [(1)+(3/4)+(2/3)]+[21/8]

D=   24
 GENUS#1;P=2; 192; [A]+[1]+[2]
 GENUS#1;P=3; 16/3; [(1)+(1)+(2)]+[3/4]
 GENUS#2;P=2; 192; [H]+[1]+[6/7]
 GENUS#2;P=3; 16/3; [(1)+(1)+(1)]+[3/2]

D=   25
 GENUS#1;P=2; 9; [A+A]
 GENUS#1;P=5; 360; [(1)+(3/4)]+[(5/3)+(5/4)]

D=   28
 GENUS#1;P=2; 96; [H]+[(1)+(1)]
 GENUS#1;P=7; 96/7; [(1)+(1)+(1)]+[7/4]
 GENUS#2;P=2; 96; [A]+[(5/3)+(7/5)]
 GENUS#2;P=7; 96/7; [(1)+(3/4)+(5/3)]+[7/5]

D=   29
 GENUS#1;P=2; 15; [A+H]
 GENUS#1;P=29; 1680/29; [(1)+(3/4)+(1)]+[29/12]

D=   32
 GENUS#1;P=2; 6144; [(1)+(1)+(1)]+[2]
 GENUS#2;P=2; 1536; [A]+[1]+[8/3]
```

```
GENUS#3;P=2;  6144;  [A]+[2/3]+[4]
GENUS#4;P=2;  2048;  [H]+[10/7]+[4/5]
GENUS#5;P=2;  1536;  [A]+[5/3]+[8/5]


D=    33
GENUS#1;P=2;  9;  [A+A]
GENUS#1;P=3;  16/3;  [(1)+(1)+(11/4)]+[3/4]
GENUS#1;P=11;  240/11;  [(1)+(3/4)+(1)]+[11/4]
GENUS#2;P=2;  9;  [H+H]
GENUS#2;P=3;  16/3;  [(1)+(1)+(7/4)]+[33/28]
GENUS#2;P=11;  240/11;  [(1)+(1)+(3/2)]+[11/8]


D=    36
GENUS#1;P=2;  576;  [A]+[2A]
GENUS#1;P=3;  16;  [(1)+(19/4)+(14/19)]+[9/14]
GENUS#2;P=2;  576;  [A]+[2A]
GENUS#2;P=3;  16;  [(1)+(1)+(1)]+[9/4]
GENUS#3;P=2;  48;  [A]+[(1)+(3)]
GENUS#3;P=3;  96;  [(1)+(1)]+[(3/4)+(3)]
GENUS#4;P=2;  576;  [A]+[2A]
GENUS#4;P=3;  24;  [(1)+(2)]+[(3/4)+(3/2)]
GENUS#5;P=2;  64;  [H]+[2H]
GENUS#5;P=3;  96;  [(1)+(1)]+[(3/2)+(3/2)]


D=    37
GENUS#1;P=2;  15;  [A+H]
GENUS#1;P=37;  2736/37;  [(1)+(3/4)+(2/3)]+[37/8]


D=    40
GENUS#1;P=2;  192;  [A]+[5]+[2/3]
GENUS#1;P=5;  48/5;  [(1)+(3/4)+(2/3)]+[5]
GENUS#2;P=2;  192;  [A]+[1]+[10/11]
GENUS#2;P=5;  48/5;  [(1)+(1)+(1)]+[5/2]


D=    41
GENUS#1;P=2;  9;  [A+A]
GENUS#1;P=41;  3360/41;  [(1)+(3/4)+(1)]+[41/12]


D=    44
GENUS#1;P=2;  96;  [A]+[(1)+(1)]
GENUS#1;P=11;  240/11;  [(1)+(1)+(1)]+[11/4]
GENUS#2;P=2;  96;  [A]+[(17/3)+(11/17)]
GENUS#2;P=11;  240/11;  [(1)+(3/4)+(2/3)]+[11/2]


D=    45
GENUS#1;P=2;  15;  [A+H]
GENUS#1;P=3;  48;  [(1)+(1)]+[(3/4)+(15/4)]
GENUS#1;P=5;  48/5;  [(1)+(3/4)+(1)]+[15/4]
GENUS#2;P=2;  15;  [A+H]
GENUS#2;P=3;  16;  [(1)+(1)+(1/2)]+[45/8]
GENUS#2;P=5;  48/5;  [(1)+(3/4)+(2/3)]+[45/8]
GENUS#3;P=2;  15;  [A+H]
GENUS#3;P=3;  48;  [(1)+(2)]+[(3/4)+(15/8)]
GENUS#3;P=5;  48/5;  [(1)+(3/4)+(2)]+[15/8]
GENUS#4;P=2;  15;  [H+A]
GENUS#4;P=3;  16;  [(1)+(1)+(7/4)]+[45/28]
GENUS#4;P=5;  48/5;  [(1)+(1)+(7/4)]+[45/28]


D=    48
GENUS#1;P=2;  3072;  [(1)+(1)+(1)+(3)]
GENUS#1;P=3;  16/3;  [(1)+(1)+(1)]+[3]
GENUS#2;P=2;  384;  [A]+[1]+[4]
GENUS#2;P=3;  16/3;  [(1)+(1)+(4)]+[3/4]
GENUS#3;P=2;  3072;  [A]+[(2/3)+(6)]
GENUS#3;P=3;  16/3;  [(1)+(1)+(1/2)]+[6]
GENUS#4;P=2;  384;  [A]+[5/3]+[12/5]
GENUS#4;P=3;  16/3;  [(1)+(7/4)+(5/7)]+[12/5]
GENUS#5;P=2;  1024;  [H]+[(6/7)+(2)]
GENUS#5;P=3;  16/3;  [(1)+(1)+(2)]+[3/2]
GENUS#6;P=2;  3072;  [(1)+(1)+(2A)]
GENUS#6;P=3;  16/3;  [(1)+(1)+(2)]+[3/2]
GENUS#7;P=2;  3072;  [A]+[(2)+(2)]
```

```
GENUS#7;P=3;  16/3;  [(1)+(2)+(2)]+[3/4]
GENUS#8;P=2;  1024;  [H]+[(10/7)+(6/5)]
GENUS#8;P=3;  16/3;  [(1)+(2)+(2)]+[3/4]

D=   49
 GENUS#1;P=2;  9;  [H+H]
 GENUS#1;P=7;  896;  [(1)+(1)]+[(7/4)+(7/4)]

D=   52
 GENUS#1;P=2;  192;  [H]+[2A]
 GENUS#1;P=13;  336/13;  [(1)+(1)+(1)]+[13/4]
 GENUS#2;P=2;  192;  [A]+[2H]
 GENUS#2;P=13;  336/13;  [(1)+(3/4)+(2/3)]+[13/2]
 GENUS#3;P=2;  48;  [H]+[(1)+(13/7)]
 GENUS#3;P=13;  336/13;  [(1)+(1)+(7/4)]+[13/7]

D=   53
 GENUS#1;P=2;  15;  [A+H]
 GENUS#1;P=53;  5616/53;  [(1)+(3/4)+(1)]+[53/12]

D=   56
 GENUS#1;P=2;  192;  [A]+[1]+[14/3]
 GENUS#1;P=7;  96/7;  [(1)+(3/4)+(1)]+[14/3]
 GENUS#2;P=2;  192;  [H]+[1]+[14/15]
 GENUS#2;P=7;  96/7;  [(1)+(1)+(1)]+[7/2]

D=   57
 GENUS#1;P=2;  9;  [A+A]
 GENUS#1;P=3;  16/3;  [(1)+(1)+(19/4)]+[3/4]
 GENUS#1;P=19;  720/19;  [(1)+(3/4)+(1)]+[19/4]
 GENUS#2;P=2;  9;  [H+H]
 GENUS#2;P=3;  16/3;  [(1)+(1)+(11/4)]+[57/44]
 GENUS#2;P=19;  720/19;  [(1)+(1)+(3/2)]+[19/8]

D=   60
 GENUS#1;P=2;  96;  [A]+[(1)+(5)]
 GENUS#1;P=3;  16/3;  [(1)+(1)+(5)]+[3/4]
 GENUS#1;P=5;  48/5;  [(1)+(3/4)+(1)]+[5]
 GENUS#2;P=2;  96;  [H]+[(1)+(1)]
 GENUS#2;P=3;  16/3;  [(1)+(1)+(1)]+[15/4]
 GENUS#2;P=5;  48/5;  [(1)+(1)+(1)]+[15/4]
 GENUS#3;P=2;  96;  [A]+[(3)+(5/3)]
 GENUS#3;P=3;  16/3;  [(1)+(2)+(5/2)]+[3/4]
 GENUS#3;P=5;  48/5;  [(1)+(3/4)+(2)]+[5/2]
 GENUS#4;P=2;  96;  [H]+[(13/7)+(15/13)]
 GENUS#4;P=3;  16/3;  [(1)+(7/4)+(10/7)]+[3/2]
 GENUS#4;P=5;  48/5;  [(1)+(7/4)+(13/7)]+[15/13]

D=   61
 GENUS#1;P=2;  15;  [A+H]
 GENUS#1;P=61;  7440/61;  [(1)+(3/4)+(2/3)]+[61/8]

D=   64
 GENUS#1;P=2;  49152;  [(1)+(1)+(1)]+[4]
 GENUS#2;P=2;  12288;  [A]+[2/3]+[8]
 GENUS#3;P=2;  32768;  [(1)+(1)]+[(2)+(2)]
 GENUS#4;P=2;  36864;  [A]+[8A]
 GENUS#5;P=2;  12288;  [A]+[2]+[8/3]
 GENUS#6;P=2;  49152;  [(1)+(2A)]+[4/3]

D=   65
 GENUS#1;P=2;  9;  [A+A]
 GENUS#1;P=5;  48/5;  [(1)+(3/4)+(1)]+[65/12]
 GENUS#1;P=13;  336/13;  [(1)+(3/4)+(1)]+[65/12]
 GENUS#2;P=2;  9;  [H+H]
 GENUS#2;P=5;  48/5;  [(1)+(1)+(3/2)]+[65/24]
 GENUS#2;P=13;  336/13;  [(1)+(1)+(3/2)]+[65/24]

D=   68
 GENUS#1;P=2;  576;  [A]+[2A]
 GENUS#1;P=17;  576/17;  [(1)+(1)+(1)]+[17/4]
 GENUS#2;P=2;  48;  [A]+[(1)+(17/3)]
```

```
 GENUS#2;P=17; 576/17; [(1)+(3/4)+(1)]+[17/3]
 GENUS#3;P=2; 64; [H]+[2H]
 GENUS#3;P=17; 576/17; [(1)+(1)+(3/2)]+[17/6]


D=   69
 GENUS#1;P=2; 15; [A+H]
 GENUS#1;P=3; 16/3; [(1)+(1)+(23/4)]+[3/4]
 GENUS#1;P=23; 1056/23; [(1)+(3/4)+(1)]+[23/4]
 GENUS#2;P=2; 15; [H+A]
 GENUS#2;P=3; 16/3; [(1)+(1)+(7/4)]+[69/28]
 GENUS#2;P=23; 1056/23; [(1)+(1)+(7/4)]+[69/28]


D=   72
 GENUS#1;P=2; 192; [A]+[1]+[6]
 GENUS#1;P=3; 48; [(1)+(1)]+[(3/4)+(6)]
 GENUS#2;P=2; 192; [A]+[9]+[2/3]
 GENUS#2;P=3; 16; [(1)+(1)+(1/2)]+[9]
 GENUS#3;P=2; 192; [A]+[1]+[18/19]
 GENUS#3;P=3; 16; [(1)+(1)+(1)]+[9/2]
 GENUS#4;P=2; 192; [A]+[3]+[2]
 GENUS#4;P=3; 48; [(1)+(2)]+[(3/4)+(3)]


D=   73
 GENUS#1;P=2; 9; [H+H]
 GENUS#1;P=73; 10656/73; [(1)+(1)+(7/4)]+[73/28]


D=   76
 GENUS#1;P=2; 96; [A]+[(29/3)+(19/29)]
 GENUS#1;P=19; 720/19; [(1)+(3/4)+(2/3)]+[19/2]
 GENUS#2;P=2; 96; [A]+[(1)+(1)]
 GENUS#2;P=19; 720/19; [(1)+(1)+(1)]+[19/4]


D=   77
 GENUS#1;P=2; 15; [A+H]
 GENUS#1;P=7; 96/7; [(1)+(3/4)+(2/3)]+[77/8]
 GENUS#1;P=11; 240/11; [(1)+(3/4)+(2/3)]+[77/8]
 GENUS#2;P=2; 15; [A+H]
 GENUS#2;P=7; 96/7; [(1)+(3/4)+(1)]+[77/12]
 GENUS#2;P=11; 240/11; [(1)+(3/4)+(1)]+[77/12]


D=   80
 GENUS#1;P=2; 1536; [A]+[(2/3)+(10)]
 GENUS#1;P=5; 48/5; [(1)+(3/4)+(2/3)]+[10]
 GENUS#2;P=2; 6144; [(1)+(1)+(1)+(5)]
 GENUS#2;P=5; 48/5; [(1)+(1)+(1)]+[5]
 GENUS#3;P=2; 384; [A]+[1]+[20/3]
 GENUS#3;P=5; 48/5; [(1)+(3/4)+(1)]+[20/3]
 GENUS#4;P=2; 1536; [A]+[4H]
 GENUS#4;P=5; 48/5; [(1)+(3/4)+(8/3)]+[5/2]
 GENUS#5;P=2; 2048; [(1)+(1)+(3)+(5/3)]
 GENUS#5;P=5; 48/5; [(1)+(1)+(2)]+[5/2]
 GENUS#6;P=2; 384; [A]+[5/3]+[4]
 GENUS#6;P=5; 48/5; [(1)+(3/4)+(4)]+[5/3]
 GENUS#7;P=2; 512; [H]+[(6/7)+(10/3)]
 GENUS#7;P=5; 48/5; [(1)+(1)+(3/2)]+[10/3]
 GENUS#8;P=2; 1536; [H]+[4A]
 GENUS#8;P=5; 48/5; [(1)+(7/4)+(12/7)]+[5/3]


D=   81
 GENUS#1;P=2; 9; [A+A]
 GENUS#1;P=3; 1296; [(1)+(1)]+[3/4]+[27/4]
 GENUS#2;P=2; 9; [H+H]
 GENUS#2;P=3; 1296; [(1)+(1)]+[3/2]+[27/8]
 GENUS#3;P=2; 9; [A+A]
 GENUS#3;P=3; 11664; [1]+[(3/4)+(3)]+[9/4]
 GENUS#4;P=2; 9; [H+H]
 GENUS#4;P=3; 11664; [2]+[(15/8)+(6/5)]+[9/8]


D=   84
 GENUS#1;P=2; 192; [A]+[2H]
 GENUS#1;P=3; 16/3; [(1)+(43/4)+(32/43)]+[21/32]
 GENUS#1;P=7; 96/7; [(1)+(3/4)+(2/3)]+[21/2]
```

```
GENUS#2;P=2;  192;  [H]+[2A]
GENUS#2;P=3;  16/3;  [(1)+(1)+(1)]+[21/4]
GENUS#2;P=7;  96/7;  [(1)+(1)+(1)]+[21/4]
GENUS#3;P=2;  48;  [A]+[(1)+(7)]
GENUS#3;P=3;  16/3;  [(1)+(1)+(7)]+[3/4]
GENUS#3;P=7;  96/7;  [(1)+(3/4)+(1)]+[7]
GENUS#4;P=2;  48;  [A]+[(21/11)+(1)]
GENUS#4;P=3;  16/3;  [(1)+(1)+(2)]+[21/8]
GENUS#4;P=7;  96/7;  [(1)+(1)+(2)]+[21/8]
GENUS#5;P=2;  192;  [H]+[2A]
GENUS#5;P=3;  16/3;  [(1)+(1)+(7/2)]+[3/2]
GENUS#5;P=7;  96/7;  [(1)+(1)+(3/2)]+[7/2]
GENUS#6;P=2;  192;  [A]+[2H]
GENUS#6;P=3;  16/3;  [(1)+(2)+(7/2)]+[3/4]
GENUS#6;P=7;  96/7;  [(1)+(3/4)+(2)]+[7/2]


D=   85
 GENUS#1;P=2;  15;  [A+H]
 GENUS#1;P=5;  48/5;  [(1)+(3/4)+(2/3)]+[85/8]
 GENUS#1;P=17;  576/17;  [(1)+(3/4)+(2/3)]+[85/8]
 GENUS#2;P=2;  15;  [A+H]
 GENUS#2;P=5;  48/5;  [(1)+(3/4)+(2)]+[85/24]
 GENUS#2;P=17;  576/17;  [(1)+(3/4)+(2)]+[85/24]


D=   88
 GENUS#1;P=2;  192;  [A]+[11]+[2/3]
 GENUS#1;P=11;  240/11;  [(1)+(3/4)+(2/3)]+[11]
 GENUS#2;P=2;  192;  [H]+[1]+[22/23]
 GENUS#2;P=11;  240/11;  [(1)+(1)+(1)]+[11/2]


D=   89
 GENUS#1;P=2;  9;  [A+A]
 GENUS#1;P=89;  15840/89;  [(1)+(3/4)+(1)]+[89/12]


D=   92
 GENUS#1;P=2;  96;  [A]+[(35/3)+(23/35)]
 GENUS#1;P=23;  1056/23;  [(1)+(3/4)+(2/3)]+[23/2]
 GENUS#2;P=2;  96;  [A]+[(5/3)+(23/5)]
 GENUS#2;P=23;  1056/23;  [(1)+(3/4)+(5/3)]+[23/5]


D=   93
 GENUS#1;P=2;  15;  [A+H]
 GENUS#1;P=3;  16/3;  [(1)+(1)+(1/2)]+[93/8]
 GENUS#1;P=31;  1920/31;  [(1)+(3/4)+(2/3)]+[93/8]
 GENUS#2;P=2;  15;  [A+H]
 GENUS#2;P=3;  16/3;  [(1)+(1)+(31/4)]+[3/4]
 GENUS#2;P=31;  1920/31;  [(1)+(3/4)+(1)]+[31/4]


D=   96
 GENUS#1;P=2;  6144;  [A]+[2/3]+[12]
 GENUS#1;P=3;  16/3;  [(1)+(1)+(1/2)]+[12]
 GENUS#2;P=2;  6144;  [(1)+(1)+(1)]+[6]
 GENUS#2;P=3;  16/3;  [(1)+(1)+(1)]+[6]
 GENUS#3;P=2;  1536;  [A]+[1]+[8]
 GENUS#3;P=3;  16/3;  [(1)+(1)+(8)]+[3/4]
 GENUS#4;P=2;  6144;  [(1)+(1)+(3)]+[2]
 GENUS#4;P=3;  16/3;  [(1)+(1)+(2)]+[3]
 GENUS#5;P=2;  1536;  [A]+[3]+[8/3]
 GENUS#5;P=3;  16/3;  [(1)+(11/4)+(8/11)]+[3]
 GENUS#6;P=2;  2048;  [H]+[26/15]+[12/13]
 GENUS#6;P=3;  16/3;  [(1)+(1)+(2)]+[3]
 GENUS#7;P=2;  2048;  [H]+[6/7]+[4]
 GENUS#7;P=3;  16/3;  [(1)+(1)+(4)]+[3/2]
 GENUS#8;P=2;  1536;  [H]+[1]+[24/7]
 GENUS#8;P=3;  16/3;  [(1)+(1)+(7/4)]+[24/7]
 GENUS#9;P=2;  6144;  [A]+[10/11]+[12/5]
 GENUS#9;P=3;  16/3;  [(1)+(1)+(5/2)]+[12/5]
 GENUS#10;P=2;  1536;  [H]+[13/7]+[24/13]
 GENUS#10;P=3;  16/3;  [(1)+(2)+(2)]+[3/2]


D=   97
 GENUS#1;P=2;  9;  [H+H]
```

```
GENUS#1;P=97; 18816/97; [(1)+(1)+(7/4)]+[97/28]

D= 100
 GENUS#1;P=2; 576; [A]+[2A]
 GENUS#1;P=5; 48; [(1)+(3/4)+(2/3)]+[25/2]
 GENUS#2;P=2; 576; [A]+[2A]
 GENUS#2;P=5; 48; [(1)+(1)+(1)]+[25/4]
 GENUS#3;P=2; 576; [A]+[2A]
 GENUS#3;P=5; 160; [(1)+(1)]+[(5/2)+(5/2)]
 GENUS#4;P=2; 48; [A]+[(5/3)+(5)]
 GENUS#4;P=5; 360; [(1)+(3/4)]+[(5/3)+(5)]
 GENUS#5;P=2; 64; [H]+[2H]
 GENUS#5;P=5; 360; [(1)+(7/4)]+[(10/7)+(5/2)]

D= 101
 GENUS#1;P=2; 15; [A+H]
 GENUS#1;P=101; 20400/101; [(1)+(3/4)+(2/3)]+[101/8]

D= 104
 GENUS#1;P=2; 192; [A]+[13]+[2/3]
 GENUS#1;P=13; 336/13; [(1)+(3/4)+(2/3)]+[13]
 GENUS#2;P=2; 192; [A]+[1]+[26/27]
 GENUS#2;P=13; 336/13; [(1)+(1)+(1)]+[13/2]

D= 105
 GENUS#1;P=2; 9; [A+A]
 GENUS#1;P=3; 16/3; [(1)+(1)+(35/4)]+[3/4]
 GENUS#1;P=5; 48/5; [(1)+(3/4)+(1)]+[35/4]
 GENUS#1;P=7; 96/7; [(1)+(3/4)+(1)]+[35/4]
 GENUS#2;P=2; 9; [H+H]
 GENUS#2;P=3; 16/3; [(1)+(1)+(7/4)]+[15/4]
 GENUS#2;P=5; 48/5; [(1)+(1)+(7/4)]+[15/4]
 GENUS#2;P=7; 96/7; [(1)+(1)+(15/4)]+[7/4]
 GENUS#3;P=2; 9; [H+H]
 GENUS#3;P=3; 16/3; [(1)+(1)+(19/4)]+[105/76]
 GENUS#3;P=5; 48/5; [(1)+(1)+(3/2)]+[35/8]
 GENUS#3;P=7; 96/7; [(1)+(1)+(3/2)]+[35/8]
 GENUS#4;P=2; 9; [A+A]
 GENUS#4;P=3; 16/3; [(1)+(7)+(5/4)]+[3/4]
 GENUS#4;P=5; 48/5; [(1)+(3/4)+(3)]+[35/12]
 GENUS#4;P=7; 96/7; [(1)+(3/4)+(3)]+[35/12]

D= 108
 GENUS#1;P=2; 96; [A]+[(41/3)+(27/41)]
 GENUS#1;P=3; 48; [(1)+(55/4)+(41/55)]+[27/41]
 GENUS#2;P=2; 96; [A]+[(1)+(9)]
 GENUS#2;P=3; 432; [(1)+(1)]+[3/4]+[9]
 GENUS#3;P=2; 96; [A]+[(5)+(9/5)]
 GENUS#3;P=3; 216; [(1)+(2)]+[3/4]+[9/2]
 GENUS#4;P=2; 96; [A]+[(1)+(1)]
 GENUS#4;P=3; 48; [(1)+(1)+(1)]+[27/4]
 GENUS#5;P=2; 96; [H]+[(33/7)+(9/11)]
 GENUS#5;P=3; 432; [(1)+(1)]+[3/2]+[9/2]
 GENUS#6;P=2; 96; [A]+[(3)+(3)]
 GENUS#6;P=3; 1296; [1]+[(3/4)+(3)+(3)]
 GENUS#7;P=2; 96; [A]+[(21/11)+(9/7)]
 GENUS#7;P=3; 216; [(1)+(2)]+[3/2]+[9/4]
 GENUS#8;P=2; 96; [H]+[(9/5)+(1)]
 GENUS#8;P=3; 1296; [2]+[(3/2)+(3/2)+(3/2)]

D= 109
 GENUS#1;P=2; 15; [A+H]
 GENUS#1;P=109; 23760/109; [(1)+(3/4)+(2/3)]+[109/8]

D= 112
 GENUS#1;P=2; 3072; [A]+[(2/3)+(14)]
 GENUS#1;P=7; 96/7; [(1)+(3/4)+(2/3)]+[14]
 GENUS#2;P=2; 3072; [(1)+(1)+(1)+(7)]
 GENUS#2;P=7; 96/7; [(1)+(1)+(1)]+[7]
 GENUS#3;P=2; 3072; [A]+[(2)+(14/3)]
 GENUS#3;P=7; 96/7; [(1)+(3/4)+(2)]+[14/3]
 GENUS#4;P=2; 384; [A]+[5/3]+[28/5]
```

```
GENUS#4;P=7;  96/7;  [(1)+(3/4)+(5/3)]+[28/5]
GENUS#5;P=2;  384;  [H]+[1]+[4]
GENUS#5;P=7;  96/7;  [(1)+(1)+(4)]+[7/4]
GENUS#6;P=2;  1024;  [H]+[(2)+(14/15)]
GENUS#6;P=7;  96/7;  [(1)+(1)+(2)]+[7/2]
GENUS#7;P=2;  3072;  [(1)+(3)+(5/3)+(7/5)]
GENUS#7;P=7;  96/7;  [(1)+(2)+(3/2)]+[7/3]
GENUS#8;P=2;  1024;  [H]+[(10/7)+(14/5)]
GENUS#8;P=7;  96/7;  [(1)+(2)+(5/4)]+[14/5]


D=  113
 GENUS#1;P=2;  9;  [A+A]
 GENUS#1;P=113;  25536/113;  [(1)+(3/4)+(1)]+[113/12]


D=  116
 GENUS#1;P=2;  48;  [A]+[(1)+(29/3)]
 GENUS#1;P=29;  1680/29;  [(1)+(3/4)+(1)]+[29/3]
 GENUS#2;P=2;  192;  [A]+[2H]
 GENUS#2;P=29;  1680/29;  [(1)+(3/4)+(2/3)]+[29/2]
 GENUS#3;P=2;  192;  [H]+[2A]
 GENUS#3;P=29;  1680/29;  [(1)+(1)+(1)]+[29/4]


D=  117
 GENUS#1;P=2;  15;  [A+H]
 GENUS#1;P=3;  96;  [(1)+(1)]+[(3/4)+(39/4)]
 GENUS#1;P=13;  336/13;  [(1)+(3/4)+(1)]+[39/4]
 GENUS#2;P=2;  15;  [A+H]
 GENUS#2;P=3;  16;  [(1)+(1)+(1/2)]+[117/8]
 GENUS#2;P=13;  336/13;  [(1)+(3/4)+(2/3)]+[117/8]
 GENUS#3;P=2;  15;  [A+H]
 GENUS#3;P=3;  24;  [(1)+(2)]+[(3/4)+(39/8)]
 GENUS#3;P=13;  336/13;  [(1)+(3/4)+(2)]+[39/8]
 GENUS#4;P=2;  15;  [H+A]
 GENUS#4;P=3;  16;  [(1)+(1)+(7/4)]+[117/28]
 GENUS#4;P=13;  336/13;  [(1)+(1)+(7/4)]+[117/28]


D=  120
 GENUS#1;P=2;  192;  [A]+[1]+[10]
 GENUS#1;P=3;  16/3;  [(1)+(1)+(10)]+[3/4]
 GENUS#1;P=5;  48/5;  [(1)+(3/4)+(1)]+[10]
 GENUS#2;P=2;  192;  [A]+[15]+[2/3]
 GENUS#2;P=3;  16/3;  [(1)+(1)+(1/2)]+[15]
 GENUS#2;P=5;  48/5;  [(1)+(3/4)+(2/3)]+[15]
 GENUS#3;P=2;  192;  [H]+[1]+[30/31]
 GENUS#3;P=3;  16/3;  [(1)+(1)+(1)]+[15/2]
 GENUS#3;P=5;  48/5;  [(1)+(1)+(1)]+[15/2]
 GENUS#4;P=2;  192;  [H]+[5]+[6/7]
 GENUS#4;P=3;  16/3;  [(1)+(1)+(5)]+[3/2]
 GENUS#4;P=5;  48/5;  [(1)+(1)+(3/2)]+[5]


D=  121
 GENUS#1;P=2;  9;  [A+A]
 GENUS#1;P=11;  3168;  [(1)+(1)]+[(11/4)+(11/4)]


D=  124
 GENUS#1;P=2;  96;  [A]+[(47/3)+(31/47)]
 GENUS#1;P=31;  1920/31;  [(1)+(3/4)+(2/3)]+[31/2]
 GENUS#2;P=2;  96;  [A]+[(17/3)+(31/17)]
 GENUS#2;P=31;  1920/31;  [(1)+(3/4)+(2)]+[31/6]


D=  125
 GENUS#1;P=2;  15;  [A+H]
 GENUS#1;P=5;  240;  [(1)+(3/4)+(1)]+[125/12]
 GENUS#2;P=2;  15;  [A+H]
 GENUS#2;P=5;  3000;  [(1)+(3/4)]+[5/3]+[25/4]
 GENUS#3;P=2;  15;  [H+A]
 GENUS#3;P=5;  30000;  [2]+[(15/8)+(5/3)+(5/4)]
 GENUS#4;P=2;  15;  [A+H]
 GENUS#4;P=5;  2000;  [(1)+(1)]+[5/2]+[25/8]


D=  128
 GENUS#1;P=2;  6144;  [A]+[1]+[32/3]
```

```
GENUS#2;P=2;  49152;  [A]+[2/3]+[16]
GENUS#3;P=2;  196608;  [(1)+(1)+(1)]+[8]
GENUS#4;P=2;  65536;  [(1)+(1)]+[2]+[4]
GENUS#5;P=2;  16384;  [H]+[6/7]+[16/3]
GENUS#6;P=2;  6144;  [A]+[5/3]+[32/5]
GENUS#7;P=2;  49152;  [A]+[4]+[8/3]
GENUS#8;P=2;  49152;  [A]+[14/3]+[16/7]
GENUS#9;P=2;  196608;  [1]+[(2)+(2)+(2)]
GENUS#10;P=2;  16384;  [H]+[10/7]+[16/5]
GENUS#11;P=2;  65536;  [(1)+(1)+(3)]+[8/3]
GENUS#12;P=2;  16384;  [H]+[12/7]+[8/3]
GENUS#13;P=2;  65536;  [(1)+(3)+(5/3)]+[8/5]
GENUS#14;P=2;  196608;  [(3)+(5/3)+(7/5)]+[8/7]


D=  129
 GENUS#1;P=2;  9;  [A+A]
 GENUS#1;P=3;  16/3;  [(1)+(1)+(43/4)]+[3/4]
 GENUS#1;P=43;  3696/43;  [(1)+(3/4)+(1)]+[43/4]
 GENUS#2;P=2;  9;  [H+H]
 GENUS#2;P=3;  16/3;  [(1)+(1)+(23/4)]+[129/92]
 GENUS#2;P=43;  3696/43;  [(1)+(1)+(3/2)]+[43/8]


D=  132
 GENUS#1;P=2;  48;  [A]+[(1)+(11)]
 GENUS#1;P=3;  16/3;  [(1)+(1)+(11)]+[3/4]
 GENUS#1;P=11;  240/11;  [(1)+(3/4)+(1)]+[11]
 GENUS#2;P=2;  576;  [A]+[2A]
 GENUS#2;P=3;  16/3;  [(1)+(67/4)+(50/67)]+[33/50]
 GENUS#2;P=11;  240/11;  [(1)+(3/4)+(2/3)]+[33/2]
 GENUS#3;P=2;  576;  [A]+[2A]
 GENUS#3;P=3;  16/3;  [(1)+(1)+(1)]+[33/4]
 GENUS#3;P=11;  240/11;  [(1)+(1)+(1)]+[33/4]
 GENUS#4;P=2;  48;  [H]+[(1)+(33/7)]
 GENUS#4;P=3;  16/3;  [(1)+(1)+(7/4)]+[33/7]
 GENUS#4;P=11;  240/11;  [(1)+(1)+(7/4)]+[33/7]
 GENUS#5;P=2;  64;  [H]+[2H]
 GENUS#5;P=3;  16/3;  [(1)+(1)+(11/2)]+[3/2]
 GENUS#5;P=11;  240/11;  [(1)+(1)+(3/2)]+[11/2]
 GENUS#6;P=2;  64;  [H]+[2H]
 GENUS#6;P=3;  16/3;  [(1)+(1)+(7/2)]+[33/14]
 GENUS#6;P=11;  240/11;  [(1)+(1)+(3)]+[11/4]


D=  133
 GENUS#1;P=2;  15;  [A+H]
 GENUS#1;P=7;  96/7;  [(1)+(3/4)+(2/3)]+[133/8]
 GENUS#1;P=19;  720/19;  [(1)+(3/4)+(2/3)]+[133/8]
 GENUS#2;P=2;  15;  [H+A]
 GENUS#2;P=7;  96/7;  [(1)+(1)+(19/4)]+[7/4]
 GENUS#2;P=19;  720/19;  [(1)+(1)+(7/4)]+[19/4]


D=  136
 GENUS#1;P=2;  192;  [A]+[17]+[2/3]
 GENUS#1;P=17;  576/17;  [(1)+(3/4)+(2/3)]+[17]
 GENUS#2;P=2;  192;  [H]+[1]+[34/7]
 GENUS#2;P=17;  576/17;  [(1)+(1)+(7/4)]+[34/7]


D=  137
 GENUS#1;P=2;  9;  [A+A]
 GENUS#1;P=137;  37536/137;  [(1)+(3/4)+(1)]+[137/12]


D=  140
 GENUS#1;P=2;  96;  [A]+[(1)+(35/3)]
 GENUS#1;P=5;  48/5;  [(1)+(3/4)+(1)]+[35/3]
 GENUS#1;P=7;  96/7;  [(1)+(3/4)+(1)]+[35/3]
 GENUS#2;P=2;  96;  [A]+[(53/3)+(35/53)]
 GENUS#2;P=5;  48/5;  [(1)+(3/4)+(2/3)]+[35/2]
 GENUS#2;P=7;  96/7;  [(1)+(3/4)+(2/3)]+[35/2]
 GENUS#3;P=2;  96;  [H]+[(41/7)+(35/41)]
 GENUS#3;P=5;  48/5;  [(1)+(1)+(3/2)]+[35/6]
 GENUS#3;P=7;  96/7;  [(1)+(1)+(3/2)]+[35/6]
 GENUS#4;P=2;  96;  [A]+[(1)+(1)]
 GENUS#4;P=5;  48/5;  [(1)+(1)+(1)]+[35/4]
```

```
GENUS#4;P=7; 96/7; [(1)+(1)+(1)]+[35/4]


D=  141
 GENUS#1;P=2; 15; [A+H]
 GENUS#1;P=3; 16/3; [(1)+(1)+(47/4)]+[3/4]
 GENUS#1;P=47; 4416/47; [(1)+(3/4)+(1)]+[47/4]
 GENUS#2;P=2; 15; [A+H]
 GENUS#2;P=3; 16/3; [(1)+(1)+(5/2)]+[141/40]
 GENUS#2;P=47; 4416/47; [(1)+(1)+(5/2)]+[141/40]


D=  144
 GENUS#1;P=2; 384; [A]+[1]+[12]
 GENUS#1;P=3; 96; [(1)+(1)]+[(3/4)+(12)]
 GENUS#2;P=2; 1536; [A]+[(2/3)+(18)]
 GENUS#2;P=3; 16; [(1)+(1)+(1/2)]+[18]
 GENUS#3;P=2; 6144; [(1)+(1)+(1)+(9)]
 GENUS#3;P=3; 16; [(1)+(1)+(1)]+[9]
 GENUS#4;P=2; 2048; [(1)+(1)+(3)+(3)]
 GENUS#4;P=3; 96; [(1)+(1)]+[(3)+(3)]
 GENUS#5;P=2; 4608; [A]+[4A]
 GENUS#5;P=3; 96; [(1)+(4)]+[(3/4)+(3)]
 GENUS#6;P=2; 512; [H]+[4H]
 GENUS#6;P=3; 96; [(1)+(7/4)]+[(12/7)+(3)]
 GENUS#7;P=2; 6144; [(1)+(1)+(5)+(9/5)]
 GENUS#7;P=3; 16; [(1)+(1)+(2)]+[9/2]
 GENUS#8;P=2; 512; [H]+[(6/7)+(6)]
 GENUS#8;P=3; 96; [(1)+(1)]+[(3/2)+(6)]
 GENUS#9;P=2; 1536; [A]+[(2)+(6)]
 GENUS#9;P=3; 24; [(1)+(2)]+[(3/4)+(6)]
 GENUS#10;P=2; 384; [A]+[3]+[4]
 GENUS#10;P=3; 96; [(1)+(4)]+[(3)+(3/4)]
 GENUS#11;P=2; 1536; [A]+[(10/11)+(18/5)]
 GENUS#11;P=3; 16; [(1)+(1)+(5/2)]+[18/5]
 GENUS#12;P=2; 6144; [(1)+(3)+(2A)]
 GENUS#12;P=3; 24; [(1)+(2)]+[(3/2)+(3)]


D=  145
 GENUS#1;P=2; 9; [H+H]
 GENUS#1;P=5; 48/5; [(1)+(1)+(7/4)]+[145/28]
 GENUS#1;P=29; 1680/29; [(1)+(1)+(7/4)]+[145/28]
 GENUS#2;P=2; 9; [A+A]
 GENUS#2;P=5; 48/5; [(1)+(1)+(11/4)]+[145/44]
 GENUS#2;P=29; 1680/29; [(1)+(1)+(11/4)]+[145/44]


D=  148
 GENUS#1;P=2; 192; [H]+[2A]
 GENUS#1;P=37; 2736/37; [(1)+(1)+(1)]+[37/4]
 GENUS#2;P=2; 192; [A]+[2H]
 GENUS#2;P=37; 2736/37; [(1)+(3/4)+(2/3)]+[37/2]
 GENUS#3;P=2; 48; [A]+[(37/19)+(1)]
 GENUS#3;P=37; 2736/37; [(1)+(1)+(2)]+[37/8]


D=  149
 GENUS#1;P=2; 15; [A+H]
 GENUS#1;P=149; 44400/149; [(1)+(3/4)+(1)]+[149/12]


D=  152
 GENUS#1;P=2; 192; [A]+[1]+[38/3]
 GENUS#1;P=19; 720/19; [(1)+(3/4)+(1)]+[38/3]
 GENUS#2;P=2; 192; [H]+[1]+[38/39]
 GENUS#2;P=19; 720/19; [(1)+(1)+(1)]+[19/2]


D=  153
 GENUS#1;P=2; 9; [A+A]
 GENUS#1;P=3; 48; [(1)+(1)]+[(3/4)+(51/4)]
 GENUS#1;P=17; 576/17; [(1)+(3/4)+(1)]+[51/4]
 GENUS#2;P=2; 9; [H+H]
 GENUS#2;P=3; 16; [(1)+(1)+(7/4)]+[153/28]
 GENUS#2;P=17; 576/17; [(1)+(1)+(7/4)]+[153/28]
 GENUS#3;P=2; 9; [A+A]
 GENUS#3;P=3; 16; [(1)+(7/4)+(5/7)]+[153/20]
 GENUS#3;P=17; 576/17; [(1)+(3/4)+(5/3)]+[153/20]
```

```
 GENUS#4;P=2;  9;  [A+A]
 GENUS#4;P=3;  48;  [(1)+(5)]+[(51/20)+(3/4)]
 GENUS#4;P=17;  576/17;  [(1)+(3/4)+(3)]+[17/4]

D=  156
 GENUS#1;P=2;  96;  [H]+[(1)+(1)]
 GENUS#1;P=3;  16/3;  [(1)+(1)+(1)]+[39/4]
 GENUS#1;P=13;  336/13;  [(1)+(1)+(1)]+[39/4]
 GENUS#2;P=2;  96;  [A]+[(59/3)+(39/59)]
 GENUS#2;P=3;  16/3;  [(1)+(79/4)+(59/79)]+[39/59]
 GENUS#2;P=13;  336/13;  [(1)+(3/4)+(2/3)]+[39/2]
 GENUS#3;P=2;  96;  [A]+[(7)+(13/7)]
 GENUS#3;P=3;  16/3;  [(1)+(2)+(13/2)]+[3/4]
 GENUS#3;P=13;  336/13;  [(1)+(3/4)+(2)]+[13/2]
 GENUS#4;P=2;  96;  [A]+[(3)+(13/3)]
 GENUS#4;P=3;  16/3;  [(1)+(19/4)+(53/19)]+[39/53]
 GENUS#4;P=13;  336/13;  [(1)+(3/4)+(3)]+[13/3]

D=  157
 GENUS#1;P=2;  15;  [A+H]
 GENUS#1;P=157;  49296/157;  [(1)+(3/4)+(2/3)]+[157/8]

D=  160
 GENUS#1;P=2;  6144;  [(1)+(1)+(1)]+[10]
 GENUS#1;P=5;  48/5;  [(1)+(1)+(1)]+[10]
 GENUS#2;P=2;  6144;  [A]+[2/3]+[20]
 GENUS#2;P=5;  48/5;  [(1)+(3/4)+(2/3)]+[20]
 GENUS#3;P=2;  6144;  [(1)+(1)+(5)]+[2]
 GENUS#3;P=5;  48/5;  [(1)+(1)+(2)]+[5]
 GENUS#4;P=2;  6144;  [A]+[2]+[20/3]
 GENUS#4;P=5;  48/5;  [(1)+(3/4)+(2)]+[20/3]
 GENUS#5;P=2;  1536;  [A]+[5/3]+[8]
 GENUS#5;P=5;  48/5;  [(1)+(3/4)+(8)]+[5/3]
 GENUS#6;P=2;  1536;  [H]+[1]+[40/7]
 GENUS#6;P=5;  48/5;  [(1)+(1)+(7/4)]+[40/7]
 GENUS#7;P=2;  1536;  [A]+[17/3]+[40/17]
 GENUS#7;P=5;  48/5;  [(1)+(3/4)+(8/3)]+[5]
 GENUS#8;P=2;  2048;  [H]+[42/23]+[20/21]
 GENUS#8;P=5;  48/5;  [(1)+(1)+(2)]+[5]
 GENUS#9;P=2;  1536;  [A]+[1]+[40/11]
 GENUS#9;P=5;  48/5;  [(1)+(1)+(11/4)]+[40/11]
 GENUS#10;P=2;  2048;  [H]+[14/15]+[20/7]
 GENUS#10;P=5;  48/5;  [(1)+(1)+(4)]+[5/2]

D=  161
 GENUS#1;P=2;  9;  [A+A]
 GENUS#1;P=7;  96/7;  [(1)+(3/4)+(1)]+[161/12]
 GENUS#1;P=23;  1056/23;  [(1)+(3/4)+(1)]+[161/12]
 GENUS#2;P=2;  9;  [H+H]
 GENUS#2;P=7;  96/7;  [(1)+(1)+(23/4)]+[7/4]
 GENUS#2;P=23;  1056/23;  [(1)+(1)+(7/4)]+[23/4]

D=  164
 GENUS#1;P=2;  576;  [A]+[2A]
 GENUS#1;P=41;  3360/41;  [(1)+(1)+(1)]+[41/4]
 GENUS#2;P=2;  48;  [A]+[(1)+(41/3)]
 GENUS#2;P=41;  3360/41;  [(1)+(3/4)+(1)]+[41/3]
 GENUS#3;P=2;  64;  [H]+[2H]
 GENUS#3;P=41;  3360/41;  [(1)+(1)+(3/2)]+[41/6]

D=  165
 GENUS#1;P=2;  15;  [A+H]
 GENUS#1;P=3;  16/3;  [(1)+(1)+(55/4)]+[3/4]
 GENUS#1;P=5;  48/5;  [(1)+(3/4)+(1)]+[55/4]
 GENUS#1;P=11;  240/11;  [(1)+(3/4)+(1)]+[55/4]
 GENUS#2;P=2;  15;  [A+H]
 GENUS#2;P=3;  16/3;  [(1)+(1)+(1/2)]+[165/8]
 GENUS#2;P=5;  48/5;  [(1)+(3/4)+(2/3)]+[165/8]
 GENUS#2;P=11;  240/11;  [(1)+(3/4)+(2/3)]+[165/8]
 GENUS#3;P=2;  15;  [A+H]
 GENUS#3;P=3;  16/3;  [(1)+(2)+(55/8)]+[3/4]
 GENUS#3;P=5;  48/5;  [(1)+(3/4)+(2)]+[55/8]
```

```
GENUS#3;P=11; 240/11; [(1)+(3/4)+(2)]+[55/8]
GENUS#4;P=2; 15; [A+H]
GENUS#4;P=3; 16/3; [(1)+(7/4)+(5/7)]+[33/4]
GENUS#4;P=5; 48/5; [(1)+(3/4)+(26/3)]+[165/104]
GENUS#4;P=11; 240/11; [(1)+(3/4)+(5/3)]+[33/4]

D=  168
 GENUS#1;P=2; 192; [A]+[1]+[14]
 GENUS#1;P=3; 16/3; [(1)+(1)+(14)]+[3/4]
 GENUS#1;P=7; 96/7; [(1)+(3/4)+(1)]+[14]
 GENUS#2;P=2; 192; [A]+[21]+[2/3]
 GENUS#2;P=3; 16/3; [(1)+(1)+(1/2)]+[21]
 GENUS#2;P=7; 96/7; [(1)+(3/4)+(2/3)]+[21]
 GENUS#3;P=2; 192; [H]+[1]+[6]
 GENUS#3;P=3; 16/3; [(1)+(1)+(7/4)]+[6]
 GENUS#3;P=7; 96/7; [(1)+(1)+(6)]+[7/4]
 GENUS#4;P=2; 192; [A]+[1]+[42/43]
 GENUS#4;P=3; 16/3; [(1)+(1)+(1)]+[21/2]
 GENUS#4;P=7; 96/7; [(1)+(1)+(1)]+[21/2]

D=  169
 GENUS#1;P=2; 9; [H+H]
 GENUS#1;P=13; 5096; [(1)+(7/4)]+[(13/7)+(13/4)]

D=  172
 GENUS#1;P=2; 96; [A]+[(65/3)+(43/65)]
 GENUS#1;P=43; 3696/43; [(1)+(3/4)+(2/3)]+[43/2]
 GENUS#2;P=2; 96; [A]+[(1)+(1)]
 GENUS#2;P=43; 3696/43; [(1)+(1)+(1)]+[43/4]

D=  173
 GENUS#1;P=2; 15; [A+H]
 GENUS#1;P=173; 59856/173; [(1)+(3/4)+(1)]+[173/12]

D=  176
 GENUS#1;P=2; 3072; [(1)+(1)+(1)+(11)]
 GENUS#1;P=11; 240/11; [(1)+(1)+(1)]+[11]
 GENUS#2;P=2; 384; [A]+[1]+[44/3]
 GENUS#2;P=11; 240/11; [(1)+(3/4)+(1)]+[44/3]
 GENUS#3;P=2; 3072; [A]+[(2/3)+(22)]
 GENUS#3;P=11; 240/11; [(1)+(3/4)+(2/3)]+[22]
 GENUS#4;P=2; 1024; [H]+[(2)+(22/23)]
 GENUS#4;P=11; 240/11; [(1)+(1)+(2)]+[11/2]
 GENUS#5;P=2; 384; [A]+[17/3]+[44/17]
 GENUS#5;P=11; 240/11; [(1)+(3/4)+(8/3)]+[11/2]
 GENUS#6;P=2; 3072; [A]+[(14/3)+(22/7)]
 GENUS#6;P=11; 240/11; [(1)+(3/4)+(4)]+[11/3]
 GENUS#7;P=2; 3072; [(1)+(1)+(2A)]
 GENUS#7;P=11; 240/11; [(1)+(1)+(2)]+[11/2]
 GENUS#8;P=2; 1024; [H]+[(26/7)+(22/13)]
 GENUS#8;P=11; 240/11; [(1)+(7/4)+(12/7)]+[11/3]

D=  177
 GENUS#1;P=2; 9; [A+A]
 GENUS#1;P=3; 16/3; [(1)+(1)+(59/4)]+[3/4]
 GENUS#1;P=59; 6960/59; [(1)+(3/4)+(1)]+[59/4]
 GENUS#2;P=2; 9; [H+H]
 GENUS#2;P=3; 16/3; [(1)+(1)+(31/4)]+[177/124]
 GENUS#2;P=59; 6960/59; [(1)+(1)+(3/2)]+[59/8]

D=  180
 GENUS#1;P=2; 192; [H]+[2A]
 GENUS#1;P=3; 16; [(1)+(1)+(1)]+[45/4]
 GENUS#1;P=5; 48/5; [(1)+(1)+(1)]+[45/4]
 GENUS#2;P=2; 48; [A]+[(1)+(15)]
 GENUS#2;P=3; 48; [(1)+(1)]+[(3/4)+(15)]
 GENUS#2;P=5; 48/5; [(1)+(3/4)+(1)]+[15]
 GENUS#3;P=2; 192; [A]+[2H]
 GENUS#3;P=3; 16; [(1)+(91/4)+(68/91)]+[45/68]
 GENUS#3;P=5; 48/5; [(1)+(3/4)+(2/3)]+[45/2]
 GENUS#4;P=2; 48; [H]+[(1)+(45/7)]
 GENUS#4;P=3; 16; [(1)+(1)+(7/4)]+[45/7]
```

```
GENUS#4;P=5;  48/5;  [(1)+(1)+(7/4)]+[45/7]
GENUS#5;P=2;  48;  [H]+[(45/23)+(1)]
GENUS#5;P=3;  16;  [(1)+(1)+(2)]+[45/8]
GENUS#5;P=5;  48/5;  [(1)+(1)+(2)]+[45/8]
GENUS#6;P=2;  48;  [A]+[(3)+(5)]
GENUS#6;P=3;  48;  [(1)+(5)]+[(3)+(3/4)]
GENUS#6;P=5;  48/5;  [(1)+(3/4)+(3)]+[5]
GENUS#7;P=2;  192;  [H]+[2A]
GENUS#7;P=3;  48;  [(1)+(2)]+[(3/2)+(15/4)]
GENUS#7;P=5;  48/5;  [(1)+(2)+(3/2)]+[15/4]
GENUS#8;P=2;  192;  [A]+[2H]
GENUS#8;P=3;  48;  [(1)+(2)]+[(3/4)+(15/2)]
GENUS#8;P=5;  48/5;  [(1)+(3/4)+(2)]+[15/2]
GENUS#9;P=2;  192;  [H]+[2A]
GENUS#9;P=3;  48;  [(1)+(1)]+[(3/2)+(15/2)]
GENUS#9;P=5;  48/5;  [(1)+(1)+(3/2)]+[15/2]
GENUS#10;P=2;  192;  [A]+[2H]
GENUS#10;P=3;  16;  [(1)+(1)+(5/2)]+[9/2]
GENUS#10;P=5;  48/5;  [(1)+(1)+(9/2)]+[5/2]
GENUS#11;P=2;  192;  [A]+[2H]
GENUS#11;P=3;  48;  [(1)+(1)]+[(3)+(15/4)]
GENUS#11;P=5;  48/5;  [(1)+(1)+(3)]+[15/4]
GENUS#12;P=2;  192;  [H]+[2A]
GENUS#12;P=3;  16;  [(1)+(1)+(7/2)]+[45/14]
GENUS#12;P=5;  48/5;  [(1)+(1)+(7/2)]+[45/14]

D=  181
GENUS#1;P=2;  15;  [A+H]
GENUS#1;P=181;  65520/181;  [(1)+(3/4)+(2/3)]+[181/8]

D=  184
GENUS#1;P=2;  192;  [H]+[1]+[46/47]
GENUS#1;P=23;  1056/23;  [(1)+(1)+(1)]+[23/2]
GENUS#2;P=2;  192;  [A]+[51/11]+[46/51]
GENUS#2;P=23;  1056/23;  [(1)+(1)+(5/2)]+[23/5]

D=  185
GENUS#1;P=2;  9;  [A+A]
GENUS#1;P=5;  48/5;  [(1)+(3/4)+(1)]+[185/12]
GENUS#1;P=37;  2736/37;  [(1)+(3/4)+(1)]+[185/12]
GENUS#2;P=2;  9;  [A+A]
GENUS#2;P=5;  48/5;  [(1)+(3/4)+(29/3)]+[185/116]
GENUS#2;P=37;  2736/37;  [(1)+(3/4)+(5/3)]+[37/4]

D=  188
GENUS#1;P=2;  96;  [H]+[(1)+(1)]
GENUS#1;P=47;  4416/47;  [(1)+(1)+(1)]+[47/4]
GENUS#2;P=2;  96;  [A]+[(5/3)+(47/5)]
GENUS#2;P=47;  4416/47;  [(1)+(3/4)+(5/3)]+[47/5]

D=  189
GENUS#1;P=2;  15;  [A+H]
GENUS#1;P=3;  432;  [(1)+(1)]+[3/4]+[63/4]
GENUS#1;P=7;  96/7;  [(1)+(3/4)+(1)]+[63/4]
GENUS#2;P=2;  15;  [A+H]
GENUS#2;P=3;  48;  [(1)+(1)+(1/2)]+[189/8]
GENUS#2;P=7;  96/7;  [(1)+(3/4)+(2/3)]+[189/8]
GENUS#3;P=2;  15;  [H+A]
GENUS#3;P=3;  48;  [(1)+(1)+(7/4)]+[27/4]
GENUS#3;P=7;  96/7;  [(1)+(1)+(27/4)]+[7/4]
GENUS#4;P=2;  15;  [A+H]
GENUS#4;P=3;  216;  [(1)+(2)]+[3/4]+[63/8]
GENUS#4;P=7;  96/7;  [(1)+(3/4)+(2)]+[63/8]
GENUS#5;P=2;  15;  [H+A]
GENUS#5;P=3;  1296;  [2]+[(15/8)+(21/5)+(3/4)]
GENUS#5;P=7;  96/7;  [(2)+(15/8)+(9/5)]+[7/4]
GENUS#6;P=2;  15;  [A+H]
GENUS#6;P=3;  1296;  [1]+[(3/4)+(3)+(21/4)]
GENUS#6;P=7;  96/7;  [(1)+(3/4)+(3)]+[21/4]
GENUS#7;P=2;  15;  [H+A]
GENUS#7;P=3;  432;  [(1)+(1)]+[15/4]+[63/20]
GENUS#7;P=7;  96/7;  [(1)+(1)+(15/4)]+[63/20]
```

```
 GENUS#8;P=2; 15; [A+H]
 GENUS#8;P=3; 216; [(1)+(2)]+[21/8]+[9/4]
 GENUS#8;P=7; 96/7; [(1)+(2)+(9/4)]+[21/8]


D=  192
 GENUS#1;P=2; 49152; [(1)+(1)+(1)]+[12]
 GENUS#1;P=3; 16/3; [(1)+(1)+(1)]+[12]
 GENUS#2;P=2; 3072; [A]+[1]+[16]
 GENUS#2;P=3; 16/3; [(1)+(1)+(16)]+[3/4]
 GENUS#3;P=2; 12288; [A]+[2/3]+[24]
 GENUS#3;P=3; 16/3; [(1)+(1)+(1/2)]+[24]
 GENUS#4;P=2; 3072; [A]+[5/3]+[48/5]
 GENUS#4;P=3; 16/3; [(1)+(7/4)+(5/7)]+[48/5]
 GENUS#5;P=2; 32768; [(1)+(1)]+[(2)+(6)]
 GENUS#5;P=3; 16/3; [(1)+(1)+(2)]+[6]
 GENUS#6;P=2; 16384; [(1)+(1)+(3)]+[4]
 GENUS#6;P=3; 16/3; [(1)+(1)+(4)]+[3]
 GENUS#7;P=2; 24576; [A]+[(20/3)+(12/5)]
 GENUS#7;P=3; 16/3; [(1)+(11/4)+(8/11)]+[6]
 GENUS#8;P=2; 3072; [H]+[1]+[48/7]
 GENUS#8;P=3; 16/3; [(1)+(1)+(7/4)]+[48/7]
 GENUS#9;P=2; 32768; [(1)+(3)]+[(2)+(2)]
 GENUS#9;P=3; 16/3; [(1)+(2)+(2)]+[3]
 GENUS#10;P=2; 4096; [H]+[6/7]+[8]
 GENUS#10;P=3; 16/3; [(1)+(1)+(8)]+[3/2]
 GENUS#11;P=2; 12288; [A]+[2]+[8]
 GENUS#11;P=3; 16/3; [(1)+(2)+(8)]+[3/4]
 GENUS#12;P=2; 3072; [A]+[3]+[16/3]
 GENUS#12;P=3; 16/3; [(1)+(23/4)+(65/23)]+[48/65]
 GENUS#13;P=2; 98304; [(1)+(1)]+[4A]
 GENUS#13;P=3; 16/3; [(1)+(1)+(4)]+[3]
 GENUS#14;P=2; 24576; [A]+[(4)+(4)]
 GENUS#14;P=3; 16/3; [(1)+(4)+(4)]+[3/4]
 GENUS#15;P=2; 8192; [H]+[(12/7)+(4)]
 GENUS#15;P=3; 16/3; [(1)+(7/4)+(4)]+[12/7]
 GENUS#16;P=2; 49152; [(1)+(2A)]+[4]
 GENUS#16;P=3; 16/3; [(1)+(2)+(4)]+[3/2]
 GENUS#17;P=2; 4096; [H]+[10/7]+[24/5]
 GENUS#17;P=3; 16/3; [(1)+(7/4)+(10/7)]+[24/5]
 GENUS#18;P=2; 8192; [H]+[(20/7)+(12/5)]
 GENUS#18;P=3; 16/3; [(1)+(7/4)+(20/7)]+[12/5]
 GENUS#19;P=2; 16384; [(1)+(3)+(5/3)]+[12/5]
 GENUS#19;P=3; 16/3; [(1)+(2)+(5/2)]+[12/5]
 GENUS#20;P=2; 98304; [(3)+(5/3)]+[4H]
 GENUS#20;P=3; 16/3; [(2)+(5/2)+(8/5)]+[3/2]


D=  193
 GENUS#1;P=2; 9; [A+A]
 GENUS#1;P=193; 74496/193; [(1)+(3/4)+(5/3)]+[193/20]


D=  196
 GENUS#1;P=2; 576; [A]+[2A]
 GENUS#1;P=7; 96; [(1)+(1)+(1)]+[49/4]
 GENUS#2;P=2; 48; [H]+[(1)+(7)]
 GENUS#2;P=7; 896; [(1)+(1)]+[(7/4)+(7)]
 GENUS#3;P=2; 576; [A]+[2A]
 GENUS#3;P=7; 96; [(1)+(3/4)+(2)]+[49/6]
 GENUS#4;P=2; 64; [H]+[2H]
 GENUS#4;P=7; 896; [(1)+(1)]+[(7/2)+(7/2)]
 GENUS#5;P=2; 576; [A]+[2A]
 GENUS#5;P=7; 504; [(1)+(3/4)]+[(14/3)+(7/2)]


D=  197
 GENUS#1;P=2; 15; [A+H]
 GENUS#1;P=197; 77616/197; [(1)+(3/4)+(1)]+[197/12]


D=  200
 GENUS#1;P=2; 192; [A]+[1]+[50/3]
 GENUS#1;P=5; 48; [(1)+(3/4)+(1)]+[50/3]
 GENUS#2;P=2; 192; [A]+[5/3]+[10]
 GENUS#2;P=5; 240; [(1)+(3/4)]+[(5/3)+(10)]
 GENUS#3;P=2; 192; [A]+[1]+[50/51]
```

```
GENUS#3;P=5;  48;  [(1)+(1)+(1)]+[25/2]
GENUS#4;P=2;  192;  [A]+[5]+[10/11]
GENUS#4;P=5;  240;  [(1)+(1)]+[(5/2)+(5)]
```

D= 201
```
GENUS#1;P=2;  9;  [A+A]
GENUS#1;P=3;  16/3;  [(1)+(1)+(67/4)]+[3/4]
GENUS#1;P=67;  8976/67;  [(1)+(3/4)+(1)]+[67/4]
GENUS#2;P=2;  9;  [H+H]
GENUS#2;P=3;  16/3;  [(1)+(1)+(35/4)]+[201/140]
GENUS#2;P=67;  8976/67;  [(1)+(1)+(3/2)]+[67/8]
```

D= 204
```
GENUS#1;P=2;  96;  [A]+[(1)+(17)]
GENUS#1;P=3;  16/3;  [(1)+(1)+(17)]+[3/4]
GENUS#1;P=17;  576/17;  [(1)+(3/4)+(1)]+[17]
GENUS#2;P=2;  96;  [A]+[(77/3)+(51/77)]
GENUS#2;P=3;  16/3;  [(1)+(103/4)+(77/103)]+[51/77]
GENUS#2;P=17;  576/17;  [(1)+(3/4)+(2/3)]+[51/2]
GENUS#3;P=2;  96;  [A]+[(1)+(1)]
GENUS#3;P=3;  16/3;  [(1)+(1)+(1)]+[51/4]
GENUS#3;P=17;  576/17;  [(1)+(1)+(1)]+[51/4]
GENUS#4;P=2;  96;  [H]+[(61/7)+(51/61)]
GENUS#4;P=3;  16/3;  [(1)+(1)+(17/2)]+[3/2]
GENUS#4;P=17;  576/17;  [(1)+(1)+(3/2)]+[17/2]
```

D= 205
```
GENUS#1;P=2;  15;  [A+H]
GENUS#1;P=5;  48/5;  [(1)+(3/4)+(2/3)]+[205/8]
GENUS#1;P=41;  3360/41;  [(1)+(3/4)+(2/3)]+[205/8]
GENUS#2;P=2;  15;  [A+H]
GENUS#2;P=5;  48/5;  [(1)+(3/4)+(2)]+[205/24]
GENUS#2;P=41;  3360/41;  [(1)+(3/4)+(2)]+[205/24]
```

D= 208
```
GENUS#1;P=2;  6144;  [(1)+(1)+(1)+(13)]
GENUS#1;P=13;  336/13;  [(1)+(1)+(1)]+[13]
GENUS#2;P=2;  1536;  [A]+[(2/3)+(26)]
GENUS#2;P=13;  336/13;  [(1)+(3/4)+(2/3)]+[26]
GENUS#3;P=2;  384;  [A]+[5/3]+[52/5]
GENUS#3;P=13;  336/13;  [(1)+(3/4)+(5/3)]+[52/5]
GENUS#4;P=2;  384;  [H]+[1]+[52/7]
GENUS#4;P=13;  336/13;  [(1)+(1)+(7/4)]+[52/7]
GENUS#5;P=2;  1536;  [A]+[4H]
GENUS#5;P=13;  336/13;  [(1)+(3/4)+(8/3)]+[13/2]
GENUS#6;P=2;  512;  [H]+[(14/15)+(26/7)]
GENUS#6;P=13;  336/13;  [(1)+(1)+(7/2)]+[26/7]
GENUS#7;P=2;  1536;  [H]+[4A]
GENUS#7;P=13;  336/13;  [(1)+(7/4)+(20/7)]+[13/5]
GENUS#8;P=2;  2048;  [(1)+(1)+(7)+(13/7)]
GENUS#8;P=13;  336/13;  [(1)+(1)+(2)]+[13/2]
```

D= 209
```
GENUS#1;P=2;  9;  [A+A]
GENUS#1;P=11;  240/11;  [(1)+(3/4)+(1)]+[209/12]
GENUS#1;P=19;  720/19;  [(1)+(3/4)+(1)]+[209/12]
GENUS#2;P=2;  9;  [H+H]
GENUS#2;P=11;  240/11;  [(1)+(1)+(7/4)]+[209/28]
GENUS#2;P=19;  720/19;  [(1)+(1)+(7/4)]+[209/28]
```

D= 212
```
GENUS#1;P=2;  192;  [H]+[2A]
GENUS#1;P=53;  5616/53;  [(1)+(1)+(1)]+[53/4]
GENUS#2;P=2;  48;  [A]+[(1)+(53/3)]
GENUS#2;P=53;  5616/53;  [(1)+(3/4)+(1)]+[53/3]
GENUS#3;P=2;  192;  [A]+[2H]
GENUS#3;P=53;  5616/53;  [(1)+(3/4)+(2/3)]+[53/2]
```

D= 213
```
GENUS#1;P=2;  15;  [A+H]
GENUS#1;P=3;  16/3;  [(1)+(1)+(71/4)]+[3/4]
GENUS#1;P=71;  10080/71;  [(1)+(3/4)+(1)]+[71/4]
```

```
 GENUS#2;P=2; 15; [H+A]
 GENUS#2;P=3; 16/3; [(1)+(1)+(7/4)]+[213/28]
 GENUS#2;P=71; 10080/71; [(1)+(1)+(7/4)]+[213/28]


D= 216
 GENUS#1;P=2; 192; [A]+[1]+[18]
 GENUS#1;P=3; 432; [(1)+(1)]+[3/4]+[18]
 GENUS#2;P=2; 192; [A]+[27]+[2/3]
 GENUS#2;P=3; 48; [(1)+(1)+(1/2)]+[27]
 GENUS#3;P=2; 192; [H]+[1]+[54/55]
 GENUS#3;P=3; 48; [(1)+(1)+(1)]+[27/2]
 GENUS#4;P=2; 192; [H]+[9]+[6/7]
 GENUS#4;P=3; 432; [(1)+(1)]+[3/2]+[9]
 GENUS#5;P=2; 192; [A]+[9]+[2]
 GENUS#5;P=3; 216; [(1)+(2)]+[3/4]+[9]
 GENUS#6;P=2; 192; [A]+[3]+[6]
 GENUS#6;P=3; 1296; [1]+[(3/4)+(3)+(6)]
 GENUS#7;P=2; 192; [H]+[1]+[18/13]
 GENUS#7;P=3; 216; [(1)+(2)]+[3/2]+[9/2]
 GENUS#8;P=2; 192; [H]+[6/5]
 GENUS#8;P=3; 1296; [2]+[(3/2)+(3/2)+(3)]


D= 217
 GENUS#1;P=2; 9; [A+A]
 GENUS#1;P=7; 96/7; [(1)+(3/4)+(5/3)]+[217/20]
 GENUS#1;P=31; 1920/31; [(1)+(3/4)+(5/3)]+[217/20]
 GENUS#2;P=2; 9; [A+A]
 GENUS#2;P=7; 96/7; [(1)+(1)+(11/4)]+[217/44]
 GENUS#2;P=31; 1920/31; [(1)+(1)+(11/4)]+[217/44]


D= 220
 GENUS#1;P=2; 96; [A]+[(83/3)+(55/83)]
 GENUS#1;P=5; 48/5; [(1)+(3/4)+(2/3)]+[55/2]
 GENUS#1;P=11; 240/11; [(1)+(3/4)+(2/3)]+[55/2]
 GENUS#2;P=2; 96; [A]+[(29/3)+(55/29)]
 GENUS#2;P=5; 48/5; [(1)+(3/4)+(2)]+[55/6]
 GENUS#2;P=11; 240/11; [(1)+(3/4)+(2)]+[55/6]
 GENUS#3;P=2; 96; [A]+[(5/3)+(11)]
 GENUS#3;P=5; 48/5; [(1)+(3/4)+(11)]+[5/3]
 GENUS#3;P=11; 240/11; [(1)+(3/4)+(5/3)]+[11]
 GENUS#4;P=2; 96; [H]+[(1)+(1)]
 GENUS#4;P=5; 48/5; [(1)+(1)+(1)]+[55/4]
 GENUS#4;P=11; 240/11; [(1)+(1)+(1)]+[55/4]


D= 221
 GENUS#1;P=2; 15; [A+H]
 GENUS#1;P=13; 336/13; [(1)+(3/4)+(1)]+[221/12]
 GENUS#1;P=17; 576/17; [(1)+(3/4)+(1)]+[221/12]
 GENUS#2;P=2; 15; [A+H]
 GENUS#2;P=13; 336/13; [(1)+(3/4)+(2/3)]+[221/8]
 GENUS#2;P=17; 576/17; [(1)+(3/4)+(2/3)]+[221/8]


D= 224
 GENUS#1;P=2; 6144; [(1)+(1)+(1)]+[14]
 GENUS#1;P=7; 96/7; [(1)+(1)+(1)]+[14]
 GENUS#2;P=2; 1536; [A]+[1]+[56/3]
 GENUS#2;P=7; 96/7; [(1)+(3/4)+(1)]+[56/3]
 GENUS#3;P=2; 6144; [A]+[2/3]+[28]
 GENUS#3;P=7; 96/7; [(1)+(3/4)+(2/3)]+[28]
 GENUS#4;P=2; 2048; [H]+[6/7]+[28/3]
 GENUS#4;P=7; 96/7; [(1)+(1)+(3/2)]+[28/3]
 GENUS#5;P=2; 6144; [A]+[10/11]+[28/5]
 GENUS#5;P=7; 96/7; [(1)+(1)+(5/2)]+[28/5]
 GENUS#6;P=2; 1536; [A]+[7]+[8/3]
 GENUS#6;P=7; 96/7; [(1)+(3/4)+(8/3)]+[7]
 GENUS#7;P=2; 1536; [H]+[1]+[8]
 GENUS#7;P=7; 96/7; [(1)+(1)+(8)]+[7/4]
 GENUS#8;P=2; 1536; [A]+[17/3]+[56/17]
 GENUS#8;P=7; 96/7; [(1)+(3/4)+(4)]+[14/3]
 GENUS#9;P=2; 2048; [H]+[58/31]+[28/29]
 GENUS#9;P=7; 96/7; [(1)+(1)+(2)]+[7]
 GENUS#10;P=2; 6144; [(1)+(1)+(3)]+[14/3]
```

GENUS#10;P=7; 96/7; [(1)+(1)+(3)]+[14/3]

D= 225
 GENUS#1;P=2; 9; [A+A]
 GENUS#1;P=3; 96; [(1)+(1)]+[(3/4)+(75/4)]
 GENUS#1;P=5; 48; [(1)+(3/4)+(1)]+[75/4]
 GENUS#2;P=2; 9; [H+H]
 GENUS#2;P=3; 96; [(1)+(1)]+[(3/2)+(75/8)]
 GENUS#2;P=5; 48; [(1)+(1)+(3/2)]+[75/8]
 GENUS#3;P=2; 9; [A+A]
 GENUS#3;P=3; 16; [(1)+(7/4)+(5/7)]+[45/4]
 GENUS#3;P=5; 360; [(1)+(3/4)]+[(5/3)+(45/4)]
 GENUS#4;P=2; 9; [H+H]
 GENUS#4;P=3; 96; [(1)+(1)]+[(15/4)+(15/4)]
 GENUS#4;P=5; 160; [(1)+(1)]+[(15/4)+(15/4)]
 GENUS#5;P=2; 9; [A+A]
 GENUS#5;P=3; 24; [(1)+(5)]+[(3/4)+(15/4)]
 GENUS#5;P=5; 360; [(1)+(3/4)]+[(5)+(15/4)]
 GENUS#6;P=2; 9; [H+H]
 GENUS#6;P=3; 16; [(1)+(7/4)+(10/7)]+[45/8]
 GENUS#6;P=5; 360; [(1)+(7/4)]+[(10/7)+(45/8)]

D= 228
 GENUS#1;P=2; 576; [A]+[2A]
 GENUS#1;P=3; 16/3; [(1)+(1)+(1)]+[57/4]
 GENUS#1;P=19; 720/19; [(1)+(1)+(1)]+[57/4]
 GENUS#2;P=2; 48; [A]+[(1)+(19)]
 GENUS#2;P=3; 16/3; [(1)+(1)+(19)]+[3/4]
 GENUS#2;P=19; 720/19; [(1)+(3/4)+(1)]+[19]
 GENUS#3;P=2; 576; [A]+[2A]
 GENUS#3;P=3; 16/3; [(1)+(115/4)+(86/115)]+[57/86]
 GENUS#3;P=19; 720/19; [(1)+(3/4)+(2/3)]+[57/2]
 GENUS#4;P=2; 48; [A]+[(5/3)+(57/5)]
 GENUS#4;P=3; 16/3; [(1)+(7/4)+(5/7)]+[57/5]
 GENUS#4;P=19; 720/19; [(1)+(3/4)+(5/3)]+[57/5]
 GENUS#5;P=2; 64; [H]+[2H]
 GENUS#5;P=3; 16/3; [(1)+(1)+(11/2)]+[57/22]
 GENUS#5;P=19; 720/19; [(1)+(1)+(3)]+[19/4]
 GENUS#6;P=2; 64; [H]+[2H]
 GENUS#6;P=3; 16/3; [(1)+(1)+(19/2)]+[3/2]
 GENUS#6;P=19; 720/19; [(1)+(1)+(3/2)]+[19/2]

D= 229
 GENUS#1;P=2; 15; [A+H]
 GENUS#1;P=229; 104880/229; [(1)+(3/4)+(2/3)]+[229/8]

D= 232
 GENUS#1;P=2; 192; [A]+[29]+[2/3]
 GENUS#1;P=29; 1680/29; [(1)+(3/4)+(2/3)]+[29]
 GENUS#2;P=2; 192; [A]+[1]+[58/59]
 GENUS#2;P=29; 1680/29; [(1)+(1)+(1)]+[29/2]

D= 233
 GENUS#1;P=2; 9; [A+A]
 GENUS#1;P=233; 108576/233; [(1)+(3/4)+(1)]+[233/12]

D= 236
 GENUS#1;P=2; 96; [A]+[(1)+(1)]
 GENUS#1;P=59; 6960/59; [(1)+(1)+(1)]+[59/4]
 GENUS#2;P=2; 96; [A]+[(89/3)+(59/89)]
 GENUS#2;P=59; 6960/59; [(1)+(3/4)+(2/3)]+[59/2]

D= 237
 GENUS#1;P=2; 15; [A+H]
 GENUS#1;P=3; 16/3; [(1)+(1)+(79/4)]+[3/4]
 GENUS#1;P=79; 12480/79; [(1)+(3/4)+(1)]+[79/4]
 GENUS#2;P=2; 15; [A+H]
 GENUS#2;P=3; 16/3; [(1)+(1)+(1/2)]+[237/8]
 GENUS#2;P=79; 12480/79; [(1)+(3/4)+(2/3)]+[237/8]

D= 240
 GENUS#1;P=2; 3072; [(1)+(1)+(1)+(15)]

```
GENUS#1;P=3;  16/3;  [(1)+(1)+(1)]+[15]
GENUS#1;P=5;  48/5;  [(1)+(1)+(1)]+[15]
GENUS#2;P=2;  384;  [A]+[1]+[20]
GENUS#2;P=3;  16/3;  [(1)+(1)+(20)]+[3/4]
GENUS#2;P=5;  48/5;  [(1)+(3/4)+(1)]+[20]
GENUS#3;P=2;  3072;  [A]+[(2/3)+(30)]
GENUS#3;P=3;  16/3;  [(1)+(1)+(1/2)]+[30]
GENUS#3;P=5;  48/5;  [(1)+(3/4)+(2/3)]+[30]
GENUS#4;P=2;  384;  [A]+[5/3]+[12]
GENUS#4;P=3;  16/3;  [(1)+(7/4)+(5/7)]+[12]
GENUS#4;P=5;  48/5;  [(1)+(3/4)+(12)]+[5/3]
GENUS#5;P=2;  3072;  [(1)+(1)+(2H)]
GENUS#5;P=3;  16/3;  [(1)+(1)+(2)]+[15/2]
GENUS#5;P=5;  48/5;  [(1)+(1)+(2)]+[15/2]
GENUS#6;P=2;  3072;  [A]+[(2)+(10)]
GENUS#6;P=3;  16/3;  [(1)+(2)+(10)]+[3/4]
GENUS#6;P=5;  48/5;  [(1)+(3/4)+(2)]+[10]
GENUS#7;P=2;  1024;  [H]+[(2)+(30/31)]
GENUS#7;P=3;  16/3;  [(1)+(1)+(2)]+[15/2]
GENUS#7;P=5;  48/5;  [(1)+(1)+(2)]+[15/2]
GENUS#8;P=2;  1024;  [H]+[(6/7)+(10)]
GENUS#8;P=3;  16/3;  [(1)+(1)+(10)]+[3/2]
GENUS#8;P=5;  48/5;  [(1)+(1)+(3/2)]+[10]
GENUS#9;P=2;  3072;  [A]+[(10/11)+(6)]
GENUS#9;P=3;  16/3;  [(1)+(1)+(5/2)]+[6]
GENUS#9;P=5;  48/5;  [(1)+(1)+(6)]+[5/2]
GENUS#10;P=2;  384;  [H]+[1]+[4]
GENUS#10;P=3;  16/3;  [(1)+(1)+(4)]+[15/4]
GENUS#10;P=5;  48/5;  [(1)+(1)+(4)]+[15/4]
GENUS#11;P=2;  384;  [H]+[13/7]+[60/13]
GENUS#11;P=3;  16/3;  [(1)+(7/4)+(13/7)]+[60/13]
GENUS#11;P=5;  48/5;  [(1)+(7/4)+(13/7)]+[60/13]
GENUS#12;P=2;  3072;  [(1)+(5)+(2A)]
GENUS#12;P=3;  16/3;  [(1)+(2)+(5)]+[3/2]
GENUS#12;P=5;  48/5;  [(1)+(2)+(3/2)]+[5]
GENUS#13;P=2;  1024;  [H]+[(10/7)+(6)]
GENUS#13;P=3;  16/3;  [(1)+(7/4)+(10/7)]+[6]
GENUS#13;P=5;  48/5;  [(1)+(7/4)+(6)]+[10/7]
GENUS#14;P=2;  3072;  [A]+[(2)+(30/11)]
GENUS#14;P=3;  16/3;  [(1)+(2)+(11/4)]+[30/11]
GENUS#14;P=5;  48/5;  [(1)+(2)+(11/4)]+[30/11]
GENUS#15;P=2;  3072;  [(1)+(3)+(3)+(5/3)]
GENUS#15;P=3;  16/3;  [(1)+(2)+(5/2)]+[3]
GENUS#15;P=5;  48/5;  [(1)+(2)+(3)]+[5/2]
GENUS#16;P=2;  1024;  [H]+[(26/15)+(30/13)]
GENUS#16;P=3;  16/3;  [(1)+(2)+(13/4)]+[30/13]
GENUS#16;P=5;  48/5;  [(1)+(2)+(3)]+[5/2]


D=  241
 GENUS#1;P=2;  9;  [H+H]
 GENUS#1;P=241;  116160/241;  [(1)+(1)+(7/4)]+[241/28]


D=  244
 GENUS#1;P=2;  192;  [H]+[2A]
 GENUS#1;P=61;  7440/61;  [(1)+(1)+(1)]+[61/4]
 GENUS#2;P=2;  192;  [A]+[2H]
 GENUS#2;P=61;  7440/61;  [(1)+(3/4)+(2/3)]+[61/2]
 GENUS#3;P=2;  48;  [H]+[(61/31)+(1)]
 GENUS#3;P=61;  7440/61;  [(1)+(1)+(2)]+[61/8]


D=  245
 GENUS#1;P=2;  15;  [A+H]
 GENUS#1;P=5;  48/5;  [(1)+(3/4)+(1)]+[245/12]
 GENUS#1;P=7;  96;  [(1)+(3/4)+(1)]+[245/12]
 GENUS#2;P=2;  15;  [A+H]
 GENUS#2;P=5;  48/5;  [(1)+(3/4)+(2/3)]+[245/8]
 GENUS#2;P=7;  96;  [(1)+(3/4)+(2/3)]+[245/8]
 GENUS#3;P=2;  15;  [H+A]
 GENUS#3;P=5;  48/5;  [(1)+(1)+(7/4)]+[35/4]
 GENUS#3;P=7;  672;  [(1)+(1)]+[(7/4)+(35/4)]
 GENUS#4;P=2;  15;  [A+H]
 GENUS#4;P=5;  48/5;  [(1)+(3/4)+(14/3)]+[35/8]
```

GENUS#4;P=7; 672; [(1)+(3/4)]+[(14/3)+(35/8)]

D= 248
GENUS#1;P=2; 192; [A]+[1]+[62/3]
GENUS#1;P=31; 1920/31; [(1)+(3/4)+(1)]+[62/3]
GENUS#2;P=2; 192; [H]+[1]+[62/63]
GENUS#2;P=31; 1920/31; [(1)+(1)+(1)]+[31/2]

D= 249
GENUS#1;P=2; 9; [A+A]
GENUS#1;P=3; 16/3; [(1)+(1)+(83/4)]+[3/4]
GENUS#1;P=83; 13776/83; [(1)+(3/4)+(1)]+[83/4]
GENUS#2;P=2; 9; [H+H]
GENUS#2;P=3; 16/3; [(1)+(1)+(43/4)]+[249/172]
GENUS#2;P=83; 13776/83; [(1)+(1)+(3/2)]+[83/8]

D= 252
GENUS#1;P=2; 96; [A]+[(1)+(21)]
GENUS#1;P=3; 96; [(1)+(1)]+[(3/4)+(21)]
GENUS#1;P=7; 96/7; [(1)+(3/4)+(1)]+[21]
GENUS#2;P=2; 96; [A]+[(95/3)+(63/95)]
GENUS#2;P=3; 16; [(1)+(127/4)+(95/127)]+[63/95]
GENUS#2;P=7; 96/7; [(1)+(3/4)+(2/3)]+[63/2]
GENUS#3;P=2; 96; [A]+[(5/3)+(63/5)]
GENUS#3;P=3; 16; [(1)+(7/4)+(5/7)]+[63/5]
GENUS#3;P=7; 96/7; [(1)+(3/4)+(5/3)]+[63/5]
GENUS#4;P=2; 96; [H]+[(1)+(1)]
GENUS#4;P=3; 16; [(1)+(1)+(1)]+[63/4]
GENUS#4;P=7; 96/7; [(1)+(1)+(1)]+[63/4]
GENUS#5;P=2; 96; [A]+[(3)+(7)]
GENUS#5;P=3; 96; [(1)+(7)]+[(3)+(3/4)]
GENUS#5;P=7; 96/7; [(1)+(3/4)+(3)]+[7]
GENUS#6;P=2; 96; [A]+[(5)+(21/5)]
GENUS#6;P=3; 24; [(1)+(5)]+[(3/4)+(21/5)]
GENUS#6;P=7; 96/7; [(1)+(3/4)+(5)]+[21/5]
GENUS#7;P=2; 96; [H]+[(13/7)+(63/13)]
GENUS#7;P=3; 16; [(1)+(7/4)+(13/7)]+[63/13]
GENUS#7;P=7; 96/7; [(1)+(2)+(13/8)]+[63/13]
GENUS#8;P=2; 96; [A]+[(11)+(21/11)]
GENUS#8;P=3; 24; [(1)+(2)]+[(3/4)+(21/2)]
GENUS#8;P=7; 96/7; [(1)+(3/4)+(2)]+[21/2]

D= 253
GENUS#1;P=2; 15; [A+H]
GENUS#1;P=11; 240/11; [(1)+(3/4)+(2/3)]+[253/8]
GENUS#1;P=23; 1056/23; [(1)+(3/4)+(2/3)]+[253/8]
GENUS#2;P=2; 15; [A+H]
GENUS#2;P=11; 240/11; [(1)+(3/4)+(5/3)]+[253/20]
GENUS#2;P=23; 1056/23; [(1)+(3/4)+(5/3)]+[253/20]

D= 256
GENUS#1;P=2; 393216; [(1)+(1)+(1)]+[16]
GENUS#2;P=2; 98304; [A]+[2/3]+[32]
GENUS#3;P=2; 98304; [A]+[2]+[32/3]
GENUS#4;P=2; 262144; [(1)+(1)]+[2]+[8]
GENUS#5;P=2; 98304; [A]+[10/11]+[32/5]
GENUS#6;P=2; 98304; [A]+[(8/3)+(8)]
GENUS#7;P=2; 1048576; [(1)+(1)]+[(4)+(4)]
GENUS#8;P=2; 393216; [(1)+(1)+(5)]+[16/5]
GENUS#9;P=2; 98304; [A]+[14/3]+[32/7]
GENUS#10;P=2; 524288; [1]+[(2)+(2)]+[4]
GENUS#11;P=2; 393216; [(1)+(2A)]+[16/3]
GENUS#12;P=2; 262144; [(1)+(3)]+[2]+[8/3]
GENUS#13;P=2; 1572864; [(1)+(3)]+[8A]
GENUS#14;P=2; 1048576; [(3)+(5/3)]+[(12/5)+(4/3)]
GENUS#15;P=2; 393216; [(3)+(5/3)+(7/5)]+[16/7]

D= 257
GENUS#1;P=2; 9; [A+A]
GENUS#1;P=257; 132096/257; [(1)+(3/4)+(1)]+[257/12]

```
D=  260
 GENUS#1;P=2;  576;  [A]+[2A]
 GENUS#1;P=5;  48/5;  [(1)+(1)+(1)]+[65/4]
 GENUS#1;P=13;  336/13;  [(1)+(1)+(1)]+[65/4]
 GENUS#2;P=2;  48;  [A]+[(1)+(65/3)]
 GENUS#2;P=5;  48/5;  [(1)+(3/4)+(1)]+[65/3]
 GENUS#2;P=13;  336/13;  [(1)+(3/4)+(1)]+[65/3]
 GENUS#3;P=2;  576;  [A]+[2A]
 GENUS#3;P=5;  48/5;  [(1)+(3/4)+(2/3)]+[65/2]
 GENUS#3;P=13;  336/13;  [(1)+(3/4)+(2/3)]+[65/2]
 GENUS#4;P=2;  64;  [H]+[2H]
 GENUS#4;P=5;  48/5;  [(1)+(1)+(3/2)]+[65/6]
 GENUS#4;P=13;  336/13;  [(1)+(1)+(3/2)]+[65/6]
 GENUS#5;P=2;  48;  [A]+[(5/3)+(13)]
 GENUS#5;P=5;  48/5;  [(1)+(3/4)+(13)]+[5/3]
 GENUS#5;P=13;  336/13;  [(1)+(3/4)+(5/3)]+[13]
 GENUS#6;P=2;  64;  [H]+[2H]
 GENUS#6;P=5;  48/5;  [(1)+(1)+(3)]+[65/12]
 GENUS#6;P=13;  336/13;  [(1)+(1)+(3)]+[65/12]

D=  261
 GENUS#1;P=2;  15;  [A+H]
 GENUS#1;P=3;  48;  [(1)+(1)]+[(3/4)+(87/4)]
 GENUS#1;P=29;  1680/29;  [(1)+(1)+(1)]+[87/4]
 GENUS#2;P=2;  15;  [A+H]
 GENUS#2;P=3;  16;  [(1)+(1)+(1/2)]+[261/8]
 GENUS#2;P=29;  1680/29;  [(1)+(3/4)+(2/3)]+[261/8]
 GENUS#3;P=2;  15;  [A+H]
 GENUS#3;P=3;  48;  [(1)+(2)]+[(3/4)+(87/8)]
 GENUS#3;P=29;  1680/29;  [(1)+(3/4)+(2)]+[87/8]
 GENUS#4;P=2;  15;  [A+H]
 GENUS#4;P=3;  16;  [(1)+(1)+(5/2)]+[261/40]
 GENUS#4;P=29;  1680/29;  [(1)+(1)+(5/2)]+[261/40]

D=  264
 GENUS#1;P=2;  192;  [A]+[1]+[66/67]
 GENUS#1;P=3;  16/3;  [(1)+(1)+(1)]+[33/2]
 GENUS#1;P=11;  240/11;  [(1)+(1)+(1)]+[33/2]
 GENUS#2;P=2;  192;  [A]+[33]+[2/3]
 GENUS#2;P=3;  16/3;  [(1)+(1)+(1/2)]+[33]
 GENUS#2;P=11;  240/11;  [(1)+(3/4)+(2/3)]+[33]
 GENUS#3;P=2;  192;  [H]+[1]+[66/7]
 GENUS#3;P=3;  16/3;  [(1)+(1)+(7/4)]+[66/7]
 GENUS#3;P=11;  240/11;  [(1)+(1)+(7/4)]+[66/7]
 GENUS#4;P=2;  192;  [A]+[3]+[22/3]
 GENUS#4;P=3;  16/3;  [(1)+(31/4)+(89/31)]+[66/89]
 GENUS#4;P=11;  240/11;  [(1)+(3/4)+(3)]+[22/3]

D=  265
 GENUS#1;P=2;  9;  [H+H]
 GENUS#1;P=5;  48/5;  [(1)+(1)+(7/4)]+[265/28]
 GENUS#1;P=53;  5616/53;  [(1)+(1)+(7/4)]+[265/28]
 GENUS#2;P=2;  9;  [A+A]
 GENUS#2;P=5;  48/5;  [(1)+(3/4)+(41/3)]+[265/164]
 GENUS#2;P=53;  5616/53;  [(1)+(3/4)+(5/3)]+[53/4]

D=  268
 GENUS#1;P=2;  96;  [A]+[(101/3)+(67/101)]
 GENUS#1;P=67;  8976/67;  [(1)+(3/4)+(2/3)]+[67/2]
 GENUS#2;P=2;  96;  [A]+[(1)+(1)]
 GENUS#2;P=67;  8976/67;  [(1)+(1)+(1)]+[67/4]

D=  269
 GENUS#1;P=2;  15;  [A+H]
 GENUS#1;P=269;  144720/269;  [(1)+(3/4)+(1)]+[269/12]

D=  272
 GENUS#1;P=2;  6144;  [(1)+(1)+(1)+(17)]
 GENUS#1;P=17;  576/17;  [(1)+(1)+(1)]+[17]
 GENUS#2;P=2;  384;  [A]+[1]+[68/3]
 GENUS#2;P=17;  576/17;  [(1)+(3/4)+(1)]+[68/3]
 GENUS#3;P=2;  1536;  [A]+[(2/3)+(34)]
```

```
GENUS#3;P=17; 576/17; [(1)+(3/4)+(2/3)]+[34]
GENUS#4;P=2;  512; [H]+[(6/7)+(34/3)]
GENUS#4;P=17; 576/17; [(1)+(1)+(3/2)]+[34/3]
GENUS#5;P=2;  384; [A]+[5/3]+[68/5]
GENUS#5;P=17; 576/17; [(1)+(3/4)+(5/3)]+[68/5]
GENUS#6;P=2;  2048; [(1)+(1)+(3)+(17/3)]
GENUS#6;P=17; 576/17; [(1)+(1)+(3)]+[17/3]
GENUS#7;P=2;  4608; [A]+[4A]
GENUS#7;P=17; 576/17; [(1)+(3/4)+(4)]+[17/3]
GENUS#8;P=2;  512; [H]+[4H]
GENUS#8;P=17; 576/17; [(1)+(7/4)+(12/7)]+[17/3]
```

D= 273
```
GENUS#1;P=2;  9; [A+A]
GENUS#1;P=3;  16/3; [(1)+(1)+(91/4)]+[3/4]
GENUS#1;P=7;  96/7; [(1)+(3/4)+(1)]+[91/4]
GENUS#1;P=13; 336/13; [(1)+(3/4)+(1)]+[91/4]
GENUS#2;P=2;  9; [H+H]
GENUS#2;P=3;  16/3; [(1)+(1)+(7/4)]+[39/4]
GENUS#2;P=7;  96/7; [(1)+(1)+(39/4)]+[7/4]
GENUS#2;P=13; 336/13; [(1)+(1)+(7/4)]+[39/4]
GENUS#3;P=2;  9; [H+H]
GENUS#3;P=3;  16/3; [(1)+(1)+(47/4)]+[273/188]
GENUS#3;P=7;  96/7; [(1)+(1)+(3/2)]+[91/8]
GENUS#3;P=13; 336/13; [(1)+(1)+(3/2)]+[91/8]
GENUS#4;P=2;  9; [A+A]
GENUS#4;P=3;  16/3; [(1)+(31/4)+(92/31)]+[273/368]
GENUS#4;P=7;  96/7; [(1)+(3/4)+(3)]+[91/12]
GENUS#4;P=13; 336/13; [(1)+(3/4)+(3)]+[91/12]
```

D= 276
```
GENUS#1;P=2;  192; [H]+[2A]
GENUS#1;P=3;  16/3; [(1)+(1)+(1)]+[69/4]
GENUS#1;P=23; 1056/23; [(1)+(1)+(1)]+[69/4]
GENUS#2;P=2;  48; [A]+[(1)+(23)]
GENUS#2;P=3;  16/3; [(1)+(1)+(23)]+[3/4]
GENUS#2;P=23; 1056/23; [(1)+(3/4)+(1)]+[23]
GENUS#3;P=2;  192; [A]+[2H]
GENUS#3;P=3;  16/3; [(1)+(139/4)+(104/139)]+[69/104]
GENUS#3;P=23; 1056/23; [(1)+(3/4)+(2/3)]+[69/2]
GENUS#4;P=2;  48; [H]+[(1)+(69/7)]
GENUS#4;P=3;  16/3; [(1)+(1)+(7/4)]+[69/7]
GENUS#4;P=23; 1056/23; [(1)+(1)+(7/4)]+[69/7]
GENUS#5;P=2;  192; [A]+[2H]
GENUS#5;P=3;  16/3; [(1)+(1)+(5/2)]+[69/10]
GENUS#5;P=23; 1056/23; [(1)+(1)+(5/2)]+[69/10]
GENUS#6;P=2;  192; [H]+[2A]
GENUS#6;P=3;  16/3; [(1)+(1)+(7/2)]+[69/14]
GENUS#6;P=23; 1056/23; [(1)+(1)+(7/2)]+[69/14]
```

D= 277
```
GENUS#1;P=2;  15; [A+H]
GENUS#1;P=277; 153456/277; [(1)+(3/4)+(2/3)]+[277/8]
```

D= 280
```
GENUS#1;P=2;  192; [A]+[35]+[2/3]
GENUS#1;P=5;  48/5; [(1)+(3/4)+(2/3)]+[35]
GENUS#1;P=7;  96/7; [(1)+(3/4)+(2/3)]+[35]
GENUS#2;P=2;  192; [H]+[1]+[10]
GENUS#2;P=5;  48/5; [(1)+(1)+(7/4)]+[10]
GENUS#2;P=7;  96/7; [(1)+(1)+(10)]+[7/4]
GENUS#3;P=2;  192; [A]+[35/3]+[2]
GENUS#3;P=5;  48/5; [(1)+(3/4)+(2)]+[35/3]
GENUS#3;P=7;  96/7; [(1)+(3/4)+(2)]+[35/3]
GENUS#4;P=2;  192; [H]+[1]+[70/71]
GENUS#4;P=5;  48/5; [(1)+(1)+(1)]+[35/2]
GENUS#4;P=7;  96/7; [(1)+(1)+(1)]+[35/2]
```

D= 281
```
GENUS#1;P=2;  9; [A+A]
GENUS#1;P=281; 157920/281; [(1)+(3/4)+(1)]+[281/12]
```

```
D=  284
 GENUS#1;P=2;  96;  [H]+[(1)+(1)]
 GENUS#1;P=71; 10080/71;  [(1)+(1)+(1)]+[71/4]
 GENUS#2;P=2;  96;  [A]+[(17/3)+(71/17)]
 GENUS#2;P=71; 10080/71;  [(1)+(3/4)+(14/3)]+[71/14]

D=  285
 GENUS#1;P=2;  15;  [A+H]
 GENUS#1;P=3;  16/3;  [(1)+(1)+(95/4)]+[3/4]
 GENUS#1;P=5;  48/5;  [(1)+(3/4)+(1)]+[95/4]
 GENUS#1;P=19; 720/19;  [(1)+(3/4)+(1)]+[95/4]
 GENUS#2;P=2;  15;  [H+A]
 GENUS#2;P=3;  16/3;  [(1)+(1)+(7/4)]+[285/28]
 GENUS#2;P=5;  48/5;  [(1)+(1)+(7/4)]+[285/28]
 GENUS#2;P=19; 720/19;  [(1)+(1)+(7/4)]+[285/28]
 GENUS#3;P=2;  15;  [A+H]
 GENUS#3;P=3;  16/3;  [(1)+(2)+(95/8)]+[3/4]
 GENUS#3;P=5;  48/5;  [(1)+(3/4)+(2)]+[95/8]
 GENUS#3;P=19; 720/19;  [(1)+(3/4)+(2)]+[95/8]
 GENUS#4;P=2;  15;  [A+H]
 GENUS#4;P=3;  16/3;  [(1)+(5/2)]+[57/8]
 GENUS#4;P=5;  48/5;  [(1)+(1)+(31/4)]+[285/124]
 GENUS#4;P=19; 720/19;  [(1)+(1)+(5/2)]+[57/8]

D=  288
 GENUS#1;P=2;  6144;  [(1)+(1)+(1)]+[18]
 GENUS#1;P=3;  16;  [(1)+(1)+(1)]+[18]
 GENUS#2;P=2;  1536;  [A]+[1]+[24]
 GENUS#2;P=3;  48;  [(1)+(1)]+[(3/4)+(24)]
 GENUS#3;P=2;  6144;  [A]+[2/3]+[36]
 GENUS#3;P=3;  16;  [(1)+(1)+(1/2)]+[36]
 GENUS#4;P=2;  2048;  [H]+[74/39]+[36/37]
 GENUS#4;P=3;  16;  [(1)+(1)+(2)]+[9]
 GENUS#5;P=2;  1536;  [A]+[5/3]+[72/5]
 GENUS#5;P=3;  16;  [(1)+(7/4)+(5/7)]+[72/5]
 GENUS#6;P=2;  6144;  [(1)+(1)+(9)]+[2]
 GENUS#6;P=3;  16;  [(1)+(1)+(2)]+[9]
 GENUS#7;P=2;  6144;  [(1)+(1)+(3)]+[6]
 GENUS#7;P=3;  48;  [(1)+(1)]+[(3)+(6)]
 GENUS#8;P=2;  1536;  [A]+[3]+[8]
 GENUS#8;P=3;  48;  [(1)+(8)]+[(3)+(3/4)]
 GENUS#9;P=2;  1536;  [A]+[1]+[72/11]
 GENUS#9;P=3;  16;  [(1)+(1)+(11/4)]+[72/11]
 GENUS#10;P=2;  1536;  [A]+[7]+[24/7]
 GENUS#10;P=3;  48;  [(1)+(4)]+[(3/4)+(6)]
 GENUS#11;P=2;  2048;  [H]+[22/23]+[36/11]
 GENUS#11;P=3;  16;  [(1)+(1)+(4)]+[9/2]
 GENUS#12;P=2;  6144;  [(1)+(2A)]+[6]
 GENUS#12;P=3;  48;  [(1)+(2)]+[(3/2)+(6)]
 GENUS#13;P=2;  6144;  [A]+[18/19]+[4]
 GENUS#13;P=3;  16;  [(1)+(1)+(4)]+[9/2]
 GENUS#14;P=2;  1536;  [A]+[5]+[24/5]
 GENUS#14;P=3;  48;  [(1)+(5)]+[(3/4)+(24/5)]
 GENUS#15;P=2;  2048;  [H]+[6/7]+[12]
 GENUS#15;P=3;  48;  [(1)+(1)]+[(3/2)+(12)]
 GENUS#16;P=2;  6144;  [A]+[2]+[12]
 GENUS#16;P=3;  48;  [(1)+(2)]+[(3/4)+(12)]
 GENUS#17;P=2;  1536;  [A]+[1]+[72/19]
 GENUS#17;P=3;  16;  [(1)+(1)+(4)]+[9/2]
 GENUS#18;P=2;  6144;  [A]+[6]+[4]
 GENUS#18;P=3;  48;  [(1)+(4)]+[(3/4)+(6)]
 GENUS#19;P=2;  1536;  [A]+[37/19]+[72/37]
 GENUS#19;P=3;  16;  [(1)+(2)+(2)]+[9/2]
 GENUS#20;P=2;  2048;  [H]+[2]+[12/5]
 GENUS#20;P=3;  48;  [(1)+(2)]+[(3)+(3)]

D=  289
 GENUS#1;P=2;  9;  [A+A]
 GENUS#1;P=17; 11016;  [(1)+(3/4)]+[(17/3)+(17/4)]

D=  292
 GENUS#1;P=2;  576;  [A]+[2A]
```

```
GENUS#1;P=73; 10656/73; [(1)+(1)+(1)]+[73/4]
GENUS#2;P=2; 48; [H]+[(1)+(73/7)]
GENUS#2;P=73; 10656/73; [(1)+(1)+(7/4)]+[73/7]
GENUS#3;P=2; 64; [H]+[2H]
GENUS#3;P=73; 10656/73; [(1)+(1)+(7/2)]+[73/14]
```

D= 293
```
 GENUS#1;P=2; 15; [A+H]
 GENUS#1;P=293; 171696/293; [(1)+(3/4)+(1)]+[293/12]
```

D= 296
```
 GENUS#1;P=2; 192; [A]+[1]+[74/75]
 GENUS#1;P=37; 2736/37; [(1)+(1)+(1)]+[37/2]
 GENUS#2;P=2; 192; [A]+[37]+[2/3]
 GENUS#2;P=37; 2736/37; [(1)+(3/4)+(2/3)]+[37]
```

D= 297
```
 GENUS#1;P=2; 9; [A+A]
 GENUS#1;P=3; 432; [(1)+(1)]+[3/4]+[99/4]
 GENUS#1;P=11; 240/11; [(1)+(3/4)+(1)]+[99/4]
 GENUS#2;P=2; 9; [H+H]
 GENUS#2;P=3; 48; [(1)+(1)+(7/4)]+[297/28]
 GENUS#2;P=11; 240/11; [(1)+(1)+(7/4)]+[297/28]
 GENUS#3;P=2; 9; [H+H]
 GENUS#3;P=3; 432; [(1)+(1)]+[3/2]+[99/8]
 GENUS#3;P=11; 240/11; [(1)+(1)+(3/2)]+[99/8]
 GENUS#4;P=2; 9; [A+A]
 GENUS#4;P=3; 48; [(1)+(7/4)+(5/7)]+[297/20]
 GENUS#4;P=11; 240/11; [(1)+(3/4)+(5/3)]+[297/20]
 GENUS#5;P=2; 9; [A+A]
 GENUS#5;P=3; 1296; [1]+[(3/4)+(3)+(33/4)]
 GENUS#5;P=11; 240/11; [(1)+(3/4)+(3)]+[33/4]
 GENUS#6;P=2; 9; [H+H]
 GENUS#6;P=3; 1296; [2]+[(3/2)+(3/2)+(33/8)]
 GENUS#6;P=11; 240/11; [(2)+(3/2)+(3/2)]+[33/8]
 GENUS#7;P=2; 9; [A+A]
 GENUS#7;P=3; 216; [(1)+(5)]+[3/4]+[99/20]
 GENUS#7;P=11; 240/11; [(1)+(3/4)+(5)]+[99/20]
 GENUS#8;P=2; 9; [A+A]
 GENUS#8;P=3; 216; [(1)+(2)]+[21/8]+[99/28]
 GENUS#8;P=11; 240/11; [(1)+(2)+(21/8)]+[99/28]
```

D= 300
```
 GENUS#1;P=2; 96; [A]+[(1)+(25)]
 GENUS#1;P=3; 16/3; [(1)+(1)+(25)]+[3/4]
 GENUS#1;P=5; 48; [(1)+(3/4)+(1)]+[25]
 GENUS#2;P=2; 96; [A]+[(113/3)+(75/113)]
 GENUS#2;P=3; 16/3; [(1)+(151/4)+(113/151)]+[75/113]
 GENUS#2;P=5; 48; [(1)+(3/4)+(2/3)]+[75/2]
 GENUS#3;P=2; 96; [A]+[(5/3)+(15)]
 GENUS#3;P=3; 16/3; [(1)+(7/4)+(5/7)]+[15]
 GENUS#3;P=5; 240; [(1)+(3/4)]+[(5/3)+(15)]
 GENUS#4;P=2; 96; [A]+[(85/11)+(15/17)]
 GENUS#4;P=3; 16/3; [(1)+(1)+(5/2)]+[15/2]
 GENUS#4;P=5; 240; [(1)+(1)]+[(5/2)+(15/2)]
 GENUS#5;P=2; 96; [H]+[(89/7)+(75/89)]
 GENUS#5;P=3; 16/3; [(1)+(1)+(25/2)]+[3/2]
 GENUS#5;P=5; 48; [(1)+(1)+(3/2)]+[25/2]
 GENUS#6;P=2; 96; [A]+[(5)+(5)]
 GENUS#6;P=3; 16/3; [(1)+(5)+(5)]+[3/4]
 GENUS#6;P=5; 240; [(1)+(3/4)]+[(5)+(5)]
 GENUS#7;P=2; 96; [H]+[(1)+(5)]
 GENUS#7;P=3; 16/3; [(1)+(1)+(5)]+[15/4]
 GENUS#7;P=5; 240; [(1)+(1)]+[(15/4)+(5)]
 GENUS#8;P=2; 96; [A]+[(1)+(1)]
 GENUS#8;P=3; 16/3; [(1)+(1)+(1)]+[75/4]
 GENUS#8;P=5; 48; [(1)+(1)+(1)]+[75/4]
```

D= 301
```
 GENUS#1;P=2; 15; [A+H]
 GENUS#1;P=7; 96/7; [(1)+(3/4)+(2/3)]+[301/8]
 GENUS#1;P=43; 3696/43; [(1)+(3/4)+(2/3)]+[301/8]
```

```
GENUS#2;P=2; 15; [A+H]
GENUS#2;P=7; 96/7; [(1)+(3/4)+(2)]+[301/24]
GENUS#2;P=43; 3696/43; [(1)+(3/4)+(2)]+[301/24]


D= 304
 GENUS#1;P=2; 3072; [(1)+(1)+(1)+(19)]
 GENUS#1;P=19; 720/19; [(1)+(1)+(1)]+[19]
 GENUS#2;P=2; 3072; [A]+[(2/3)+(38)]
 GENUS#2;P=19; 720/19; [(1)+(3/4)+(2/3)]+[38]
 GENUS#3;P=2; 384; [H]+[1]+[76/7]
 GENUS#3;P=19; 720/19; [(1)+(1)+(7/4)]+[76/7]
 GENUS#4;P=2; 3072; [A]+[(2)+(38/3)]
 GENUS#4;P=19; 720/19; [(1)+(3/4)+(2)]+[38/3]
 GENUS#5;P=2; 384; [A]+[29/3]+[76/29]
 GENUS#5;P=19; 720/19; [(1)+(3/4)+(8/3)]+[19/2]
 GENUS#6;P=2; 3072; [(1)+(1)+(2A)]
 GENUS#6;P=19; 720/19; [(1)+(1)+(2)]+[19/2]
 GENUS#7;P=2; 1024; [H]+[(2)+(38/39)]
 GENUS#7;P=19; 720/19; [(1)+(1)+(2)]+[19/2]
 GENUS#8;P=2; 1024; [H]+[(42/23)+(38/21)]
 GENUS#8;P=19; 720/19; [(1)+(2)+(2)]+[19/4]


D= 305
 GENUS#1;P=2; 9; [A+A]
 GENUS#1;P=5; 48/5; [(1)+(3/4)+(1)]+[305/12]
 GENUS#1;P=61; 7440/61; [(1)+(3/4)+(1)]+[305/12]
 GENUS#2;P=2; 9; [H+H]
 GENUS#2;P=5; 48/5; [(1)+(1)+(3/2)]+[305/24]
 GENUS#2;P=61; 7440/61; [(1)+(1)+(3/2)]+[305/24]


D= 308
 GENUS#1;P=2; 192; [H]+[2A]
 GENUS#1;P=7; 96/7; [(1)+(1)+(1)]+[77/4]
 GENUS#1;P=11; 240/11; [(1)+(1)+(1)]+[77/4]
 GENUS#2;P=2; 48; [A]+[(1)+(77/3)]
 GENUS#2;P=7; 96/7; [(1)+(3/4)+(1)]+[77/3]
 GENUS#2;P=11; 240/11; [(1)+(3/4)+(1)]+[77/3]
 GENUS#3;P=2; 192; [A]+[2H]
 GENUS#3;P=7; 96/7; [(1)+(3/4)+(2/3)]+[77/2]
 GENUS#3;P=11; 240/11; [(1)+(3/4)+(2/3)]+[77/2]
 GENUS#4;P=2; 48; [H]+[(77/39)+(1)]
 GENUS#4;P=7; 96/7; [(1)+(1)+(2)]+[77/8]
 GENUS#4;P=11; 240/11; [(1)+(1)+(2)]+[77/8]
 GENUS#5;P=2; 192; [H]+[2A]
 GENUS#5;P=7; 96/7; [(1)+(1)+(3/2)]+[77/6]
 GENUS#5;P=11; 240/11; [(1)+(1)+(3/2)]+[77/6]
 GENUS#6;P=2; 192; [A]+[2H]
 GENUS#6;P=7; 96/7; [(1)+(1)+(3)]+[77/12]
 GENUS#6;P=11; 240/11; [(1)+(1)+(3)]+[77/12]


D= 309
 GENUS#1;P=2; 15; [A+H]
 GENUS#1;P=3; 16/3; [(1)+(1)+(103/4)]+[3/4]
 GENUS#1;P=103; 21216/103; [(1)+(3/4)+(1)]+[103/4]
 GENUS#2;P=2; 15; [A+H]
 GENUS#2;P=3; 16/3; [(1)+(1)+(1/2)]+[309/8]
 GENUS#2;P=103; 21216/103; [(1)+(3/4)+(2/3)]+[309/8]


D= 312
 GENUS#1;P=2; 192; [A]+[1]+[26]
 GENUS#1;P=3; 16/3; [(1)+(1)+(26)]+[3/4]
 GENUS#1;P=13; 336/13; [(1)+(3/4)+(1)]+[26]
 GENUS#2;P=2; 192; [A]+[39]+[2/3]
 GENUS#2;P=3; 16/3; [(1)+(1)+(1/2)]+[39]
 GENUS#2;P=13; 336/13; [(1)+(3/4)+(2/3)]+[39]
 GENUS#3;P=2; 192; [H]+[13]+[6/7]
 GENUS#3;P=3; 16/3; [(1)+(1)+(13)]+[3/2]
 GENUS#3;P=13; 336/13; [(1)+(1)+(3/2)]+[13]
 GENUS#4;P=2; 192; [H]+[1]+[78/79]
 GENUS#4;P=3; 16/3; [(1)+(1)+(1)]+[39/2]
 GENUS#4;P=13; 336/13; [(1)+(1)+(1)]+[39/2]
```

```
D=  313
 GENUS#1;P=2;  9;  [H+H]
 GENUS#1;P=313;  195936/313;  [(1)+(1)+(7/4)]+[313/28]

D=  316
 GENUS#1;P=2;  96;  [H]+[(1)+(1)]
 GENUS#1;P=79;  12480/79;  [(1)+(1)+(1)]+[79/4]
 GENUS#2;P=2;  96;  [A]+[(41/3)+(79/41)]
 GENUS#2;P=79;  12480/79;  [(1)+(3/4)+(2)]+[79/6]

D=  317
 GENUS#1;P=2;  15;  [A+H]
 GENUS#1;P=317;  200976/317;  [(1)+(3/4)+(1)]+[317/12]

D=  320
 GENUS#1;P=2;  49152;  [(1)+(1)+(1)]+[20]
 GENUS#1;P=5;  48/5;  [(1)+(1)+(1)]+[20]
 GENUS#2;P=2;  3072;  [A]+[1]+[80/3]
 GENUS#2;P=5;  48/5;  [(1)+(3/4)+(1)]+[80/3]
 GENUS#3;P=2;  12288;  [A]+[2/3]+[40]
 GENUS#3;P=5;  48/5;  [(1)+(3/4)+(2/3)]+[40]
 GENUS#4;P=2;  32768;  [(1)+(1)]+[(2)+(10)]
 GENUS#4;P=5;  48/5;  [(1)+(1)+(2)]+[10]
 GENUS#5;P=2;  3072;  [H]+[1]+[80/7]
 GENUS#5;P=5;  48/5;  [(1)+(1)+(7/4)]+[80/7]
 GENUS#6;P=2;  4096;  [H]+[6/7]+[40/3]
 GENUS#6;P=5;  48/5;  [(1)+(1)+(3/2)]+[40/3]
 GENUS#7;P=2;  3072;  [A]+[5/3]+[16]
 GENUS#7;P=5;  48/5;  [(1)+(3/4)+(16)]+[5/3]
 GENUS#8;P=2;  12288;  [A]+[10]+[8/3]
 GENUS#8;P=5;  48/5;  [(1)+(3/4)+(8/3)]+[10]
 GENUS#9;P=2;  12288;  [A]+[8H]
 GENUS#9;P=5;  48/5;  [(1)+(3/4)+(8/3)]+[10]
 GENUS#10;P=2;  12288;  [A]+[(4)+(20/3)]
 GENUS#10;P=5;  48/5;  [(1)+(3/4)+(4)]+[20/3]
 GENUS#11;P=2;  3072;  [A]+[17/3]+[80/17]
 GENUS#11;P=5;  48/5;  [(1)+(3/4)+(17/3)]+[80/17]
 GENUS#12;P=2;  32768;  [(1)+(1)]+[(6)+(10/3)]
 GENUS#12;P=5;  48/5;  [(1)+(1)+(4)]+[5]
 GENUS#13;P=2;  4096;  [H]+[10/7]+[8]
 GENUS#13;P=5;  48/5;  [(1)+(7/4)+(8)]+[10/7]
 GENUS#14;P=2;  16384;  [(1)+(1)+(3)]+[20/3]
 GENUS#14;P=5;  48/5;  [(1)+(1)+(3)]+[20/3]
 GENUS#15;P=2;  49152;  [(1)+(2A)]+[20/11]
 GENUS#15;P=5;  48/5;  [(1)+(2)+(2)]+[5]
 GENUS#16;P=2;  4096;  [H]+[(12/7)+(20/3)]
 GENUS#16;P=5;  48/5;  [(1)+(7/4)+(12/7)]+[20/3]
 GENUS#17;P=2;  16384;  [(1)+(3)+(5/3)]+[4]
 GENUS#17;P=5;  48/5;  [(1)+(2)+(4)]+[5/2]
 GENUS#18;P=2;  12288;  [H]+[8A]
 GENUS#18;P=5;  48/5;  [(1)+(7/4)+(24/7)]+[10/3]
 GENUS#19;P=2;  49152;  [(1)+(3)]+[4H]
 GENUS#19;P=5;  48/5;  [(1)+(3)+(8/3)]+[5/2]

D=  321
 GENUS#1;P=2;  9;  [A+A]
 GENUS#1;P=3;  16/3;  [(1)+(1)+(107/4)]+[3/4]
 GENUS#1;P=107;  22896/107;  [(1)+(3/4)+(1)]+[107/4]
 GENUS#2;P=2;  9;  [H+H]
 GENUS#2;P=3;  16/3;  [(1)+(1)+(7/4)]+[321/28]
 GENUS#2;P=107;  22896/107;  [(1)+(1)+(7/4)]+[321/28]

D=  324
 GENUS#1;P=2;  576;  [A]+[2A]
 GENUS#1;P=3;  144;  [(1)+(1)+(1)]+[81/4]
 GENUS#2;P=2;  48;  [A]+[(1)+(27)]
 GENUS#2;P=3;  1296;  [(1)+(1)]+[3/4]+[27]
 GENUS#3;P=2;  576;  [A]+[2A]
 GENUS#3;P=3;  144;  [(1)+(163/4)+(122/163)]+[81/122]
 GENUS#4;P=2;  64;  [H]+[2H]
 GENUS#4;P=3;  1296;  [(1)+(1)]+[3]+[27/4]
 GENUS#5;P=2;  48;  [H]+[(1)+(27/5)]
```

```
GENUS#5;P=3;  1296;  [(1)+(1)]+[15/4]+[27/5]
GENUS#6;P=2;  576;  [A]+[2A]
GENUS#6;P=3;  5832;  [1]+[(3/4)+(6)]+[9/2]
GENUS#7;P=2;  48;  [H]+[(9/5)+(3)]
GENUS#7;P=3;  11664;  [2]+[(15/8)+(3)]+[9/5]
GENUS#8;P=2;  64;  [H]+[2H]
GENUS#8;P=3;  11664;  [2]+[(3/2)+(3/2)]+[9/2]
GENUS#9;P=2;  576;  [A]+[2A]
GENUS#9;P=3;  648;  [(1)+(2)]+[3/4]+[27/2]
GENUS#10;P=2;  64;  [H]+[2H]
GENUS#10;P=3;  1296;  [(1)+(1)]+[3/2]+[27/2]
GENUS#11;P=2;  48;  [A]+[(3)+(9)]
GENUS#11;P=3;  11664;  [1]+[(3/4)+(3)]+[9]
GENUS#12;P=2;  576;  [A]+[2A]
GENUS#12;P=3;  2592;  [(1)+(1)]+[(9/2)+(9/2)]
GENUS#13;P=2;  576;  [A]+[2A]
GENUS#13;P=3;  648;  [(1)+(2)]+[3/2]+[27/4]
GENUS#14;P=2;  576;  [A]+[2A]
GENUS#14;P=3;  648;  [(1)+(2)]+[(9/4)+(9/2)]
GENUS#15;P=2;  64;  [H]+[2H]
GENUS#15;P=3;  11664;  [1]+[(3)+(3)]+[9/4]
GENUS#16;P=2;  576;  [A]+[2A]
GENUS#16;P=3;  5832;  [2]+[(3/2)+(3)]+[9/4]


D=  325
 GENUS#1;P=2;  15;  [A+H]
 GENUS#1;P=5;  48;  [(1)+(3/4)+(2/3)]+[325/8]
 GENUS#1;P=13;  336/13;  [(1)+(3/4)+(2/3)]+[325/8]
 GENUS#2;P=2;  15;  [A+H]
 GENUS#2;P=5;  240;  [(1)+(3/4)]+[(5/3)+(65/4)]
 GENUS#2;P=13;  336/13;  [(1)+(3/4)+(5/3)]+[65/4]
 GENUS#3;P=2;  15;  [A+H]
 GENUS#3;P=5;  48;  [(1)+(3/4)+(2)]+[325/24]
 GENUS#3;P=13;  336/13;  [(1)+(3/4)+(2)]+[325/24]
 GENUS#4;P=2;  15;  [A+H]
 GENUS#4;P=5;  240;  [(1)+(1)]+[(5/2)+(65/8)]
 GENUS#4;P=13;  336/13;  [(1)+(1)+(5/2)]+[65/8]


D=  328
 GENUS#1;P=2;  192;  [A]+[1]+[82/83]
 GENUS#1;P=41;  3360/41;  [(1)+(1)+(1)]+[41/2]
 GENUS#2;P=2;  192;  [H]+[1]+[82/7]
 GENUS#2;P=41;  3360/41;  [(1)+(1)+(7/4)]+[82/7]


D=  329
 GENUS#1;P=2;  9;  [A+A]
 GENUS#1;P=7;  96/7;  [(1)+(3/4)+(1)]+[329/12]
 GENUS#1;P=47;  4416/47;  [(1)+(3/4)+(1)]+[329/12]
 GENUS#2;P=2;  9;  [A+A]
 GENUS#2;P=7;  96/7;  [(1)+(1)+(11/4)]+[329/44]
 GENUS#2;P=47;  4416/47;  [(1)+(1)+(11/4)]+[329/44]


D=  332
 GENUS#1;P=2;  96;  [A]+[(1)+(1)]
 GENUS#1;P=83;  13776/83;  [(1)+(1)+(1)]+[83/4]
 GENUS#2;P=2;  96;  [A]+[(125/3)+(83/125)]
 GENUS#2;P=83;  13776/83;  [(1)+(3/4)+(2/3)]+[83/2]


D=  333
 GENUS#1;P=2;  15;  [A+H]
 GENUS#1;P=3;  96;  [(1)+(1)]+[(3/4)+(111/4)]
 GENUS#1;P=37;  2736/37;  [(1)+(3/4)+(1)]+[111/4]
 GENUS#2;P=2;  15;  [A+H]
 GENUS#2;P=3;  16;  [(1)+(1)+(1/2)]+[333/8]
 GENUS#2;P=37;  2736/37;  [(1)+(3/4)+(2/3)]+[333/8]
 GENUS#3;P=2;  15;  [A+H]
 GENUS#3;P=3;  24;  [(1)+(2)]+[(3/4)+(111/8)]
 GENUS#3;P=37;  2736/37;  [(1)+(3/4)+(2)]+[111/8]
 GENUS#4;P=2;  15;  [A+H]
 GENUS#4;P=3;  16;  [(1)+(1)+(19/4)]+[333/76]
 GENUS#4;P=37;  2736/37;  [(1)+(1)+(19/4)]+[333/76]
```

```
D=  336
 GENUS#1;P=2;  6144;  [(1)+(1)+(1)+(21)]
 GENUS#1;P=3;  16/3;  [(1)+(1)+(1)]+[21]
 GENUS#1;P=7;  96/7;  [(1)+(1)+(1)]+[21]
 GENUS#2;P=2;  384;  [A]+[1]+[28]
 GENUS#2;P=3;  16/3;  [(1)+(1)+(28)]+[3/4]
 GENUS#2;P=7;  96/7;  [(1)+(3/4)+(1)]+[28]
 GENUS#3;P=2;  1536;  [A]+[(2/3)+(42)]
 GENUS#3;P=3;  16/3;  [(1)+(1)+(1/2)]+[42]
 GENUS#3;P=7;  96/7;  [(1)+(3/4)+(2/3)]+[42]
 GENUS#4;P=2;  1536;  [A]+[4H]
 GENUS#4;P=3;  16/3;  [(1)+(11/4)+(8/11)]+[21/2]
 GENUS#4;P=7;  96/7;  [(1)+(3/4)+(8/3)]+[21/2]
 GENUS#5;P=2;  2048;  [(1)+(1)+(3)+(7)]
 GENUS#5;P=3;  16/3;  [(1)+(1)+(7)]+[3]
 GENUS#5;P=7;  96/7;  [(1)+(1)+(3)]+[7]
 GENUS#6;P=2;  1536;  [A]+[(2)+(14)]
 GENUS#6;P=3;  16/3;  [(1)+(2)+(14)]+[3/4]
 GENUS#6;P=7;  96/7;  [(1)+(3/4)+(2)]+[14]
 GENUS#7;P=2;  384;  [A]+[7]+[4]
 GENUS#7;P=3;  16/3;  [(1)+(4)+(7)]+[3/4]
 GENUS#7;P=7;  96/7;  [(1)+(3/4)+(4)]+[7]
 GENUS#8;P=2;  1536;  [H]+[4A]
 GENUS#8;P=3;  16/3;  [(1)+(7/4)+(52/7)]+[21/13]
 GENUS#8;P=7;  96/7;  [(1)+(31/4)+(52/31)]+[21/13]
 GENUS#9;P=2;  2048;  [(1)+(1)+(11)+(21/11)]
 GENUS#9;P=3;  16/3;  [(1)+(1)+(2)]+[21/2]
 GENUS#9;P=7;  96/7;  [(1)+(1)+(2)]+[21/2]
 GENUS#10;P=2;  512;  [H]+[(6/7)+(14)]
 GENUS#10;P=3;  16/3;  [(1)+(1)+(14)]+[3/2]
 GENUS#10;P=7;  96/7;  [(1)+(1)+(3/2)]+[14]
 GENUS#11;P=2;  384;  [A]+[3]+[28/3]
 GENUS#11;P=3;  16/3;  [(1)+(11)+(32/11)]+[21/32]
 GENUS#11;P=7;  96/7;  [(1)+(3/4)+(3)]+[28/3]
 GENUS#12;P=2;  384;  [A]+[1]+[84/11]
 GENUS#12;P=3;  16/3;  [(1)+(1)+(11/4)]+[84/11]
 GENUS#12;P=7;  96/7;  [(1)+(1)+(11/4)]+[84/11]
 GENUS#13;P=2;  512;  [H]+[(22/23)+(42/11)]
 GENUS#13;P=3;  16/3;  [(1)+(1)+(4)]+[21/4]
 GENUS#13;P=7;  96/7;  [(1)+(1)+(4)]+[21/4]
 GENUS#14;P=2;  1536;  [A]+[4H]
 GENUS#14;P=3;  16/3;  [(1)+(4)+(7)]+[3/4]
 GENUS#14;P=7;  96/7;  [(1)+(3/4)+(4)]+[7]
 GENUS#15;P=2;  6144;  [(1)+(1)+(5)+(21/5)]
 GENUS#15;P=3;  16/3;  [(1)+(1)+(5)]+[21/5]
 GENUS#15;P=7;  96/7;  [(1)+(1)+(5)]+[21/5]
 GENUS#16;P=2;  1536;  [H]+[4A]
 GENUS#16;P=3;  16/3;  [(1)+(8)+(1)]+[21/8]
 GENUS#16;P=7;  96/7;  [(1)+(3)+(8/3)]+[21/8]

D=  337
 GENUS#1;P=2;  9;  [A+A]
 GENUS#1;P=337;  227136/337;  [(1)+(3/4)+(5/3)]+[337/20]

D=  340
 GENUS#1;P=2;  192;  [H]+[2A]
 GENUS#1;P=5;  48/5;  [(1)+(1)+(1)]+[85/4]
 GENUS#1;P=17;  576/17;  [(1)+(1)+(1)]+[85/4]
 GENUS#2;P=2;  192;  [A]+[2H]
 GENUS#2;P=5;  48/5;  [(1)+(3/4)+(2/3)]+[85/2]
 GENUS#2;P=17;  576/17;  [(1)+(3/4)+(2/3)]+[85/2]
 GENUS#3;P=2;  192;  [A]+[2H]
 GENUS#3;P=5;  48/5;  [(1)+(3/4)+(2)]+[85/6]
 GENUS#3;P=17;  576/17;  [(1)+(3/4)+(2)]+[85/6]
 GENUS#4;P=2;  48;  [A]+[(5/3)+(17)]
 GENUS#4;P=5;  48/5;  [(1)+(3/4)+(17)]+[5/3]
 GENUS#4;P=17;  576/17;  [(1)+(3/4)+(5/3)]+[17]
 GENUS#5;P=2;  48;  [A]+[(85/43)+(1)]
 GENUS#5;P=5;  48/5;  [(1)+(1)+(2)]+[85/8]
 GENUS#5;P=17;  576/17;  [(1)+(1)+(2)]+[85/8]
 GENUS#6;P=2;  192;  [H]+[2A]
 GENUS#6;P=5;  48/5;  [(1)+(1)+(11/2)]+[85/22]
```

```
GENUS#6;P=17; 576/17; [(1)+(1)+(5)]+[17/4]


D=   341
 GENUS#1;P=2; 15; [A+H]
 GENUS#1;P=11; 240/11; [(1)+(3/4)+(1)]+[341/12]
 GENUS#1;P=31; 1920/31; [(1)+(3/4)+(1)]+[341/12]
 GENUS#2;P=2; 15; [A+H]
 GENUS#2;P=11; 240/11; [(1)+(3/4)+(2/3)]+[341/8]
 GENUS#2;P=31; 1920/31; [(1)+(3/4)+(2/3)]+[341/8]


D=   344
 GENUS#1;P=2; 192; [A]+[1]+[86/3]
 GENUS#1;P=43; 3696/43; [(1)+(3/4)+(1)]+[86/3]
 GENUS#2;P=2; 192; [H]+[1]+[86/87]
 GENUS#2;P=43; 3696/43; [(1)+(1)+(1)]+[43/2]


D=   345
 GENUS#1;P=2; 9; [A+A]
 GENUS#1;P=3; 16/3; [(1)+(1)+(115/4)]+[3/4]
 GENUS#1;P=5; 48/5; [(1)+(3/4)+(1)]+[115/4]
 GENUS#1;P=23; 1056/23; [(1)+(3/4)+(1)]+[115/4]
 GENUS#2;P=2; 9; [H+H]
 GENUS#2;P=3; 16/3; [(1)+(1)+(59/4)]+[345/236]
 GENUS#2;P=5; 48/5; [(1)+(1)+(3/2)]+[115/8]
 GENUS#2;P=23; 1056/23; [(1)+(1)+(3/2)]+[115/8]
 GENUS#3;P=2; 9; [A+A]
 GENUS#3;P=3; 16/3; [(1)+(7/4)+(5/7)]+[69/4]
 GENUS#3;P=5; 48/5; [(1)+(3/4)+(53/3)]+[345/212]
 GENUS#3;P=23; 1056/23; [(1)+(3/4)+(5/3)]+[69/4]
 GENUS#4;P=2; 9; [A+A]
 GENUS#4;P=3; 16/3; [(1)+(1)+(19/4)]+[345/76]
 GENUS#4;P=5; 48/5; [(1)+(1)+(19/4)]+[345/76]
 GENUS#4;P=23; 1056/23; [(1)+(1)+(19/4)]+[345/76]


D=   348
 GENUS#1;P=2; 96; [A]+[(1)+(29)]
 GENUS#1;P=3; 16/3; [(1)+(1)+(29)]+[3/4]
 GENUS#1;P=29; 1680/29; [(1)+(3/4)+(1)]+[29]
 GENUS#2;P=2; 96; [H]+[(1)+(1)]
 GENUS#2;P=3; 16/3; [(1)+(1)+(1)]+[87/4]
 GENUS#2;P=29; 1680/29; [(1)+(1)+(1)]+[87/4]
 GENUS#3;P=2; 96; [A]+[(15)+(29/15)]
 GENUS#3;P=3; 16/3; [(1)+(2)+(29/2)]+[3/4]
 GENUS#3;P=29; 1680/29; [(1)+(3/4)+(2)]+[29/2]
 GENUS#4;P=2; 96; [H]+[(61/31)+(87/61)]
 GENUS#4;P=3; 16/3; [(1)+(2)+(61/8)]+[87/61]
 GENUS#4;P=29; 1680/29; [(1)+(2)+(3/2)]+[29/4]


D=   349
 GENUS#1;P=2; 15; [A+H]
 GENUS#1;P=349; 243600/349; [(1)+(3/4)+(2/3)]+[349/8]


D=   352
 GENUS#1;P=2; 6144; [(1)+(1)+(1)]+[22]
 GENUS#1;P=11; 240/11; [(1)+(1)+(1)]+[22]
 GENUS#2;P=2; 6144; [A]+[2/3]+[44]
 GENUS#2;P=11; 240/11; [(1)+(3/4)+(2/3)]+[44]
 GENUS#3;P=2; 6144; [(1)+(1)+(11)]+[2]
 GENUS#3;P=11; 240/11; [(1)+(1)+(2)]+[11]
 GENUS#4;P=2; 2048; [H]+[90/47]+[44/45]
 GENUS#4;P=11; 240/11; [(1)+(1)+(2)]+[11]
 GENUS#5;P=2; 1536; [A]+[5/3]+[88/5]
 GENUS#5;P=11; 240/11; [(1)+(3/4)+(5/3)]+[88/5]
 GENUS#6;P=2; 1536; [A]+[11]+[8/3]
 GENUS#6;P=11; 240/11; [(1)+(3/4)+(8/3)]+[11]
 GENUS#7;P=2; 1536; [A]+[35/3]+[88/35]
 GENUS#7;P=11; 240/11; [(1)+(3/4)+(8/3)]+[11]
 GENUS#8;P=2; 6144; [A]+[26/27]+[44/13]
 GENUS#8;P=11; 240/11; [(1)+(1)+(4)]+[11/2]
 GENUS#9;P=2; 2048; [H]+[22/23]+[4]
 GENUS#9;P=11; 240/11; [(1)+(1)+(4)]+[11/2]
 GENUS#10;P=2; 1536; [H]+[1]+[88/23]
```

```
GENUS#10;P=11; 240/11; [(1)+(1)+(4)]+[11/2]
```

D= 353
```
GENUS#1;P=2; 9; [A+A]
GENUS#1;P=353; 249216/353; [(1)+(3/4)+(1)]+[353/12]
```

D= 356
```
GENUS#1;P=2; 576; [A]+[2A]
GENUS#1;P=89; 15840/89; [(1)+(1)+(1)]+[89/4]
GENUS#2;P=2; 48; [A]+[(1)+(89/3)]
GENUS#2;P=89; 15840/89; [(1)+(3/4)+(1)]+[89/3]
GENUS#3;P=2; 64; [H]+[2H]
GENUS#3;P=89; 15840/89; [(1)+(1)+(3/2)]+[89/6]
```

D= 357
```
GENUS#1;P=2; 15; [A+H]
GENUS#1;P=3; 16/3; [(1)+(1)+(119/4)]+[3/4]
GENUS#1;P=7; 96/7; [(1)+(3/4)+(1)]+[119/4]
GENUS#1;P=17; 576/17; [(1)+(3/4)+(1)]+[119/4]
GENUS#2;P=2; 15; [A+H]
GENUS#2;P=3; 16/3; [(1)+(1)+(1/2)]+[357/8]
GENUS#2;P=7; 96/7; [(1)+(3/4)+(2/3)]+[357/8]
GENUS#2;P=17; 576/17; [(1)+(3/4)+(2/3)]+[357/8]
GENUS#3;P=2; 15; [H+A]
GENUS#3;P=3; 16/3; [(1)+(1)+(7/4)]+[51/4]
GENUS#3;P=7; 96/7; [(1)+(1)+(51/4)]+[7/4]
GENUS#3;P=17; 576/17; [(1)+(1)+(7/4)]+[51/4]
GENUS#4;P=2; 15; [A+H]
GENUS#4;P=3; 16/3; [(1)+(1)+(19/4)]+[357/76]
GENUS#4;P=7; 96/7; [(1)+(1)+(19/4)]+[357/76]
GENUS#4;P=17; 576/17; [(1)+(1)+(19/4)]+[357/76]
```

D= 360
```
GENUS#1;P=2; 192; [A]+[1]+[30]
GENUS#1;P=3; 96; [(1)+(1)]+[(3/4)+(30)]
GENUS#1;P=5; 48/5; [(1)+(3/4)+(1)]+[30]
GENUS#2;P=2; 192; [A]+[45]+[2/3]
GENUS#2;P=3; 16; [(1)+(1)+(1/2)]+[45]
GENUS#2;P=5; 48/5; [(1)+(3/4)+(2/3)]+[45]
GENUS#3;P=2; 192; [H]+[1]+[90/7]
GENUS#3;P=3; 16; [(1)+(1)+(7/4)]+[90/7]
GENUS#3;P=5; 48/5; [(1)+(1)+(7/4)]+[90/7]
GENUS#4;P=2; 192; [A]+[5/3]+[18]
GENUS#4;P=3; 16; [(1)+(7/4)+(5/7)]+[18]
GENUS#4;P=5; 48/5; [(1)+(3/4)+(18)]+[5/3]
GENUS#5;P=2; 192; [H]+[15]+[6/7]
GENUS#5;P=3; 96; [(1)+(1)]+[(3/2)+(15)]
GENUS#5;P=5; 48/5; [(1)+(1)+(3/2)]+[15]
GENUS#6;P=2; 192; [A]+[1]+[90/91]
GENUS#6;P=3; 16; [(1)+(1)+(1)]+[45/2]
GENUS#6;P=5; 48/5; [(1)+(1)+(1)]+[45/2]
GENUS#7;P=2; 192; [A]+[15]+[2]
GENUS#7;P=3; 24; [(1)+(2)]+[(3/4)+(15)]
GENUS#7;P=5; 48/5; [(1)+(3/4)+(2)]+[15]
GENUS#8;P=2; 192; [A]+[5]+[6]
GENUS#8;P=3; 24; [(1)+(5)]+[(3/4)+(6)]
GENUS#8;P=5; 48/5; [(1)+(3/4)+(6)]+[5]
```

D= 361
```
GENUS#1;P=2; 9; [A+A]
GENUS#1;P=19; 15200; [(1)+(1)]+[(19/4)+(19/4)]
```

D= 364
```
GENUS#1;P=2; 96; [A]+[(137/3)+(91/137)]
GENUS#1;P=7; 96/7; [(1)+(3/4)+(2/3)]+[91/2]
GENUS#1;P=13; 336/13; [(1)+(3/4)+(2/3)]+[91/2]
GENUS#2;P=2; 96; [H]+[(1)+(13)]
GENUS#2;P=7; 96/7; [(1)+(1)+(13)]+[7/4]
GENUS#2;P=13; 336/13; [(1)+(1)+(7/4)]+[13]
GENUS#3;P=2; 96; [A]+[(101/11)+(91/101)]
GENUS#3;P=7; 96/7; [(1)+(1)+(5/2)]+[91/10]
GENUS#3;P=13; 336/13; [(1)+(1)+(5/2)]+[91/10]
```

```
GENUS#4;P=2;  96;  [A]+[(1)+(1)]
GENUS#4;P=7;  96/7;  [(1)+(1)+(1)]+[91/4]
GENUS#4;P=13;  336/13;  [(1)+(1)+(1)]+[91/4]


D=  365
 GENUS#1;P=2;  15;  [A+H]
 GENUS#1;P=5;  48/5;  [(1)+(3/4)+(1)]+[365/12]
 GENUS#1;P=73;  10656/73;  [(1)+(3/4)+(1)]+[365/12]
 GENUS#2;P=2;  15;  [A+H]
 GENUS#2;P=5;  48/5;  [(1)+(3/4)+(56/3)]+[365/224]
 GENUS#2;P=73;  10656/73;  [(1)+(3/4)+(5/3)]+[73/4]


D=  368
 GENUS#1;P=2;  3072;  [(1)+(1)+(1)+(23)]
 GENUS#1;P=23;  1056/23;  [(1)+(1)+(1)]+[23]
 GENUS#2;P=2;  384;  [A]+[1]+[92/3]
 GENUS#2;P=23;  1056/23;  [(1)+(3/4)+(1)]+[92/3]
 GENUS#3;P=2;  3072;  [A]+[(2/3)+(46)]
 GENUS#3;P=23;  1056/23;  [(1)+(3/4)+(2/3)]+[46]
 GENUS#4;P=2;  1024;  [H]+[(2)+(46/47)]
 GENUS#4;P=23;  1056/23;  [(1)+(1)+(2)]+[23/2]
 GENUS#5;P=2;  384;  [A]+[5/3]+[92/5]
 GENUS#5;P=23;  1056/23;  [(1)+(3/4)+(5/3)]+[92/5]
 GENUS#6;P=2;  3072;  [A]+[(14/3)+(46/7)]
 GENUS#6;P=23;  1056/23;  [(1)+(3/4)+(14/3)]+[46/7]
 GENUS#7;P=2;  1024;  [H]+[(10/7)+(46/5)]
 GENUS#7;P=23;  1056/23;  [(1)+(7/4)+(10/7)]+[46/5]
 GENUS#8;P=2;  3072;  [(1)+(3)+(5/3)+(23/5)]
 GENUS#8;P=23;  1056/23;  [(1)+(2)+(5/2)]+[23/5]


D=  369
 GENUS#1;P=2;  9;  [A+A]
 GENUS#1;P=3;  48;  [(1)+(1)]+[(3/4)+(123/4)]
 GENUS#1;P=41;  3360/41;  [(1)+(3/4)+(1)]+[123/4]
 GENUS#2;P=2;  9;  [H+H]
 GENUS#2;P=3;  16;  [(1)+(1)+(7/4)]+[369/28]
 GENUS#2;P=41;  3360/41;  [(1)+(1)+(7/4)]+[369/28]
 GENUS#3;P=2;  9;  [A+A]
 GENUS#3;P=3;  48;  [(1)+(11)]+[(123/44)+(3/4)]
 GENUS#3;P=41;  3360/41;  [(1)+(3/4)+(3)]+[41/4]
 GENUS#4;P=2;  9;  [A+A]
 GENUS#4;P=3;  16;  [(1)+(1)+(11/4)]+[369/44]
 GENUS#4;P=41;  3360/41;  [(1)+(1)+(11/4)]+[369/44]


D=  372
 GENUS#1;P=2;  192;  [H]+[2A]
 GENUS#1;P=3;  16/3;  [(1)+(1)+(1)]+[93/4]
 GENUS#1;P=31;  1920/31;  [(1)+(1)+(1)]+[93/4]
 GENUS#2;P=2;  48;  [A]+[(1)+(31)]
 GENUS#2;P=3;  16/3;  [(1)+(1)+(31)]+[3/4]
 GENUS#2;P=31;  1920/31;  [(1)+(3/4)+(1)]+[31]
 GENUS#3;P=2;  192;  [A]+[2H]
 GENUS#3;P=3;  16/3;  [(1)+(187/4)+(140/187)]+[93/140]
 GENUS#3;P=31;  1920/31;  [(1)+(3/4)+(2/3)]+[93/2]
 GENUS#4;P=2;  48;  [H]+[(93/47)+(1)]
 GENUS#4;P=3;  16/3;  [(1)+(1)+(2)]+[93/8]
 GENUS#4;P=31;  1920/31;  [(1)+(1)+(2)]+[93/8]
 GENUS#5;P=2;  192;  [A]+[2H]
 GENUS#5;P=3;  16/3;  [(1)+(2)+(31/2)]+[3/4]
 GENUS#5;P=31;  1920/31;  [(1)+(3/4)+(2)]+[31/2]
 GENUS#6;P=2;  192;  [H]+[2A]
 GENUS#6;P=3;  16/3;  [(1)+(1)+(31/2)]+[3/2]
 GENUS#6;P=31;  1920/31;  [(1)+(1)+(3/2)]+[31/2]


D=  373
 GENUS#1;P=2;  15;  [A+H]
 GENUS#1;P=373;  278256/373;  [(1)+(3/4)+(2/3)]+[373/8]


D=  376
 GENUS#1;P=2;  192;  [H]+[1]+[94/95]
 GENUS#1;P=47;  4416/47;  [(1)+(1)+(1)]+[47/2]
 GENUS#2;P=2;  192;  [A]+[107/11]+[94/107]
```

```
GENUS#2;P=47;  4416/47;  [(1)+(1)+(5/2)]+[47/5]
```

D= 377
```
 GENUS#1;P=2;  9;  [A+A]
 GENUS#1;P=13;  336/13;  [(1)+(3/4)+(1)]+[377/12]
 GENUS#1;P=29;  1680/29;  [(1)+(3/4)+(1)]+[377/12]
 GENUS#2;P=2;  9;  [H+H]
 GENUS#2;P=13;  336/13;  [(1)+(1)+(7/4)]+[377/28]
 GENUS#2;P=29;  1680/29;  [(1)+(1)+(7/4)]+[377/28]
```

D= 380
```
 GENUS#1;P=2;  96;  [A]+[(1)+(95/3)]
 GENUS#1;P=5;  48/5;  [(1)+(3/4)+(1)]+[95/3]
 GENUS#1;P=19;  720/19;  [(1)+(3/4)+(1)]+[95/3]
 GENUS#2;P=2;  96;  [H]+[(1)+(1)]
 GENUS#2;P=5;  48/5;  [(1)+(1)+(1)]+[95/4]
 GENUS#2;P=19;  720/19;  [(1)+(1)+(1)]+[95/4]
 GENUS#3;P=2;  96;  [A]+[(5/3)+(19)]
 GENUS#3;P=5;  48/5;  [(1)+(3/4)+(19)]+[5/3]
 GENUS#3;P=19;  720/19;  [(1)+(3/4)+(5/3)]+[19]
 GENUS#4;P=2;  96;  [A]+[(7)+(95/21)]
 GENUS#4;P=5;  48/5;  [(1)+(3/4)+(14/3)]+[95/14]
 GENUS#4;P=19;  720/19;  [(1)+(3/4)+(14/3)]+[95/14]
```

D= 381
```
 GENUS#1;P=2;  15;  [A+H]
 GENUS#1;P=3;  16/3;  [(1)+(1)+(127/4)]+[3/4]
 GENUS#1;P=127;  32256/127;  [(1)+(3/4)+(1)]+[127/4]
 GENUS#2;P=2;  15;  [A+H]
 GENUS#2;P=3;  16/3;  [(1)+(1)+(1/2)]+[381/8]
 GENUS#2;P=127;  32256/127;  [(1)+(3/4)+(2/3)]+[381/8]
```

D= 384
```
 GENUS#1;P=2;  196608;  [(1)+(1)+(1)]+[24]
 GENUS#1;P=3;  16/3;  [(1)+(1)+(1)]+[24]
 GENUS#2;P=2;  6144;  [A]+[1]+[32]
 GENUS#2;P=3;  16/3;  [(1)+(1)+(32)]+[3/4]
 GENUS#3;P=2;  49152;  [A]+[2/3]+[48]
 GENUS#3;P=3;  16/3;  [(1)+(1)+(1/2)]+[48]
 GENUS#4;P=2;  65536;  [(1)+(1)]+[2]+[12]
 GENUS#4;P=3;  16/3;  [(1)+(1)+(2)]+[12]
 GENUS#5;P=2;  6144;  [H]+[1]+[96/7]
 GENUS#5;P=3;  16/3;  [(1)+(1)+(7/4)]+[96/7]
 GENUS#6;P=2;  49152;  [A]+[10/11]+[48/5]
 GENUS#6;P=3;  16/3;  [(1)+(1)+(5/2)]+[48/5]
 GENUS#7;P=2;  49152;  [A]+[12]+[8/3]
 GENUS#7;P=3;  16/3;  [(1)+(11/4)+(8/11)]+[12]
 GENUS#8;P=2;  49152;  [A]+[2]+[16]
 GENUS#8;P=3;  16/3;  [(1)+(2)+(16)]+[3/4]
 GENUS#9;P=2;  65536;  [(1)+(1)+(3)]+[8]
 GENUS#9;P=3;  16/3;  [(1)+(1)+(8)]+[3]
 GENUS#10;P=2;  65536;  [(1)+(1)]+[6]+[4]
 GENUS#10;P=3;  16/3;  [(1)+(1)+(4)]+[6]
 GENUS#11;P=2;  196608;  [(1)+(1)+(5)]+[24/5]
 GENUS#11;P=3;  16/3;  [(1)+(1)+(5)]+[24/5]
 GENUS#12;P=2;  196608;  [1]+[(2)+(2)+(6)]
 GENUS#12;P=3;  16/3;  [(1)+(2)+(2)]+[6]
 GENUS#13;P=2;  6144;  [A]+[3]+[32/3]
 GENUS#13;P=3;  16/3;  [(1)+(43/4)+(32/43)]+[3]
 GENUS#14;P=2;  16384;  [H]+[6/7]+[16]
 GENUS#14;P=3;  16/3;  [(1)+(1)+(16)]+[3/2]
 GENUS#15;P=2;  65536;  [(1)+(1)+(7)]+[24/7]
 GENUS#15;P=3;  16/3;  [(1)+(1)+(4)]+[6]
 GENUS#16;P=2;  16384;  [H]+[14/15]+[48/7]
 GENUS#16;P=3;  16/3;  [(1)+(1)+(7/2)]+[48/7]
 GENUS#17;P=2;  65536;  [(1)+(7)+(13/7)]+[24/13]
 GENUS#17;P=3;  16/3;  [(1)+(2)+(2)]+[6]
 GENUS#18;P=2;  6144;  [H]+[13/7]+[96/13]
 GENUS#18;P=3;  16/3;  [(1)+(7/4)+(13/7)]+[96/13]
 GENUS#19;P=2;  196608;  [(1)+(2A)]+[8]
 GENUS#19;P=3;  16/3;  [(1)+(2)+(8)]+[3/2]
 GENUS#20;P=2;  16384;  [H]+[12/7]+[8]
```

```
GENUS#20;P=3;  16/3;  [(1)+(7/4)+(8)]+[12/7]
GENUS#21;P=2;  16384;  [H]+[26/7]+[48/13]
GENUS#21;P=3;  16/3;  [(1)+(7/4)+(26/7)]+[48/13]
GENUS#22;P=2;  196608;  [1]+[(2)+(4A)]
GENUS#22;P=3;  16/3;  [(1)+(2)+(4)]+[3]
GENUS#23;P=2;  16384;  [H]+[4]+[24/7]
GENUS#23;P=3;  16/3;  [(1)+(7/4)+(4)]+[24/7]
GENUS#24;P=2;  16384;  [H]+[34/7]+[48/17]
GENUS#24;P=3;  16/3;  [(1)+(7/4)+(34/7)]+[48/17]
GENUS#25;P=2;  65536;  [(1)+(3)+(3)]+[8/3]
GENUS#25;P=3;  16/3;  [(1)+(8)+(1)]+[3]
GENUS#26;P=2;  49152;  [A]+[30/11]+[16/5]
GENUS#26;P=3;  16/3;  [(1)+(11/4)+(40/11)]+[12/5]
GENUS#27;P=2;  49152;  [A]+[28/11]+[24/7]
GENUS#27;P=3;  16/3;  [(1)+(11/4)+(28/11)]+[24/7]
GENUS#28;P=2;  196608;  [(3)+(3)+(5/3)]+[8/5]
GENUS#28;P=3;  16/3;  [(2)+(2)+(2)]+[3]


D=  385
 GENUS#1;P=2;  9;  [H+H]
 GENUS#1;P=5;  48/5;  [(1)+(1)+(7/4)]+[55/4]
 GENUS#1;P=7;  96/7;  [(1)+(1)+(55/4)]+[7/4]
 GENUS#1;P=11;  240/11;  [(1)+(1)+(7/4)]+[55/4]
 GENUS#2;P=2;  9;  [A+A]
 GENUS#2;P=5;  48/5;  [(1)+(3/4)+(59/3)]+[385/236]
 GENUS#2;P=7;  96/7;  [(1)+(3/4)+(5/3)]+[77/4]
 GENUS#2;P=11;  240/11;  [(1)+(3/4)+(5/3)]+[77/4]
 GENUS#3;P=2;  9;  [A+A]
 GENUS#3;P=5;  48/5;  [(1)+(1)+(11/4)]+[35/4]
 GENUS#3;P=7;  96/7;  [(1)+(1)+(11/4)]+[35/4]
 GENUS#3;P=11;  240/11;  [(1)+(1)+(35/4)]+[11/4]
 GENUS#4;P=2;  9;  [A+A]
 GENUS#4;P=5;  48/5;  [(1)+(3/4)+(23/3)]+[385/92]
 GENUS#4;P=7;  96/7;  [(1)+(3/4)+(5)]+[77/12]
 GENUS#4;P=11;  240/11;  [(1)+(3/4)+(5)]+[77/12]


D=  388
 GENUS#1;P=2;  576;  [A]+[2A]
 GENUS#1;P=97;  18816/97;  [(1)+(1)+(1)]+[97/4]
 GENUS#2;P=2;  48;  [H]+[(1)+(97/7)]
 GENUS#2;P=97;  18816/97;  [(1)+(1)+(7/4)]+[97/7]
 GENUS#3;P=2;  64;  [H]+[2H]
 GENUS#3;P=97;  18816/97;  [(1)+(1)+(7/2)]+[97/14]


D=  389
 GENUS#1;P=2;  15;  [A+H]
 GENUS#1;P=389;  302640/389;  [(1)+(3/4)+(1)]+[389/12]


D=  392
 GENUS#1;P=2;  192;  [A]+[1]+[98/3]
 GENUS#1;P=7;  96;  [(1)+(3/4)+(1)]+[98/3]
 GENUS#2;P=2;  192;  [A]+[1]+[98/99]
 GENUS#2;P=7;  96;  [(1)+(1)+(1)]+[49/2]
 GENUS#3;P=2;  192;  [H]+[1]+[14]
 GENUS#3;P=7;  896;  [(1)+(1)]+[(7/4)+(14)]
 GENUS#4;P=2;  192;  [A]+[7]+[14/3]
 GENUS#4;P=7;  504;  [(1)+(3/4)]+[(14/3)+(7)]


D=  393
 GENUS#1;P=2;  9;  [A+A]
 GENUS#1;P=3;  16/3;  [(1)+(1)+(131/4)]+[3/4]
 GENUS#1;P=131;  34320/131;  [(1)+(3/4)+(1)]+[131/4]
 GENUS#2;P=2;  9;  [H+H]
 GENUS#2;P=3;  16/3;  [(1)+(1)+(67/4)]+[393/268]
 GENUS#2;P=131;  34320/131;  [(1)+(1)+(3/2)]+[131/8]


D=  396
 GENUS#1;P=2;  96;  [A]+[(1)+(33)]
 GENUS#1;P=3;  48;  [(1)+(1)]+[(3/4)+(33)]
 GENUS#1;P=11;  240/11;  [(1)+(3/4)+(1)]+[33]
 GENUS#2;P=2;  96;  [A]+[(149/3)+(99/149)]
 GENUS#2;P=3;  16;  [(1)+(199/4)+(149/199)]+[99/149]
```

```
GENUS#2;P=11;  240/11;  [(1)+(3/4)+(2/3)]+[99/2]
GENUS#3;P=2;  96;  [A]+[(109/11)+(99/109)]
GENUS#3;P=3;  16;  [(1)+(1)+(5/2)]+[99/10]
GENUS#3;P=11;  240/11;  [(1)+(1)+(5/2)]+[99/10]
GENUS#4;P=2;  96;  [A]+[(3)+(11)]
GENUS#4;P=3;  48;  [(1)+(11)]+[(3)+(3/4)]
GENUS#4;P=11;  240/11;  [(1)+(3/4)+(3)]+[11]
GENUS#5;P=2;  96;  [A]+[(1)+(9)]
GENUS#5;P=3;  16;  [(1)+(1)+(11/4)]+[9]
GENUS#5;P=11;  240/11;  [(1)+(1)+(9)]+[11/4]
GENUS#6;P=2;  96;  [H]+[(117/7)+(11/13)]
GENUS#6;P=3;  48;  [(1)+(1)]+[(3/2)+(33/2)]
GENUS#6;P=11;  240/11;  [(1)+(1)+(3/2)]+[33/2]
GENUS#7;P=2;  96;  [A]+[(17)+(33/17)]
GENUS#7;P=3;  48;  [(1)+(2)]+[(3/4)+(33/2)]
GENUS#7;P=11;  240/11;  [(1)+(3/4)+(2)]+[33/2]
GENUS#8;P=2;  96;  [A]+[(1)+(1)]
GENUS#8;P=3;  16;  [(1)+(1)+(1)]+[99/4]
GENUS#8;P=11;  240/11;  [(1)+(1)+(1)]+[99/4]
```

D= 397
```
GENUS#1;P=2;  15;  [A+H]
GENUS#1;P=397;  315216/397;  [(1)+(3/4)+(2/3)]+[397/8]
```

D= 400
```
GENUS#1;P=2;  6144;  [(1)+(1)+(1)+(25)]
GENUS#1;P=5;  48;  [(1)+(1)+(1)]+[25]
GENUS#2;P=2;  1536;  [A]+[(2/3)+(50)]
GENUS#2;P=5;  48;  [(1)+(3/4)+(2/3)]+[50]
GENUS#3;P=2;  1536;  [A]+[(2)+(50/3)]
GENUS#3;P=5;  48;  [(1)+(3/4)+(2)]+[50/3]
GENUS#4;P=2;  384;  [A]+[5/3]+[20]
GENUS#4;P=5;  360;  [(1)+(3/4)]+[(5/3)+(20)]
GENUS#5;P=2;  1536;  [A]+[(10/11)+(10)]
GENUS#5;P=5;  160;  [(1)+(1)]+[(5/2)+(10)]
GENUS#6;P=2;  6144;  [(1)+(1)+(5)+(5)]
GENUS#6;P=5;  160;  [(1)+(1)]+[(5)+(5)]
GENUS#7;P=2;  384;  [A]+[5]+[20/3]
GENUS#7;P=5;  360;  [(1)+(3/4)]+[(5)+(20/3)]
GENUS#8;P=2;  4608;  [A]+[4A]
GENUS#8;P=5;  360;  [(1)+(3/4)]+[(20/3)+(5)]
GENUS#9;P=2;  512;  [H]+[(10/7)+(10)]
GENUS#9;P=5;  360;  [(1)+(7/4)]+[(10/7)+(10)]
GENUS#10;P=2;  6144;  [(1)+(1)+(13)+(25/13)]
GENUS#10;P=5;  48;  [(1)+(1)+(2)]+[25/2]
GENUS#11;P=2;  2048;  [(1)+(3)+(5/3)+(5)]
GENUS#11;P=5;  360;  [(1)+(2)]+[(5/2)+(5)]
GENUS#12;P=2;  512;  [H]+[4H]
GENUS#12;P=5;  360;  [(1)+(7/4)]+[(20/7)+(5)]
```

D= 401
```
GENUS#1;P=2;  9;  [A+A]
GENUS#1;P=401;  321600/401;  [(1)+(3/4)+(1)]+[401/12]
```

D= 404
```
GENUS#1;P=2;  192;  [H]+[2A]
GENUS#1;P=101;  20400/101;  [(1)+(1)+(1)]+[101/4]
GENUS#2;P=2;  48;  [A]+[(1)+(101/3)]
GENUS#2;P=101;  20400/101;  [(1)+(3/4)+(1)]+[101/3]
GENUS#3;P=2;  192;  [A]+[2H]
GENUS#3;P=101;  20400/101;  [(1)+(3/4)+(2/3)]+[101/2]
```

D= 405
```
GENUS#1;P=2;  15;  [A+H]
GENUS#1;P=3;  1296;  [(1)+(1)]+[3/4]+[135/4]
GENUS#1;P=5;  48/5;  [(1)+(3/4)+(1)]+[135/4]
GENUS#2;P=2;  15;  [A+H]
GENUS#2;P=3;  144;  [(1)+(1)+(1/2)]+[405/8]
GENUS#2;P=5;  48/5;  [(1)+(3/4)+(2/3)]+[405/8]
GENUS#3;P=2;  15;  [H+A]
GENUS#3;P=3;  144;  [(1)+(1)+(7/4)]+[405/28]
GENUS#3;P=5;  48/5;  [(1)+(1)+(7/4)]+[405/28]
```

```
GENUS#4;P=2;  15;  [A+H]
GENUS#4;P=3;  648;  [(1)+(2)]+[3/4]+[135/8]
GENUS#4;P=5;  48/5;  [(1)+(3/4)+(2)]+[135/8]
GENUS#5;P=2;  15;  [A+H]
GENUS#5;P=3;  11664;  [1]+[(3/4)+(3)]+[45/4]
GENUS#5;P=5;  48/5;  [(1)+(3/4)+(3)]+[45/4]
GENUS#6;P=2;  15;  [H+A]
GENUS#6;P=3;  1296;  [(1)+(1)]+[15/4]+[27/4]
GENUS#6;P=5;  48/5;  [(1)+(1)+(27/4)]+[15/4]
GENUS#7;P=2;  15;  [A+H]
GENUS#7;P=3;  1296;  [(1)+(1)]+[(9/2)+(45/8)]
GENUS#7;P=5;  48/5;  [(1)+(1)+(9/2)]+[45/8]
GENUS#8;P=2;  15;  [A+H]
GENUS#8;P=3;  5832;  [1]+[(3/4)+(6)]+[45/8]
GENUS#8;P=5;  48/5;  [(1)+(3/4)+(6)]+[45/8]
GENUS#9;P=2;  15;  [A+H]
GENUS#9;P=3;  648;  [(1)+(2)]+[21/8]+[135/28]
GENUS#9;P=5;  48/5;  [(1)+(2)+(21/8)]+[135/28]
GENUS#10;P=2;  15;  [A+H]
GENUS#10;P=3;  1296;  [(1)+(2)]+[(9/4)+(45/8)]
GENUS#10;P=5;  48/5;  [(1)+(2)+(9/4)]+[45/8]
GENUS#11;P=2;  15;  [H+A]
GENUS#11;P=3;  5832;  [2]+[(15/8)+(21/5)]+[45/28]
GENUS#11;P=5;  48/5;  [(2)+(9/2)+(7/4)]+[45/28]
GENUS#12;P=2;  15;  [H+A]
GENUS#12;P=3;  11664;  [2]+[(15/8)+(3)]+[9/4]
GENUS#12;P=5;  48/5;  [(2)+(3)+(9/4)]+[15/8]


D=  408
 GENUS#1;P=2;  192;  [A]+[1]+[34]
 GENUS#1;P=3;  16/3;  [(1)+(1)+(34)]+[3/4]
 GENUS#1;P=17;  576/17;  [(1)+(3/4)+(1)]+[34]
 GENUS#2;P=2;  192;  [A]+[51]+[2/3]
 GENUS#2;P=3;  16/3;  [(1)+(1)+(1/2)]+[51]
 GENUS#2;P=17;  576/17;  [(1)+(3/4)+(2/3)]+[51]
 GENUS#3;P=2;  192;  [H]+[17]+[6/7]
 GENUS#3;P=3;  16/3;  [(1)+(1)+(17)]+[3/2]
 GENUS#3;P=17;  576/17;  [(1)+(1)+(3/2)]+[17]
 GENUS#4;P=2;  192;  [H]+[1]+[102/103]
 GENUS#4;P=3;  16/3;  [(1)+(1)+(1)]+[51/2]
 GENUS#4;P=17;  576/17;  [(1)+(1)+(1)]+[51/2]


D=  409
 GENUS#1;P=2;  9;  [H+H]
 GENUS#1;P=409;  334560/409;  [(1)+(1)+(7/4)]+[409/28]


D=  412
 GENUS#1;P=2;  96;  [H]+[(1)+(1)]
 GENUS#1;P=103;  21216/103;  [(1)+(1)+(1)]+[103/4]
 GENUS#2;P=2;  96;  [A]+[(53/3)+(103/53)]
 GENUS#2;P=103;  21216/103;  [(1)+(3/4)+(2)]+[103/6]


D=  413
 GENUS#1;P=2;  15;  [A+H]
 GENUS#1;P=7;  96/7;  [(1)+(3/4)+(1)]+[413/12]
 GENUS#1;P=59;  6960/59;  [(1)+(3/4)+(1)]+[413/12]
 GENUS#2;P=2;  15;  [A+H]
 GENUS#2;P=7;  96/7;  [(1)+(3/4)+(2/3)]+[413/8]
 GENUS#2;P=59;  6960/59;  [(1)+(3/4)+(2/3)]+[413/8]


D=  416
 GENUS#1;P=2;  1536;  [A]+[29/3]+[104/29]
 GENUS#1;P=13;  336/13;  [(1)+(3/4)+(4)]+[26/3]
 GENUS#2;P=2;  6144;  [(1)+(1)+(1)]+[26]
 GENUS#2;P=13;  336/13;  [(1)+(1)+(1)]+[26]
 GENUS#3;P=2;  1536;  [A]+[1]+[104/3]
 GENUS#3;P=13;  336/13;  [(1)+(3/4)+(1)]+[104/3]
 GENUS#4;P=2;  6144;  [A]+[2/3]+[52]
 GENUS#4;P=13;  336/13;  [(1)+(3/4)+(2/3)]+[52]
 GENUS#5;P=2;  6144;  [(1)+(1)+(13)]+[2]
 GENUS#5;P=13;  336/13;  [(1)+(1)+(2)]+[13]
 GENUS#6;P=2;  2048;  [H]+[106/55]+[52/53]
```

```
GENUS#6;P=13;  336/13;  [(1)+(1)+(2)]+[13]
GENUS#7;P=2;  1536;  [H]+[1]+[104/7]
GENUS#7;P=13;  336/13;  [(1)+(1)+(7/4)]+[104/7]
GENUS#8;P=2;  6144;  [A]+[10/11]+[52/5]
GENUS#8;P=13;  336/13;  [(1)+(1)+(5/2)]+[52/5]
GENUS#9;P=2;  1536;  [A]+[41/3]+[104/41]
GENUS#9;P=13;  336/13;  [(1)+(3/4)+(8/3)]+[13]
GENUS#10;P=2;  2048;  [H]+[30/31]+[52/15]
GENUS#10;P=13;  336/13;  [(1)+(1)+(4)]+[13/2]
```

```
D=  417
 GENUS#1;P=2;  9;  [A+A]
 GENUS#1;P=3;  16/3;  [(1)+(1)+(139/4)]+[3/4]
 GENUS#1;P=139;  38640/139;  [(1)+(3/4)+(1)]+[139/4]
 GENUS#2;P=2;  9;  [H+H]
 GENUS#2;P=3;  16/3;  [(1)+(1)+(71/4)]+[417/284]
 GENUS#2;P=139;  38640/139;  [(1)+(1)+(3/2)]+[139/8]
```

```
D=  420
 GENUS#1;P=2;  576;  [A]+[2A]
 GENUS#1;P=3;  16/3;  [(1)+(1)+(1)]+[105/4]
 GENUS#1;P=5;  48/5;  [(1)+(1)+(1)]+[105/4]
 GENUS#1;P=7;  96/7;  [(1)+(1)+(1)]+[105/4]
 GENUS#2;P=2;  48;  [A]+[(1)+(35)]
 GENUS#2;P=3;  16/3;  [(1)+(1)+(35)]+[3/4]
 GENUS#2;P=5;  48/5;  [(1)+(3/4)+(1)]+[35]
 GENUS#2;P=7;  96/7;  [(1)+(3/4)+(1)]+[35]
 GENUS#3;P=2;  576;  [A]+[2A]
 GENUS#3;P=3;  16/3;  [(1)+(211/4)+(158/211)]+[105/158]
 GENUS#3;P=5;  48/5;  [(1)+(3/4)+(2/3)]+[105/2]
 GENUS#3;P=7;  96/7;  [(1)+(3/4)+(2/3)]+[105/2]
 GENUS#4;P=2;  48;  [H]+[(1)+(15)]
 GENUS#4;P=3;  16/3;  [(1)+(1)+(7/4)]+[15]
 GENUS#4;P=5;  48/5;  [(1)+(1)+(7/4)]+[15]
 GENUS#4;P=7;  96/7;  [(1)+(1)+(15)]+[7/4]
 GENUS#5;P=2;  576;  [A]+[2A]
 GENUS#5;P=3;  16/3;  [(1)+(1)+(5/2)]+[21/2]
 GENUS#5;P=5;  48/5;  [(1)+(1)+(21/2)]+[5/2]
 GENUS#5;P=7;  96/7;  [(1)+(1)+(5/2)]+[21/2]
 GENUS#6;P=2;  48;  [A]+[(1)+(105/11)]
 GENUS#6;P=3;  16/3;  [(1)+(1)+(11/4)]+[105/11]
 GENUS#6;P=5;  48/5;  [(1)+(1)+(11/4)]+[105/11]
 GENUS#6;P=7;  96/7;  [(1)+(1)+(11/4)]+[105/11]
 GENUS#7;P=2;  64;  [H]+[2H]
 GENUS#7;P=3;  16/3;  [(1)+(1)+(7/2)]+[15/2]
 GENUS#7;P=5;  48/5;  [(1)+(1)+(7/2)]+[15/2]
 GENUS#7;P=7;  96/7;  [(1)+(1)+(15/2)]+[7/2]
 GENUS#8;P=2;  48;  [A]+[(1)+(105/19)]
 GENUS#8;P=3;  16/3;  [(1)+(1)+(19/4)]+[105/19]
 GENUS#8;P=5;  48/5;  [(1)+(1)+(19/4)]+[105/19]
 GENUS#8;P=7;  96/7;  [(1)+(1)+(19/4)]+[105/19]
 GENUS#9;P=2;  576;  [A]+[2A]
 GENUS#9;P=3;  16/3;  [(1)+(2)+(35/2)]+[3/4]
 GENUS#9;P=5;  48/5;  [(1)+(3/4)+(2)]+[35/2]
 GENUS#9;P=7;  96/7;  [(1)+(3/4)+(2)]+[35/2]
 GENUS#10;P=2;  64;  [H]+[2H]
 GENUS#10;P=3;  16/3;  [(1)+(1)+(35/2)]+[3/2]
 GENUS#10;P=5;  48/5;  [(1)+(1)+(3/2)]+[35/2]
 GENUS#10;P=7;  96/7;  [(1)+(1)+(3/2)]+[35/2]
 GENUS#11;P=2;  64;  [H]+[2H]
 GENUS#11;P=3;  16/3;  [(1)+(1)+(19/2)]+[105/38]
 GENUS#11;P=5;  48/5;  [(1)+(1)+(3)]+[35/4]
 GENUS#11;P=7;  96/7;  [(1)+(1)+(3)]+[35/4]
 GENUS#12;P=2;  64;  [H]+[2H]
 GENUS#12;P=3;  16/3;  [(1)+(1)+(11/2)]+[105/22]
 GENUS#12;P=5;  48/5;  [(1)+(1)+(11/2)]+[105/22]
 GENUS#12;P=7;  96/7;  [(1)+(1)+(11/2)]+[105/22]
```

```
D=  421
 GENUS#1;P=2;  15;  [A+H]
 GENUS#1;P=421;  354480/421;  [(1)+(3/4)+(2/3)]+[421/8]
```

```
D=  424
 GENUS#1;P=2;  192;  [A]+[53]+[2/3]
 GENUS#1;P=53;  5616/53;  [(1)+(3/4)+(2/3)]+[53]
 GENUS#2;P=2;  192;  [A]+[1]+[106/107]
 GENUS#2;P=53;  5616/53;  [(1)+(1)+(1)]+[53/2]

D=  425
 GENUS#1;P=2;  9;  [A+A]
 GENUS#1;P=5;  48;  [(1)+(3/4)+(1)]+[425/12]
 GENUS#1;P=17;  576/17;  [(1)+(3/4)+(1)]+[425/12]
 GENUS#2;P=2;  9;  [A+A]
 GENUS#2;P=5;  240;  [(1)+(3/4)]+[(5/3)+(85/4)]
 GENUS#2;P=17;  576/17;  [(1)+(3/4)+(5/3)]+[85/4]
 GENUS#3;P=2;  9;  [H+H]
 GENUS#3;P=5;  48;  [(1)+(1)+(3/2)]+[425/24]
 GENUS#3;P=17;  576/17;  [(1)+(1)+(3/2)]+[425/24]
 GENUS#4;P=2;  9;  [H+H]
 GENUS#4;P=5;  240;  [(1)+(1)]+[(15/4)+(85/12)]
 GENUS#4;P=17;  576/17;  [(1)+(1)+(15/4)]+[85/12]

D=  428
 GENUS#1;P=2;  96;  [A]+[(1)+(1)]
 GENUS#1;P=107;  22896/107;  [(1)+(1)+(1)]+[107/4]
 GENUS#2;P=2;  96;  [A]+[(161/3)+(107/161)]
 GENUS#2;P=107;  22896/107;  [(1)+(3/4)+(2/3)]+[107/2]

D=  429
 GENUS#1;P=2;  15;  [A+H]
 GENUS#1;P=3;  16/3;  [(1)+(1)+(143/4)]+[3/4]
 GENUS#1;P=11;  240/11;  [(1)+(3/4)+(1)]+[143/4]
 GENUS#1;P=13;  336/13;  [(1)+(3/4)+(1)]+[143/4]
 GENUS#2;P=2;  15;  [A+H]
 GENUS#2;P=3;  16/3;  [(1)+(1)+(1/2)]+[429/8]
 GENUS#2;P=11;  240/11;  [(1)+(3/4)+(2/3)]+[429/8]
 GENUS#2;P=13;  336/13;  [(1)+(3/4)+(2/3)]+[429/8]
 GENUS#3;P=2;  15;  [H+A]
 GENUS#3;P=3;  16/3;  ·[(1)+(1)+(31/4)]+[429/124]
 GENUS#3;P=11;  240/11;  [(1)+(1)+(15/4)]+[143/20]
 GENUS#3;P=13;  336/13;  [(1)+(1)+(15/4)]+[143/20]
 GENUS#4;P=2;  15;  [A+H]
 GENUS#4;P=3;  16/3;  [(1)+(1)+(5/2)]+[429/40]
 GENUS#4;P=11;  240/11;  [(1)+(1)+(5/2)]+[429/40]
 GENUS#4;P=13;  336/13;  [(1)+(1)+(5/2)]+[429/40]

D=  432
 GENUS#1;P=2;  3072;  [(1)+(1)+(1)+(27)]
 GENUS#1;P=3;  48;  [(1)+(1)+(1)]+[27]
 GENUS#2;P=2;  384;  [A]+[1]+[36]
 GENUS#2;P=3;  432;  [(1)+(1)]+[3/4]+[36]
 GENUS#3;P=2;  3072;  [A]+[(2/3)+(54)]
 GENUS#3;P=3;  48;  [(1)+(1)+(1/2)]+[54]
 GENUS#4;P=2;  384;  [H]+[1]+[108/7]
 GENUS#4;P=3;  48;  [(1)+(1)+(7/4)]+[108/7]
 GENUS#5;P=2;  384;  [A]+[5/3]+[108/5]
 GENUS#5;P=3;  48;  [(1)+(7/4)+(5/7)]+[108/5]
 GENUS#6;P=2;  384;  [A]+[3]+[12]
 GENUS#6;P=3;  1296;  [1]+[(3/4)+(3)+(12)]
 GENUS#7;P=2;  3072;  [(1)+(1)+(3)+(9)]
 GENUS#7;P=3;  432;  [(1)+(1)]+[3]+[9]
 GENUS#8;P=2;  3072;  [A]+[(10)+(18/5)]
 GENUS#8;P=3;  432;  [(1)+(4)]+[3/4]+[9]
 GENUS#9;P=2;  3072;  [A]+[(18/19)+(6)]
 GENUS#9;P=3;  432;  [(1)+(1)]+[6]+[9/2]
 GENUS#10;P=2;  384;  [A]+[5]+[36/5]
 GENUS#10;P=3;  216;  [(1)+(5)]+[3/4]+[36/5]
 GENUS#11;P=2;  1024;  [H]+[(66/7)+(18/11)]
 GENUS#11;P=3;  432;  [(1)+(7/4)]+[12/7]+[9]
 GENUS#12;P=2;  1024;  [H]+[(10/7)+(54/5)]
 GENUS#12;P=3;  48;  [(1)+(7/4)+(10/7)]+[54/5]
 GENUS#13;P=2;  3072;  [(1)+(3)+(3)]
 GENUS#13;P=3;  1296;  [1]+[(3)+(3)+(3)]
 GENUS#14;P=2;  1024;  [H]+[(6/5)+(6)]
```

```
GENUS#14;P=3;  1296;  [2]+[(3/2)+(3/2)+(6)]
GENUS#15;P=2;  3072;  [(1)+(1)+(2A)]
GENUS#15;P=3;  48;  [(1)+(1)+(2)]+[27/2]
GENUS#16;P=2;  1024;  [H]+[(2)+(54/55)]
GENUS#16;P=3;  48;  [(1)+(1)+(2)]+[27/2]
GENUS#17;P=2;  1024;  [H]+[(6/7)+(18)]
GENUS#17;P=3;  432;  [(1)+(1)]+[3/2]+[18]
GENUS#18;P=2;  3072;  [A]+[(2)+(18)]
GENUS#18;P=3;  216;  [(1)+(2)]+[3/4]+[18]
GENUS#19;P=2;  3072;  [(1)+(1)+(2A)]
GENUS#19;P=3;  432;  [(1)+(1)]+[6]+[9/2]
GENUS#20;P=2;  3072;  [A]+[(6)+(6)]
GENUS#20;P=3;  1296;  [1]+[(3/4)+(6)+(6)]
GENUS#21;P=2;  3072;  [A]+[(2)+(2)]
GENUS#21;P=3;  48;  [(1)+(2)+(2)]+[27/4]
GENUS#22;P=2;  3072;  [(1)+(9)+(2A)]
GENUS#22;P=3;  216;  [(1)+(2)]+[3/2]+[9]
GENUS#23;P=2;  3072;  [(1)+(3)+(5)+(9/5)]
GENUS#23;P=3;  216;  [(1)+(2)]+[3]+[9/2]
GENUS#24;P=2;  1024;  [H]+[(42/23)+(18/7)]
GENUS#24;P=3;  216;  [(1)+(2)]+[3]+[9/2]
GENUS#25;P=2;  3072;  [A]+[(18/11)+(6)]
GENUS#25;P=3;  216;  [(1)+(2)]+[6]+[9/4]
GENUS#26;P=2;  384;  [A]+[21/11]+[36/7]
GENUS#26;P=3;  216;  [(1)+(2)]+[21/8]+[36/7]
GENUS#27;P=2;  1024;  [H]+[(2)+(18/5)]
GENUS#27;P=3;  216;  [(1)+(2)]+[15/4]+[18/5]
GENUS#28;P=2;  384;  [H]+[33/7]+[36/11]
GENUS#28;P=3;  432;  [(1)+(7/4)]+[24/7]+[9/2]
GENUS#29;P=2;  1024;  [H]+[(18/5)+(2)]
GENUS#29;P=3;  1296;  [1]+[(3)+(3)+(3)]
GENUS#30;P=2;  384;  [H]+[9/5]+[4]
GENUS#30;P=3;  1296;  [2]+[(15/8)+(21/5)+(12/7)]
GENUS#31;P=2;  3072;  [(3)+(3)+(2A)]
GENUS#31;P=3;  1296;  [2]+[(3)+(3/2)+(3)]
GENUS#32;P=2;  3072;  [A]+[(2)+(2)]
GENUS#32;P=3;  1296;  [2]+[(3)+(3)+(3/2)]


D=  433
 GENUS#1;P=2;  9;  [H+H]
 GENUS#1;P=433;  374976/433;  [(1)+(1)+(7/4)]+[433/28]


D=  436
 GENUS#1;P=2;  192;  [H]+[2A]
 GENUS#1;P=109;  23760/109;  [(1)+(1)+(1)]+[109/4]
 GENUS#2;P=2;  192;  [A]+[2H]
 GENUS#2;P=109;  23760/109;  [(1)+(3/4)+(2/3)]+[109/2]
 GENUS#3;P=2;  48;  [H]+[(109/55)+(1)]
 GENUS#3;P=109;  23760/109;  [(1)+(1)+(2)]+[109/8]


D=  437
 GENUS#1;P=2;  15;  [A+H]
 GENUS#1;P=19;  720/19;  [(1)+(3/4)+(1)]+[437/12]
 GENUS#1;P=23;  1056/23;  [(1)+(3/4)+(1)]+[437/12]
 GENUS#2;P=2;  15;  [H+A]
 GENUS#2;P=19;  720/19;  [(1)+(1)+(7/4)]+[437/28]
 GENUS#2;P=23;  1056/23;  [(1)+(1)+(7/4)]+[437/28]


D=  440
 GENUS#1;P=2;  192;  [A]+[1]+[110/3]
 GENUS#1;P=5;  48/5;  [(1)+(3/4)+(1)]+[110/3]
 GENUS#1;P=11;  240/11;  [(1)+(3/4)+(1)]+[110/3]
 GENUS#2;P=2;  192;  [A]+[55]+[2/3]
 GENUS#2;P=5;  48/5;  [(1)+(3/4)+(2/3)]+[55]
 GENUS#2;P=11;  240/11;  [(1)+(3/4)+(2/3)]+[55]
 GENUS#3;P=2;  192;  [H]+[131/7]+[110/131]
 GENUS#3;P=5;  48/5;  [(1)+(1)+(3/2)]+[55/3]
 GENUS#3;P=11;  240/11;  [(1)+(1)+(3/2)]+[55/3]
 GENUS#4;P=2;  192;  [H]+[1]+[110/111]
 GENUS#4;P=5;  48/5;  [(1)+(1)+(1)]+[55/2]
 GENUS#4;P=11;  240/11;  [(1)+(1)+(1)]+[55/2]
```

```
D=  441
 GENUS#1;P=2;  9;  [A+A]
 GENUS#1;P=3;  96;  [(1)+(1)]+[(3/4)+(147/4)]
 GENUS#1;P=7;  96;  [(1)+(3/4)+(1)]+[147/4]
 GENUS#2;P=2;  9;  [H+H]
 GENUS#2;P=3;  16;  [(1)+(1)+(7/4)]+[63/4]
 GENUS#2;P=7;  896;  [(1)+(1)]+[(7/4)+(63/4)]
 GENUS#3;P=2;  9;  [A+A]
 GENUS#3;P=3;  96;  [(1)+(13)]+[(147/52)+(3/4)]
 GENUS#3;P=7;  96;  [(1)+(3/4)+(3)]+[49/4]
 GENUS#4;P=2;  9;  [H+H]
 GENUS#4;P=3;  16;  [(1)+(1)+(7/2)]+[63/8]
 GENUS#4;P=7;  896;  [(1)+(1)]+[(7/2)+(63/8)]
 GENUS#5;P=2;  9;  [A+A]
 GENUS#5;P=3;  96;  [(1)+(7)]+[(3/4)+(21/4)]
 GENUS#5;P=7;  504;  [(1)+(3/4)]+[(7)+(21/4)]
 GENUS#6;P=2;  9;  [A+A]
 GENUS#6;P=3;  24;  [(1)+(2)]+[(21/8)+(21/4)]
 GENUS#6;P=7;  896;  [(1)+(2)]+[(21/8)+(21/4)]


D=  444
 GENUS#1;P=2;  96;  [H]+[(1)+(1)]
 GENUS#1;P=3;  16/3;  [(1)+(1)+(1)]+[111/4]
 GENUS#1;P=37;  2736/37;  [(1)+(1)+(1)]+[111/4]
 GENUS#2;P=2;  96;  [A]+[(167/3)+(111/167)]
 GENUS#2;P=3;  16/3;  [(1)+(223/4)+(167/223)]+[111/167]
 GENUS#2;P=37;  2736/37;  [(1)+(3/4)+(2/3)]+[111/2]
 GENUS#3;P=2;  96;  [A]+[(19)+(37/19)]
 GENUS#3;P=3;  16/3;  [(1)+(2)+(37/2)]+[3/4]
 GENUS#3;P=37;  2736/37;  [(1)+(3/4)+(2)]+[37/2]
 GENUS#4;P=2;  96;  [A]+[(3)+(37/3)]
 GENUS#4;P=3;  16/3;  [(1)+(13)+(38/13)]+[111/152]
 GENUS#4;P=37;  2736/37;  [(1)+(3/4)+(3)]+[37/3]


D=  445
 GENUS#1;P=2;  15;  [A+H]
 GENUS#1;P=5;  48/5;  [(1)+(3/4)+(2/3)]+[445/8]
 GENUS#1;P=89;  15840/89;  [(1)+(3/4)+(2/3)]+[445/8]
 GENUS#2;P=2;  15;  [A+H]
 GENUS#2;P=5;  48/5;  [(1)+(3/4)+(2)]+[445/24]
 GENUS#2;P=89;  15840/89;  [(1)+(3/4)+(2)]+[445/24]


D=  448
 GENUS#1;P=?;  49152;  [(1)+(1)+(1)]+[28]
 GENUS#1;P=7;  96/7;  [(1)+(1)+(1)]+[28]
 GENUS#2;P=2;  12288;  [A]+[2/3]+[56]
 GENUS#2;P=7;  96/7;  [(1)+(3/4)+(2/3)]+[56]
 GENUS#3;P=2;  32768;  [(1)+(1)]+[(2)+(14)]
 GENUS#3;P=7;  96/7;  [(1)+(1)+(2)]+[14]
 GENUS#4;P=2;  3072;  [H]+[1]+[16]
 GENUS#4;P=7;  96/7;  [(1)+(1)+(16)]+[7/4]
 GENUS#5;P=2;  12288;  [A]+[2]+[56/3]
 GENUS#5;P=7;  96/7;  [(1)+(3/4)+(2)]+[56/3]
 GENUS#6;P=2;  3072;  [A]+[5/3]+[112/5]
 GENUS#6;P=7;  96/7;  [(1)+(3/4)+(5/3)]+[112/5]
 GENUS#7;P=2;  3072;  [A]+[1]+[112/11]
 GENUS#7;P=7;  96/7;  [(1)+(1)+(11/4)]+[112/11]
 GENUS#8;P=2;  24576;  [A]+[(44/3)+(28/11)]
 GENUS#8;P=7;  96/7;  [(1)+(3/4)+(8/3)]+[14]
 GENUS#9;P=2;  16384;  [(1)+(1)+(7)]+[4]
 GENUS#9;P=7;  96/7;  [(1)+(1)+(4)]+[7]
 GENUS#10;P=2;  3072;  [A]+[17/3]+[112/17]
 GENUS#10;P=7;  96/7;  [(1)+(3/4)+(17/3)]+[112/17]
 GENUS#11;P=2;  24576;  [A]+[(20/3)+(28/5)]
 GENUS#11;P=7;  96/7;  [(1)+(3/4)+(20/3)]+[28/5]
 GENUS#12;P=2;  49152;  [(1)+(2A)]+[28/3]
 GENUS#12;P=7;  96/7;  [(1)+(2)+(3/2)]+[28/3]
 GENUS#13;P=2;  4096;  [H]+[10/7]+[56/5]
 GENUS#13;P=7;  96/7;  [(1)+(2)+(5/4)]+[56/5]
 GENUS#14;P=2;  8192;  [H]+[(4)+(4)]
 GENUS#14;P=7;  96/7;  [(1)+(4)+(4)]+[7/4]
 GENUS#15;P=2;  4096;  [H]+[14/15]+[8]
```

```
GENUS#15;P=7;  96/7;  [(1)+(1)+(8)]+[7/2]
GENUS#16;P=2;  98304;  [(1)+(1)]+[4H]
GENUS#16;P=7;  96/7;  [(1)+(1)+(4)]+[7]
GENUS#17;P=2;  32768;  [(1)+(3)]+[(2)+(14/3)]
GENUS#17;P=7;  96/7;  [(1)+(2)+(3)]+[14/3]
GENUS#18;P=2;  16384;  [(1)+(3)+(5/3)]+[28/5]
GENUS#18;P=7;  96/7;  [(1)+(2)+(5/2)]+[28/5]
GENUS#19;P=2;  8192;  [H]+[(20/7)+(28/5)]
GENUS#19;P=7;  96/7;  [(1)+(3)+(5/3)]+[28/5]
GENUS#20;P=2;  98304;  [(3)+(5/3)]+[4A]
GENUS#20;P=7;  96/7;  [(3)+(5/3)+(12/5)]+[7/3]

D=  449
GENUS#1;P=2;  9;  [A+A]
GENUS#1;P=449;  403200/449;  [(1)+(3/4)+(1)]+[449/12]

D=  452
GENUS#1;P=2;  576;  [A]+[2A]
GENUS#1;P=113;  25536/113;  [(1)+(1)+(1)]+[113/4]
GENUS#2;P=2;  48;  [A]+[(1)+(113/3)]
GENUS#2;P=113;  25536/113;  [(1)+(3/4)+(1)]+[113/3]
GENUS#3;P=2;  64;  [H]+[2H]
GENUS#3;P=113;  25536/113;  [(1)+(1)+(3/2)]+[113/6]

D=  453
GENUS#1;P=2;  15;  [A+H]
GENUS#1;P=3;  16/3;  [(1)+(1)+(151/4)]+[3/4]
GENUS#1;P=151;  45600/151;  [(1)+(3/4)+(1)]+[151/4]
GENUS#2;P=2;  15;  [A+H]
GENUS#2;P=3;  16/3;  [(1)+(1)+(1/2)]+[453/8]
GENUS#2;P=151;  45600/151;  [(1)+(3/4)+(2/3)]+[453/8]

D=  456
GENUS#1;P=2;  192;  [A]+[1]+[38]
GENUS#1;P=3;  16/3;  [(1)+(1)+(38)]+[3/4]
GENUS#1;P=19;  720/19;  [(1)+(3/4)+(1)]+[38]
GENUS#2;P=2;  192;  [A]+[129/11]+[38/43]
GENUS#2;P=3;  16/3;  [(1)+(1)+(5/2)]+[57/5]
GENUS#2;P=19;  720/19;  [(1)+(1)+(5/2)]+[57/5]
GENUS#3;P=2;  192;  [A]+[19]+[2]
GENUS#3;P=3;  16/3;  [(1)+(2)+(19)]+[3/4]
GENUS#3;P=19;  720/19;  [(1)+(3/4)+(2)]+[19]
GENUS#4;P=2;  192;  [A]+[1]+[114/115]
GENUS#4;P=3;  16/3;  [(1)+(1)+(1)]+[57/2]
GENUS#4;P=19;  720/19;  [(1)+(1)+(1)]+[57/2]

D=  457
GENUS#1;P=2;  9;  [A+A]
GENUS#1;P=457;  417696/457;  [(1)+(3/4)+(5/3)]+[457/20]

D=  460
GENUS#1;P=2;  96;  [A]+[(173/3)+(115/173)]
GENUS#1;P=5;  48/5;  [(1)+(3/4)+(2/3)]+[115/2]
GENUS#1;P=23;  1056/23;  [(1)+(3/4)+(2/3)]+[115/2]
GENUS#2;P=2;  96;  [H]+[(1)+(115/7)]
GENUS#2;P=5;  48/5;  [(1)+(1)+(7/4)]+[115/7]
GENUS#2;P=23;  1056/23;  [(1)+(1)+(7/4)]+[115/7]
GENUS#3;P=2;  96;  [A]+[(1)+(1)]
GENUS#3;P=5;  48/5;  [(1)+(1)+(1)]+[115/4]
GENUS#3;P=23;  1056/23;  [(1)+(1)+(1)]+[115/4]
GENUS#4;P=2;  96;  [A]+[(5/3)+(23)]
GENUS#4;P=5;  48/5;  [(1)+(3/4)+(23)]+[5/3]
GENUS#4;P=23;  1056/23;  [(1)+(3/4)+(5/3)]+[23]

D=  461
GENUS#1;P=2;  15;  [A+H]
GENUS#1;P=461;  425040/461;  [(1)+(3/4)+(1)]+[461/12]

D=  464
GENUS#1;P=2;  6144;  [(1)+(1)+(1)+(29)]
GENUS#1;P=29;  1680/29;  [(1)+(1)+(1)]+[29]
GENUS#2;P=2;  384;  [A]+[1]+[116/3]
```

```
GENUS#2;P=29; 1680/29; [(1)+(3/4)+(1)]+[116/3]
GENUS#3;P=2; 1536; [A]+[(2/3)+(58)]
GENUS#3;P=29; 1680/29; [(1)+(3/4)+(2/3)]+[58]
GENUS#4;P=2; 512; [H]+[(6/7)+(58/3)]
GENUS#4;P=29; 1680/29; [(1)+(1)+(3/2)]+[58/3]
GENUS#5;P=2; 1536; [A]+[4H]
GENUS#5;P=29; 1680/29; [(1)+(3/4)+(8/3)]+[29/2]
GENUS#6;P=2; 384; [A]+[29/3]+[4]
GENUS#6;P=29; 1680/29; [(1)+(3/4)+(4)]+[29/3]
GENUS#7;P=2; 2048; [(1)+(1)+(15)+(29/15)]
GENUS#7;P=29; 1680/29; [(1)+(1)+(2)]+[29/2]
GENUS#8;P=2; 1536; [H]+[4A]
GENUS#8;P=29; 1680/29; [(1)+(7/4)+(12/7)]+[29/3]


D=  465
 GENUS#1;P=2; 9; [A+A]
 GENUS#1;P=3; 16/3; [(1)+(1)+(155/4)]+[3/4]
 GENUS#1;P=5; 48/5; [(1)+(3/4)+(1)]+[155/4]
 GENUS#1;P=31; 1920/31; [(1)+(3/4)+(1)]+[155/4]
 GENUS#2;P=2; 9; [H+H]
 GENUS#2;P=3; 16/3; [(1)+(1)+(7/4)]+[465/28]
 GENUS#2;P=5; 48/5; [(1)+(1)+(7/4)]+[465/28]
 GENUS#2;P=31; 1920/31; [(1)+(1)+(7/4)]+[465/28]
 GENUS#3;P=2; 9; [H+H]
 GENUS#3;P=3; 16/3; [(1)+(1)+(79/4)]+[465/316]
 GENUS#3;P=5; 48/5; [(1)+(1)+(3/2)]+[155/8]
 GENUS#3;P=31; 1920/31; [(1)+(1)+(3/2)]+[155/8]
 GENUS#4;P=2; 9; [A+A]
 GENUS#4;P=3; 16/3; [(1)+(7/4)+(5/7)]+[93/4]
 GENUS#4;P=5; 48/5; [(1)+(3/4)+(71/3)]+[465/284]
 GENUS#4;P=31; 1920/31; [(1)+(3/4)+(5/3)]+[93/4]


D=  468
 GENUS#1;P=2; 192; [H]+[2A]
 GENUS#1;P=3; 16; [(1)+(1)+(1)]+[117/4]
 GENUS#1;P=13; 336/13; [(1)+(1)+(1)]+[117/4]
 GENUS#2;P=2; 48; [A]+[(1)+(39)]
 GENUS#2;P=3; 96; [(1)+(1)]+[(3/4)+(39)]
 GENUS#2;P=13; 336/13; [(1)+(3/4)+(1)]+[39]
 GENUS#3;P=2; 192; [A]+[2H]
 GENUS#3;P=3; 16; [(1)+(235/4)+(176/235)]+[117/176]
 GENUS#3;P=13; 336/13; [(1)+(3/4)+(2/3)]+[117/2]
 GENUS#4;P=2; 48; [H]+[(1)+(117/7)]
 GENUS#4;P=3; 16; [(1)+(1)+(7/4)]+[117/7]
 GENUS#4;P=13; 336/13; [(1)+(1)+(7/4)]+[117/7]
 GENUS#5;P=2; 48; [A]+[(117/59)+(1)]
 GENUS#5;P=3; 16; [(1)+(1)+(2)]+[117/8]
 GENUS#5;P=13; 336/13; [(1)+(1)+(2)]+[117/8]
 GENUS#6;P=2; 192; [A]+[2H]
 GENUS#6;P=3; 96; [(1)+(1)]+[(3)+(39/4)]
 GENUS#6;P=13; 336/13; [(1)+(1)+(3)]+[39/4]
 GENUS#7;P=2; 192; [H]+[2A]
 GENUS#7;P=3; 16; [(1)+(1)+(7/2)]+[117/14]
 GENUS#7;P=13; 336/13; [(1)+(1)+(7/2)]+[117/14]
 GENUS#8;P=2; 48; [A]+[(5)+(39/5)]
 GENUS#8;P=3; 24; [(1)+(5)]+[(3/4)+(39/5)]
 GENUS#8;P=13; 336/13; [(1)+(3/4)+(5)]+[39/5]
 GENUS#9;P=2; 192; [A]+[2H]
 GENUS#9;P=3; 24; [(1)+(2)]+[(3/4)+(39/2)]
 GENUS#9;P=13; 336/13; [(1)+(3/4)+(2)]+[39/2]
 GENUS#10;P=2; 192; [H]+[2A]
 GENUS#10;P=3; 96; [(1)+(1)]+[(3/2)+(39/2)]
 GENUS#10;P=13; 336/13; [(1)+(1)+(3/2)]+[39/2]
 GENUS#11;P=2; 192; [A]+[2H]
 GENUS#11;P=3; 16; [(1)+(1)+(7)]+[117/28]
 GENUS#11;P=13; 336/13; [(1)+(1)+(9/2)]+[13/2]
 GENUS#12;P=2; 192; [H]+[2A]
 GENUS#12;P=3; 24; [(1)+(2)]+[(3/2)+(39/4)]
 GENUS#12;P=13; 336/13; [(1)+(2)+(3/2)]+[39/4]


D=  469
 GENUS#1;P=2; 15; [A+H]
```

```
GENUS#1;P=7;  96/7;  [(1)+(3/4)+(2/3)]+[469/8]
GENUS#1;P=67; 8976/67;  [(1)+(3/4)+(2/3)]+[469/8]
GENUS#2;P=2; 15; [A+H]
GENUS#2;P=7; 96/7; [(1)+(3/4)+(2)]+[469/24]
GENUS#2;P=67; 8976/67; [(1)+(3/4)+(2)]+[469/24]
```

```
D=  472
GENUS#1;P=2; 192; [A]+[59]+[2/3]
GENUS#1;P=59; 6960/59; [(1)+(3/4)+(2/3)]+[59]
GENUS#2;P=2; 192; [H]+[1]+[118/119]
GENUS#2;P=59; 6960/59; [(1)+(1)+(1)]+[59/2]
```

```
D=  473
GENUS#1;P=2;  9; [A+A]
GENUS#1;P=11; 240/11; [(1)+(3/4)+(1)]+[473/12]
GENUS#1;P=43; 3696/43; [(1)+(3/4)+(1)]+[473/12]
GENUS#2;P=2;  9; [H+H]
GENUS#2;P=11; 240/11; [(1)+(1)+(3/2)]+[473/24]
GENUS#2;P=43; 3696/43; [(1)+(1)+(3/2)]+[473/24]
```

```
D=  476
GENUS#1;P=2; 96; [A]+[(1)+(119/3)]
GENUS#1;P=7; 96/7; [(1)+(3/4)+(1)]+[119/3]
GENUS#1;P=17; 576/17; [(1)+(3/4)+(1)]+[119/3]
GENUS#2;P=2; 96; [H]+[(1)+(1)]
GENUS#2;P=7; 96/7; [(1)+(1)+(1)]+[119/4]
GENUS#2;P=17; 576/17; [(1)+(1)+(1)]+[119/4]
GENUS#3;P=2; 96; [A]+[(29/3)+(119/29)]
GENUS#3;P=7; 96/7; [(1)+(3/4)+(29/3)]+[119/29]
GENUS#3;P=17; 576/17; [(1)+(3/4)+(14/3)]+[17/2]
GENUS#4;P=2; 96; [A]+[(17/3)+(7)]
GENUS#4;P=7; 96/7; [(1)+(3/4)+(17/3)]+[7]
GENUS#4;P=17; 576/17; [(1)+(3/4)+(7)]+[17/3]
```

```
D=  477
GENUS#1;P=2; 15; [A+H]
GENUS#1;P=3; 48; [(1)+(1)]+[(3/4)+(159/4)]
GENUS#1;P=53; 5616/53; [(1)+(3/4)+(1)]+[159/4]
GENUS#2;P=2; 15; [A+H]
GENUS#2;P=3; 16; [(1)+(1)+(1/2)]+[477/8]
GENUS#2;P=53; 5616/53; [(1)+(3/4)+(2/3)]+[477/8]
GENUS#3;P=2; 15; [A+H]
GENUS#3;P=3; 48; [(1)+(2)]+[(3/4)+(159/8)]
GENUS#3;P=53; 5616/53; [(1)+(3/4)+(2)]+[159/8]
GENUS#4;P=2; 15; [A+H]
GENUS#4;P=3; 16; [(1)+(1)+(19/4)]+[477/76]
GENUS#4;P=53; 5616/53; [(1)+(1)+(19/4)]+[477/76]
```

```
D=  480
GENUS#1;P=2; 6144; [(1)+(1)+(1)]+[30]
GENUS#1;P=3; 16/3; [(1)+(1)+(1)]+[30]
GENUS#1;P=5; 48/5; [(1)+(1)+(1)]+[30]
GENUS#2;P=2; 1536; [A]+[1]+[40]
GENUS#2;P=3; 16/3; [(1)+(1)+(40)]+[3/4]
GENUS#2;P=5; 48/5; [(1)+(3/4)+(1)]+[40]
GENUS#3;P=2; 6144; [A]+[2/3]+[60]
GENUS#3;P=3; 16/3; [(1)+(1)+(1/2)]+[60]
GENUS#3;P=5; 48/5; [(1)+(3/4)+(2/3)]+[60]
GENUS#4;P=2; 6144; [(1)+(1)+(15)]+[2]
GENUS#4;P=3; 16/3; [(1)+(1)+(2)]+[15]
GENUS#4;P=5; 48/5; [(1)+(1)+(2)]+[15]
GENUS#5;P=2; 2048; [H]+[122/63]+[60/61]
GENUS#5;P=3; 16/3; [(1)+(1)+(2)]+[15]
GENUS#5;P=5; 48/5; [(1)+(1)+(2)]+[15]
GENUS#6;P=2; 1536; [A]+[5/3]+[24]
GENUS#6;P=3; 16/3; [(1)+(7/4)+(5/7)]+[24]
GENUS#6;P=5; 48/5; [(1)+(3/4)+(24)]+[5/3]
GENUS#7;P=2; 6144; [(1)+(1)+(3)]+[10]
GENUS#7;P=3; 16/3; [(1)+(1)+(10)]+[3]
GENUS#7;P=5; 48/5; [(1)+(1)+(3)]+[10]
GENUS#8;P=2; 6144; [A]+[10/11]+[12]
GENUS#8;P=3; 16/3; [(1)+(1)+(5/2)]+[12]
```

```
GENUS#8;P=5;  48/5;  [(1)+(1)+(12)]+[5/2]
GENUS#9;P=2;  1536;  [A]+[47/3]+[120/47]
GENUS#9;P=3;  16/3;  [(1)+(11/4)+(8/11)]+[15]
GENUS#9;P=5;  48/5;  [(1)+(3/4)+(8/3)]+[15]
GENUS#10;P=2;  1536;  [A]+[11]+[40/11]
GENUS#10;P=3;  16/3;  [(1)+(4)+(10)]+[3/4]
GENUS#10;P=5;  48/5;  [(1)+(3/4)+(4)]+[10]
GENUS#11;P=2;  1536;  [A]+[1]+[120/11]
GENUS#11;P=3;  16/3;  [(1)+(1)+(11/4)]+[120/11]
GENUS#11;P=5;  48/5;  [(1)+(1)+(11/4)]+[120/11]
GENUS#12;P=2;  6144;  [A]+[2]+[20]
GENUS#12;P=3;  16/3;  [(1)+(2)+(20)]+[3/4]
GENUS#12;P=5;  48/5;  [(1)+(3/4)+(2)]+[20]
GENUS#13;P=2;  6144;  [(1)+(1)+(5)]+[6]
GENUS#13;P=3;  16/3;  [(1)+(1)+(5)]+[6]
GENUS#13;P=5;  48/5;  [(1)+(1)+(6)]+[5]
GENUS#14;P=2;  1536;  [H]+[1]+[120/31]
GENUS#14;P=3;  16/3;  [(1)+(1)+(4)]+[15/2]
GENUS#14;P=5;  48/5;  [(1)+(1)+(4)]+[15/2]
GENUS#15;P=2;  6144;  [A]+[14/3]+[60/7]
GENUS#15;P=3;  16/3;  [(1)+(19/4)+(14/19)]+[60/7]
GENUS#15;P=5;  48/5;  [(1)+(3/4)+(14/3)]+[60/7]
GENUS#16;P=2;  1536;  [A]+[7]+[40/7]
GENUS#16;P=3;  16/3;  [(1)+(7)+(40/7)]+[3/4]
GENUS#16;P=5;  48/5;  [(1)+(3/4)+(7)]+[40/7]
GENUS#17;P=2;  2048;  [H]+[30/31]+[4]
GENUS#17;P=3;  16/3;  [(1)+(1)+(4)]+[15/2]
GENUS#17;P=5;  48/5;  [(1)+(1)+(4)]+[15/2]
GENUS#18;P=2;  1536;  [A]+[3]+[40/3]
GENUS#18;P=3;  16/3;  [(1)+(55/4)+(161/55)]+[120/161]
GENUS#18;P=5;  48/5;  [(1)+(3/4)+(3)]+[40/3]
GENUS#19;P=2;  2048;  [H]+[22/23]+[60/11]
GENUS#19;P=3;  16/3;  [(1)+(1)+(11/2)]+[60/11]
GENUS#19;P=5;  48/5;  [(1)+(1)+(11/2)]+[60/11]
GENUS#20;P=2;  2048;  [H]+[6/7]+[20]
GENUS#20;P=3;  16/3;  [(1)+(1)+(20)]+[3/2]
GENUS#20;P=5;  48/5;  [(1)+(1)+(3/2)]+[20]


D=  481
 GENUS#1;P=2;  9;  [H+H]
 GENUS#1;P=13;  336/13;  [(1)+(1)+(7/4)]+[481/28]
 GENUS#1;P=37;  2736/37;  [(1)+(1)+(7/4)]+[481/28]
 GENUS#2;P=2;  9;  [H+H]
 GENUS#2;P=13;  336/13;  [(1)+(1)+(7/2)]+[481/56]
 GENUS#2;P=37;  2736/37;  [(1)+(1)+(7/2)]+[481/56]


D=  484
 GENUS#1;P=2;  576;  [A]+[2A]
 GENUS#1;P=11;  240;  [(1)+(1)+(1)]+[121/4]
 GENUS#2;P=2;  576;  [A]+[2A]
 GENUS#2;P=11;  240;  [(1)+(3/4)+(2/3)]+[121/2]
 GENUS#3;P=2;  48;  [A]+[(1)+(11)]
 GENUS#3;P=11;  3168;  [(1)+(1)]+[(11/4)+(11)]
 GENUS#4;P=2;  64;  [H]+[2H]
 GENUS#4;P=11;  3168;  [(1)+(1)]+[(11/2)+(11/2)]
 GENUS#5;P=2;  576;  [A]+[2A]
 GENUS#5;P=11;  2200;  [(1)+(2)]+[(11/4)+(11/2)]


D=  485
 GENUS#1;P=2;  15;  [A+H]
 GENUS#1;P=5;  48/5;  [(1)+(3/4)+(1)]+[485/12]
 GENUS#1;P=97;  18816/97;  [(1)+(3/4)+(1)]+[485/12]
 GENUS#2;P=2;  15;  [A+H]
 GENUS#2;P=5;  48/5;  [(1)+(3/4)+(74/3)]+[485/296]
 GENUS#2;P=97;  18816/97;  [(1)+(3/4)+(5/3)]+[97/4]


D=  488
 GENUS#1;P=2;  192;  [A]+[1]+[122/123]
 GENUS#1;P=61;  7440/61;  [(1)+(1)+(1)]+[61/2]
 GENUS#2;P=2;  192;  [A]+[61]+[2/3]
 GENUS#2;P=61;  7440/61;  [(1)+(3/4)+(2/3)]+[61]
```

```
D= 489
 GENUS#1;P=2; 9; [A+A]
 GENUS#1;P=3; 16/3; [(1)+(1)+(163/4)]+[3/4]
 GENUS#1;P=163; 53136/163; [(1)+(3/4)+(1)]+[163/4]
 GENUS#2;P=2; 9; [H+H]
 GENUS#2;P=3; 16/3; [(1)+(1)+(83/4)]+[489/332]
 GENUS#2;P=163; 53136/163; [(1)+(1)+(3/2)]+[163/8]

D= 492
 GENUS#1;P=2; 96; [A]+[(1)+(41)]
 GENUS#1;P=3; 16/3; [(1)+(1)+(41)]+[3/4]
 GENUS#1;P=41; 3360/41; [(1)+(3/4)+(1)]+[41]
 GENUS#2;P=2; 96; [A]+[(185/3)+(123/185)]
 GENUS#2;P=3; 16/3; [(1)+(247/4)+(185/247)]+[123/185]
 GENUS#2;P=41; 3360/41; [(1)+(3/4)+(2/3)]+[123/2]
 GENUS#3;P=2; 96; [H]+[(145/7)+(123/145)]
 GENUS#3;P=3; 16/3; [(1)+(1)+(41/2)]+[3/2]
 GENUS#3;P=41; 3360/41; [(1)+(1)+(3/2)]+[41/2]
 GENUS#4;P=2; 96; [A]+[(1)+(1)]
 GENUS#4;P=3; 16/3; [(1)+(1)+(1)]+[123/4]
 GENUS#4;P=41; 3360/41; [(1)+(1)+(1)]+[123/4]

D= 493
 GENUS#1;P=2; 15; [A+H]
 GENUS#1;P=17; 576/17; [(1)+(3/4)+(2/3)]+[493/8]
 GENUS#1;P=29; 1680/29; [(1)+(3/4)+(2/3)]+[493/8]
 GENUS#2;P=2; 15; [H+A]
 GENUS#2;P=17; 576/17; [(1)+(1)+(7/4)]+[493/28]
 GENUS#2;P=29; 1680/29; [(1)+(1)+(7/4)]+[493/28]

D= 496
 GENUS#1;P=2; 3072; [(1)+(1)+(1)+(31)]
 GENUS#1;P=31; 1920/31; [(1)+(1)+(1)]+[31]
 GENUS#2;P=2; 3072; [A]+[(2/3)+(62)]
 GENUS#2;P=31; 1920/31; [(1)+(3/4)+(2/3)]+[62]
 GENUS#3;P=2; 384; [H]+[1]+[124/7]
 GENUS#3;P=31; 1920/31; [(1)+(1)+(7/4)]+[124/7]
 GENUS#4;P=2; 3072; [A]+[(2)+(62/3)]
 GENUS#4;P=31; 1920/31; [(1)+(3/4)+(2)]+[62/3]
 GENUS#5;P=2; 1024; [H]+[(2)+(62/63)]
 GENUS#5;P=31; 1920/31; [(1)+(1)+(2)]+[31/2]
 GENUS#6;P=2; 384; [A]+[17/3]+[124/17]
 GENUS#6;P=31; 1920/31; [(1)+(3/4)+(17/3)]+[124/17]
 GENUS#7;P=2; 3072; [(1)+(11)+(21/11)+(31/21)]
 GENUS#7;P=31; 1920/31; [(1)+(2)+(3/2)]+[31/3]
 GENUS#8;P=2; 1024; [H]+[(42/23)+(62/21)]
 GENUS#8;P=31; 1920/31; [(1)+(2)+(3)]+[31/6]

D= 497
 GENUS#1;P=2; 9; [A+A]
 GENUS#1;P=7; 96/7; [(1)+(3/4)+(1)]+[497/12]
 GENUS#1;P=71; 10080/71; [(1)+(3/4)+(1)]+[497/12]
 GENUS#2;P=2; 9; [H+H]
 GENUS#2;P=7; 96/7; [(1)+(1)+(71/4)]+[7/4]
 GENUS#2;P=71; 10080/71; [(1)+(1)+(7/4)]+[71/4]

D= 500
 GENUS#1;P=2; 192; [H]+[2A]
 GENUS#1;P=5; 240; [(1)+(1)+(1)]+[125/4]
 GENUS#2;P=2; 48; [A]+[(1)+(125/3)]
 GENUS#2;P=5; 240; [(1)+(3/4)+(1)]+[125/3]
 GENUS#3;P=2; 192; [A]+[2H]
 GENUS#3;P=5; 240; [(1)+(3/4)+(2/3)]+[125/2]
 GENUS#4;P=2; 48; [A]+[(5/3)+(25)]
 GENUS#4;P=5; 3000; [(1)+(3/4)]+[5/3]+[25]
 GENUS#5;P=2; 192; [A]+[2H]
 GENUS#5;P=5; 2000; [(1)+(1)]+[5/2]+[25/2]
 GENUS#6;P=2; 192; [H]+[2A]
 GENUS#6;P=5; 2000; [(1)+(1)]+[5]+[25/4]
 GENUS#7;P=2; 192; [A]+[2H]
 GENUS#7;P=5; 3000; [(1)+(3/4)]+[20/3]+[25/4]
 GENUS#8;P=2; 192; [H]+[2A]
```

```
GENUS#8;P=5;  3000;  [(1)+(7/4)]+[10/7]+[25/2]
GENUS#9;P=2;  192;  [H]+[2A]
GENUS#9;P=5;  30000;  [1]+[(15/4)+(10/3)+(5/2)]
GENUS#10;P=2;  48;  [H]+[(1)+(25/3)]
GENUS#10;P=5;  2000;  [(1)+(1)]+[15/4]+[25/3]
GENUS#11;P=2;  48;  [H]+[(5/3)+(5)]
GENUS#11;P=5;  30000;  [2]+[(15/8)+(5/3)+(5)]
GENUS#12;P=2;  192;  [A]+[2H]
GENUS#12;P=5;  30000;  [2]+[(5/2)+(5/2)+(5/2)]


D=  729
GENUS#1;P=2;  9;  [A+A]
GENUS#1;P=3;  11664;  [(1)+(1)]+[3/4]+[243/4]
GENUS#2;P=2;  9;  [H+H]
GENUS#2;P=3;  11664;  [(1)+(1)]+[3/2]+[243/8]
GENUS#3;P=2;  9;  [A+A]
GENUS#3;P=3;  104976;  [1]+[(3/4)+(3)]+[81/4]
GENUS#4;P=2;  9;  [A+A]
GENUS#4;P=3;  69984;  [(1)+(1)]+[(27/4)+(27/4)]
GENUS#5;P=2;  9;  [A+A]
GENUS#5;P=3;  472392;  [1]+[3/4]+[9]+[27/4]
GENUS#6;P=2;  9;  [H+H]
GENUS#6;P=3;  104976;  [2]+[(3/2)+(3/2)]+[81/8]
GENUS#7;P=2;  9;  [H+H]
GENUS#7;P=3;  472392;  [2]+[15/8]+[9/5]+[27/4]
GENUS#8;P=2;  9;  [H+H]
GENUS#8;P=3;  472392;  [2]+[3/2]+[9/2]+[27/8]
GENUS#9;P=2;  9;  [H+H]
GENUS#9;P=3;  472392;  [1]+[15/4]+[18/5]+[27/8]


D= 1729
GENUS#1;P=2;  9;  [H+H]
GENUS#1;P=7;  96/7;  [(1)+(1)+(19/2)]+[91/8]
GENUS#1;P=13;  336/13;  [(1)+(1)+(19/2)]+[91/8]
GENUS#1;P=19;  720/19;  [(1)+(1)+(55/4)]+[1729/220]
GENUS#2;P=2;  9;  [A+A]
GENUS#2;P=7;  96/7;  [(1)+(3/4)+(29/3)]+[1729/116]
GENUS#2;P=13;  336/13;  [(1)+(3/4)+(29/3)]+[1729/116]
GENUS#2;P=19;  720/19;  [(1)+(3/4)+(29/3)]+[1729/116]
GENUS#3;P=2;  9;  [H+H]
GENUS#3;P=7;  96/7;  [(1)+(1)+(247/4)]+[7/4]
GENUS#3;P=13;  336/13;  [(1)+(1)+(7/4)]+[247/4]
GENUS#3;P=19;  720/19;  [(1)+(1)+(7/4)]+[247/4]
GENUS#4;P=2;  9;  [A+A]
GENUS#4;P=7;  96/7;  [(1)+(3/4)+(11)]+[1729/132]
GENUS#4;P=13;  336/13;  [(1)+(3/4)+(11)]+[1729/132]
GENUS#4;P=19;  720/19;  [(1)+(3/4)+(11)]+[1729/132]
```